D0218930

INDUCTION AND ANALOGY IN MATHEMATICS

MATHEMATICS
AND PLAUSIBLE REASONING

VOL. I. INDUCTION AND ANALOGY IN MATHEMATICS

VOL. II. PATTERNS OF PLAUSIBLE INFERENCE

Also by G. Polya:

HOW TO SOLVE IT

INDUCTION AND ANALOGY IN MATHEMATICS

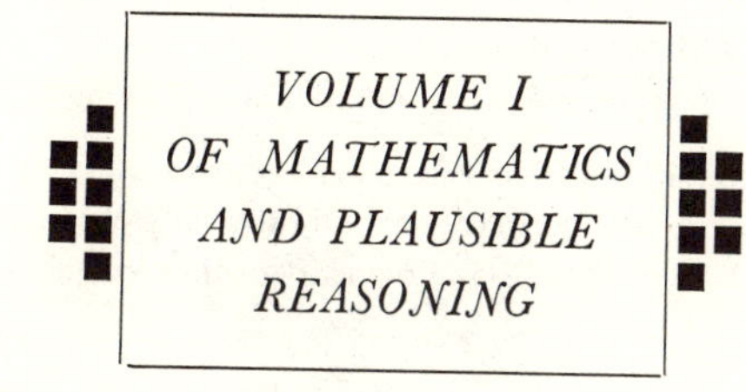

By G. POLYA

PRINCETON UNIVERSITY PRESS
PRINCETON, NEW JERSEY

Published by Princeton University Press, 41 William Street, Princeton, New Jersey 08540

Copyright © 1954 by Princeton University Press
All Rights Reserved

Library of Congress Card No. 53-6388
ISBN 0-691-08005-4
ISBN 0-691-02509-6, pbk.
Twelfth printing and first Princeton Paperback printing, 1990

Princeton University Press books are printed on acid-free paper, and meet the guidelines for permanence and durability of the Committee on Production Guidelines for Book Longevity of the Council on Library Resources

15 14 13 12 (cloth)
10 9 8 7 6 5 4 3 2 (paper)

Printed in the United States of America by Princeton University Press, Princeton, New Jersey

For information about our audio products, write us at:
Newbridge Book Clubs, 3000 Cindel Drive, Delran, NJ 08370

PREFACE

This book has various aims, closely connected with each other. In the first place, this book intends to serve students and teachers of mathematics in an important but usually neglected way. Yet in a sense the book is also a philosophical essay. It is also a continuation and requires a continuation. I shall touch upon these points, one after the other.

1. Strictly speaking, all our knowledge outside mathematics and demonstrative logic (which is, in fact, a branch of mathematics) consists of conjectures. There are, of course, conjectures and conjectures. There are highly respectable and reliable conjectures as those expressed in certain general laws of physical science. There are other conjectures, neither reliable nor respectable, some of which may make you angry when you read them in a newspaper. And in between there are all sorts of conjectures, hunches, and guesses.

We secure our mathematical knowledge by *demonstrative reasoning*, but we support our conjectures by *plausible reasoning*. A mathematical proof is demonstrative reasoning, but the inductive evidence of the physicist, the circumstantial evidence of the lawyer, the documentary evidence of the historian, and the statistical evidence of the economist belong to plausible reasoning.

The difference between the two kinds of reasoning is great and manifold. Demonstrative reasoning is safe, beyond controversy, and final. Plausible reasoning is hazardous, controversial, and provisional. Demonstrative reasoning penetrates the sciences just as far as mathematics does, but it is in itself (as mathematics is in itself) incapable of yielding essentially new knowledge about the world around us. Anything new that we learn about the world involves plausible reasoning, which is the only kind of reasoning for which we care in everyday affairs. Demonstrative reasoning has rigid standards, codified and clarified by logic (formal or demonstrative logic), which is the theory of demonstrative reasoning. The standards of plausible reasoning are fluid, and there is no theory of such reasoning that could be compared to demonstrative logic in clarity or would command comparable consensus.

2. Another point concerning the two kinds of reasoning deserves our attention. Everyone knows that mathematics offers an excellent opportunity to learn demonstrative reasoning, but I contend also that there is no other subject in the usual curricula of the schools that affords a comparable opportunity to learn plausible reasoning. I address myself to all interested students of

mathematics of all grades and I say: Certainly, let us learn proving, but also *let us learn guessing*.

This sounds a little paradoxical and I must emphasize a few points to avoid possible misunderstandings.

Mathematics is regarded as a demonstrative science. Yet this is only one of its aspects. Finished mathematics presented in a finished form appears as purely demonstrative, consisting of proofs only. Yet mathematics in the making resembles any other human knowledge in the making. You have to guess a mathematical theorem before you prove it; you have to guess the idea of the proof before you carry through the details. You have to combine observations and follow analogies; you have to try and try again. The result of the mathematician's creative work is demonstrative reasoning, a proof; but the proof is discovered by plausible reasoning, by guessing. If the learning of mathematics reflects to any degree the invention of mathematics, it must have a place for guessing, for plausible inference.

There are two kinds of reasoning, as we said: demonstrative reasoning and plausible reasoning. Let me observe that they do not contradict each other; on the contrary, they complete each other. In strict reasoning the principal thing is to distinguish a proof from a guess, a valid demonstration from an invalid attempt. In plausible reasoning the principal thing is to distinguish a guess from a guess, a more reasonable guess from a less reasonable guess. If you direct your attention to both distinctions, both may become clearer.

A serious student of mathematics, intending to make it his life's work, must learn demonstrative reasoning; it is his profession and the distinctive mark of his science. Yet for real success he must also learn plausible reasoning; this is the kind of reasoning on which his creative work will depend. The general or amateur student should also get a taste of demonstrative reasoning: he may have little opportunity to use it directly, but he should acquire a standard with which he can compare alleged evidence of all sorts aimed at him in modern life. But in all his endeavors he will need plausible reasoning. At any rate, an ambitious student of mathematics, whatever his further interests may be, should try to learn both kinds of reasoning, demonstrative and plausible.

3. I do not believe that there is a foolproof method to learn guessing. At any rate, if there is such a method, I do not know it, and quite certainly I do not pretend to offer it on the following pages. The efficient use of plausible reasoning is a practical skill and it is learned, as any other practical skill, by imitation and practice. I shall try to do my best for the reader who is anxious to learn plausible reasoning, but what I can offer are only examples for imitation and opportunity for practice.

In what follows, I shall often discuss mathematical discoveries, great and small. I cannot tell the true story how the discovery did happen, because nobody really knows that. Yet I shall try to make up a likely story how the discovery could have happened. I shall try to emphasize the motives underlying the discovery, the plausible inferences that led to it, in short, everything that deserves imitation. Of course, I shall try to impress the reader; this is my duty as teacher and author. Yet I shall be perfectly honest with the reader in the point that really matters: I shall try to impress him only with things which seem genuine and helpful to me.

Each chapter will be followed by examples and comments. The comments deal with points too technical or too subtle for the text of the chapter, or with points somewhat aside of the main line of argument. Some of the exercises give an opportunity to the reader to reconsider details only sketched in the text. Yet the majority of the exercises give an opportunity to the reader to draw plausible conclusions of his own. Before attacking a more difficult problem proposed at the end of a chapter, the reader should carefully read the relevant parts of the chapter and should also glance at the neighboring problems; one or the other may contain a clue. In order to provide (or hide) such clues with the greatest benefit to the instruction of the reader, much care has been expended not only on the contents and the form of the proposed problems, but also on their *disposition*. In fact, much more time and care went into the arrangement of these problems than an outsider could imagine or would think necessary.

In order to reach a wide circle of readers I tried to illustrate each important point by an example as elementary as possible. Yet in several cases I was obliged to take a not too elementary example to support the point impressively enough. In fact, I felt that I should present also examples of historic interest, examples of real mathematical beauty, and examples illustrating the parallelism of the procedures in other sciences, or in everyday life.

I should add that for many of the stories told the final form resulted from a sort of informal psychological experiment. I discussed the subject with several different classes, interrupting my exposition frequently with such questions as: "Well, what would you do in such a situation?" Several passages incorporated in the following text have been suggested by the answers of my students, or my original version has been modified in some other manner by the reaction of my audience.

In short, I tried to use all my experience in research and teaching to give an appropriate opportunity to the reader for intelligent imitation and for doing things by himself.

4. The examples of plausible reasoning collected in this book may be put to another use: they may throw some light upon a much agitated philosophical problem: the problem of induction. The crucial question is: Are there rules for induction? Some philosophers say Yes, most scientists think No. In order to be discussed profitably, the question should be put differently. It should be treated differently, too, with less reliance on traditional verbalisms, or on new-fangled formalisms, but in closer touch with the practice of scientists. Now, observe that inductive reasoning is a

particular case of plausible reasoning. Observe also (what modern writers almost forgot, but some older writers, such as Euler and Laplace, clearly perceived) that the role of inductive evidence in mathematical investigation is similar to its role in physical research. Then you may notice the possibility of obtaining some information about inductive reasoning by observing and comparing examples of plausible reasoning in mathematical matters. And so the door opens to *investigating induction inductively*.

When a biologist attempts to investigate some general problem, let us say, of genetics, it is very important that he should choose some particular species of plants or animals that lends itself well to an experimental study of his problem. When a chemist intends to investigate some general problem about, let us say, the velocity of chemical reactions, it is very important that he should choose some particular substances on which experiments relevant to his problem can be conveniently made. The choice of appropriate experimental material is of great importance in the inductive investigation of any problem. It seems to me that mathematics is, in several respects, the most appropriate experimental material for the study of inductive reasoning. This study involves psychological experiments of a sort: you have to experience how your confidence in a conjecture is swayed by various kinds of evidence. Thanks to their inherent simplicity and clarity, mathematical subjects lend themselves to this sort of psychological experiment much better than subjects in any other field. On the following pages the reader may find ample opportunity to convince himself of this.

It is more philosophical, I think, to consider the more general idea of plausible reasoning instead of the particular case of inductive reasoning. It seems to me that the examples collected in this book lead up to a definite and fairly satisfactory aspect of plausible reasoning. Yet I do not wish to force my views upon the reader. In fact, I do not even state them in Vol. I; I want the examples to speak for themselves. The first four chapters of Vol. II, however, are devoted to a more explicit general discussion of plausible reasoning. There I state formally the patterns of plausible inference suggested by the foregoing examples, try to systematize these patterns, and survey some of their relations to each other and to the idea of probability.

I do not know whether the contents of these four chapters deserve to be called philosophy. If this is philosophy, it is certainly a pretty low-brow kind of philosophy, more concerned with understanding concrete examples and the concrete behavior of people than with expounding generalities. I know still less, of course, how the final judgement on my views will turn out. Yet I feel pretty confident that my examples can be useful to any reasonably unprejudiced student of induction or of plausible reasoning, who wishes to form his views in close touch with the observable facts.

5. This work on *Mathematics and Plausible Reasoning*, which I have always regarded as a unit, falls naturally into two parts: *Induction and Analogy in Mathematics* (Vol. I), and *Patterns of Plausible Inference* (Vol. II). For the

convenience of the student they have been issued as separate volumes. Vol. I is entirely independent of Vol. II, and I think many students will want to go through it carefully before reading Vol. II. It has more of the mathematical "meat" of the work, and it supplies "data" for the inductive investigation of induction in Vol. II. Some readers, who should be fairly sophisticated and experienced in mathematics, will want to go directly to Vol. II, and for these it will be a convenience to have it separately. For ease of reference the chapter numbering is continuous through both volumes. I have not provided an index, since an index would tend to render the terminology more rigid than it is desirable in this kind of work. I believe the table of contents will provide a satisfactory guide to the book.

The present work is a continuation of my earlier book *How to Solve It*. The reader interested in the subject should read both, but the order does not matter much. The present text is so arranged that it can be read independently of the former work. In fact, there are only few direct references in the present book to the former and they can be disregarded in a first reading. Yet there are indirect references to the former book on almost every page, and in almost every sentence on some pages. In fact, the present work provides numerous exercises and some more advanced illustrations to the former which, in view of its size and its elementary character, had no space for them.

The present book is also related to a collection of problems in Analysis by G. Szegö and the author (see Bibliography). The problems in that collection are carefully arranged in series so that they support each other mutually, provide cues to each other, cover a certain subject-matter jointly, and give the reader an opportunity to practice various moves important in problem-solving. In the treatment of problems the present book follows the method of presentation initiated by that former work, and this link is not unimportant.

Two chapters in Vol. II of the present book deal with the theory of probability. The first of these chapters is somewhat related to an elementary exposition of the calculus of probability written by the author several years ago (see the Bibliography). The underlying views on probability and the starting points are the same, but otherwise there is little contact.

Some of the views offered in this book have been expressed before in my papers quoted in the Bibliography. Extensive passages of papers no. 4, 6, 8, 9, and 10 have been incorporated in the following text. Acknowledgment and my best thanks are due to the editors of the *American Mathematical Monthly*, *Etudes de Philosophie des Sciences en Hommage à Ferdinand Gonseth*, and *Proceedings of the International Congress of Mathematicians 1950*, who kindly gave permission to reprint these passages.

Most parts of this book have been presented in my lectures, some parts several times. In some parts and in some respects, I preserved the tone of oral presentation. I do not think that such a tone is advisable in printed presentation of mathematics in general, but in the present case it may be appropriate, or at least excusable.

6. The last chapter of Vol. II of the present book, dealing with Invention and Teaching, links the contents more explicitly to the former work of the author and points to a possible sequel.

The efficient use of plausible reasoning plays an essential role in problem-solving. The present book tries to illustrate this role by many examples, but there remain other aspects of problem-solving that need similar illustration.

Many points touched upon here need further work. My views on plausible reasoning should be confronted with the views of other authors, the historical examples should be more thoroughly explored, the views on invention and teaching should be investigated as far as possible with the methods of experimental psychology,[1] and so on. Several such tasks remain, but some of them may be thankless.

The present book is not a textbook. Yet I hope that in time it will influence the usual presentation of the textbooks and the choice of their problems. The task of rewriting the textbooks of the more usual subjects along these lines need not be thankless.

7. I wish to express my thanks to the Princeton University Press for the careful printing, and especially to Mr. Herbert S. Bailey, Jr., Director of the Press, for understanding help in several points. I am much indebted also to Mrs. Priscilla Feigen for the preparation of the typescript, and to Dr. Julius G. Baron for his kind help in reading the proofs.

George Polya

Stanford University
May 1953

[1] Exploratory work in this direction has been undertaken in the Department of Psychology of Stanford University, within the framework of a project directed by E. R. Hilgard, under O.N.R. sponsorship.

HINTS TO THE READER

THE section 2 of chapter VII is quoted as sect. 2 in chapter VII, but as sect. 7.2 in any other chapter. The subsection (3) of section 5 of chapter XIV is quoted as sect. 5 (3) in chapter XIV, but as sect. 14.5 (3) in any other chapter. We refer to example 26 of chapter XIV as ex. 26 in the same chapter, but as ex. 14.26 in any other chapter.

Some knowledge of elementary algebra and geometry may be enough to read substantial parts of the text. Thorough knowledge of elementary algebra and geometry and some knowledge of analytic geometry and calculus, including limits and infinite series, is sufficient for almost the whole text and the majority of the examples and comments. Yet more advanced knowledge is supposed in a few incidental remarks of the text, in some proposed problems, and in several comments. Usually some warning is given when more advanced knowledge is assumed.

The advanced reader who skips parts that appear to him too elementary may miss more than the less advanced reader who skips parts that appear to him too complex.

Some details of (not very difficult) demonstrations are often omitted without warning. Duly prepared for this eventuality, a reader with good critical habits need not spoil them.

Some of the problems proposed for solution are very easy, but a few are pretty hard. Hints that may facilitate the solution are enclosed in square brackets [ ]. The surrounding problems may provide hints. Especial attention should be paid to the introductory lines prefixed to the examples in some chapters, or prefixed to the First Part, or Second Part, of such examples.

The solutions are sometimes very short: they suppose that the reader has earnestly tried to solve the problem by his own means before looking at the printed solution.

A reader who spent serious effort on a problem may profit by it even if he does not succeed in solving it. For example, he may look at the solution, try to isolate what appears to him the key idea, put the book aside, and then try to work out the solution.

At some places, this book is lavish of figures or in giving small intermediate steps of a derivation. The aim is to render visible the *evolution* of a figure or a formula; see, for instance, Fig. 16.1–16.5. Yet no book can have enough figures or formulas. A reader may want to read a passage "in first approximation" or more thoroughly. If he wants to read more thoroughly, he should have paper and pencil at hand: he should be prepared to write or draw any formula or figure given in, or only indicated by, the text. Doing so, he has a better chance to see the evolution of the figure or formula, to understand how the various details contribute to the final product, and to remember the whole thing.

CONTENTS

PREFACE v

HINTS TO THE READER xi

CHAPTER I. INDUCTION 3

1. Experience and belief. 2. Suggestive contacts. 3. Supporting contacts. 4. The inductive attitude

Examples and Comments on Chapter I, 1–14. [12. Yes and No. 13. Experience and behavior. 14. The logician, the mathematician, the physicist, and the engineer.]

CHAPTER II. GENERALIZATION, SPECIALIZATION, ANALOGY 12

1. Generalization, specialization, analogy, and induction. 2. Generalization. 3. Specialization. 4. Analogy. 5. Generalization, specialization, and analogy. 6. Discovery by analogy. 7. Analogy and induction

Examples and Comments on Chapter II, 1–46; [First Part, 1–20; Second Part, 21–46]. [1. The right generalization. 5. An extreme special case. 7. A leading special case. 10. A representative special case. 11. An analogous case. 18. Great analogies. 19. Clarified analogies. 20. Quotations. 21. The conjecture *E*. 44. An objection and a first approach to a proof. 45. A second approach to a proof. 46. Dangers of analogy.]

CHAPTER III. INDUCTION IN SOLID GEOMETRY 35

1. Polyhedra. 2. First supporting contacts. 3. More supporting contacts. 4. A severe test. 5. Verifications and verifications. 6. A very different case. 7. Analogy. 8. The partition of space. 9. Modifying the problem. 10. Generalization, specialization, analogy. 11. An analogous problem. 12. An array of analogous problems. 13. Many problems may be easier than just one. 14. A conjecture. 15. Prediction and verification. 16. Again and better. 17. Induction suggests deduction, the particular case suggests the general proof. 18. More conjectures

Examples and Comments on Chapter III, 1–41. [21. Induction: adaptation of the mind, adaptation of the language. 31. Descartes' work on polyhedra. 36. Supplementary solid angles, supplementary spherical polygons.]

CHAPTER IV. INDUCTION IN THE THEORY OF NUMBERS 59

1. Right triangles in integers. 2. Sums of squares. 3. On the sum of four odd squares. 4. Examining an example. 5. Tabulating the observations. 6. What is the rule? 7. On the nature of inductive discovery. 8. On the nature of inductive evidence

Examples and Comments on Chapter IV, 1–26. [1. Notation. 26. Dangers of induction.]

CHAPTER V. MISCELLANEOUS EXAMPLES OF INDUCTION 76

1. Expansions. 2. Approximations. 3. Limits. 4. Trying to disprove it. 5. Trying to prove it. 6. The role of the inductive phase

Examples and Comments on Chapter V, 1–18. [15. Explain the observed regularities. 16. Classify the observed facts. 18. What is the difference?]

CHAPTER VI. A MORE GENERAL STATEMENT 90

1. Euler. 2. Euler's memoir. 3. Transition to a more general viewpoint. 4. Schematic outline of Euler's memoir

Examples and Comments on Chapter VI, 1–25. [1. Generating functions. 7. A combinatorial problem in plane geometry. 10. Sums of squares. 19. Another recursion formula. 20. Another Most Extraordinary Law of the Numbers concerning the Sum of their Divisors. 24. How Euler missed a discovery. 25. A generalization of Euler's theorem on $\sigma(n)$.]

CHAPTER VII. MATHEMATICAL INDUCTION 108

1. The inductive phase. 2. The demonstrative phase. 3. Examining transitions. 4. The technique of mathematical induction

Examples and Comments on Chapter VII, 1–18. [12. To prove more may be less trouble. 14. Balance your theorem. 15. Outlook. 17. Are any n numbers equal?]

CHAPTER VIII. MAXIMA AND MINIMA 121

1. Patterns. 2. Example. 3. The pattern of the tangent level line. 4. Examples. 5. The pattern of partial variation. 6. The theorem of the arithmetic and geometric means and its first consequences

Examples and Comments on Chapter VIII, 1–63; [First Part, 1–32; Second Part, 33–63]. [1. Minimum and maximum distances in plane geometry. 2. Minimum and maximum distances in solid geometry. 3. Level lines in a plane. 4. Level surfaces in space. 11. The principle of the crossing level line. 22. The principle of partial variation. 23. Existence of the extremum. 24. A modification of the pattern of partial variation: An infinite process. 25. Another modification of the pattern of partial variation: A finite process. 26. Graphic comparison. 33. Polygons and polyhedra. Area and perimeter. Volume and surface. 34. Right prism with square base. 35. Right cylinder. 36. General right prism. 37. Right double pyramid with square base. 38. Right double cone. 39. General right double pyramid. 43. Applying geometry to algebra. 45. Applying algebra to geometry. 51. Right pyramid with square base. 52. Right cone. 53. General right pyramid. 55. The box with the lid off. 56. The trough. 57. A fragment. 62. A post office problem. 63. A problem of Kepler.]

Chapter IX. Physical Mathematics 142

1. Optical interpretation. 2. Mechanical interpretation. 3. Reinterpretation. 4. Jean Bernoulli's discovery of the brachistochrone. 5. Archimedes' discovery of the integral calculus

Examples and Comments on Chapter IX, 1–38. [3. Triangle with minimum perimeter inscribed in a given triangle. 9. Traffic center of four points in space. 10. Traffic center of four points in a plane. 11. Traffic network for four points. 12. Unfold and straighten. 13. Billiards. 14. Geophysical exploration. 23. Shortest lines on a polyhedral surface. 24. Shortest lines (geodesics) on a curved surface. 26. A construction by paper-folding. 27. The die is cast. 28. The Deluge. 29. Not so deep as a well. 30. A useful extreme case. 32. The Calculus of Variations. 33. From the equilibrium of cross-sections to the equilibrium of the solids. 38. Archimedes' Method in retrospect.]

Chapter X. The Isoperimetric Problem 168

1. Descartes' inductive reasons. 2. Latent reasons. 3. Physical reasons. 4. Lord Rayleigh's inductive reasons. 5. Deriving consequences. 6. Verifying consequences. 7. Very close. 8. Three forms of the Isoperimetric Theorem. 9. Applications and questions

Examples and Comments on Chapter X, 1–43; [First Part, 1–15; Second Part, 16–43]. [1. Looking back. 2. Could you derive some part of the result differently? 3. Restate with more detail. 7. Can you use the method for some other problem? 8. Sharper form of the Isoperimetric Theorem. 16. The stick and the string. 21. Two sticks and two strings. 25. Dido's problem in solid geometry. 27. Bisectors of a plane region. 34. Bisectors of a closed surface. 40. A figure of many perfections. 41. An analogous case. 42. The regular solids. 43. Inductive reasons.]

CHAPTER XI. FURTHER KINDS OF PLAUSIBLE REASONS 190

1. Conjectures and conjectures. 2. Judging by a related case. 3. Judging by the general case. 4. Preferring the simpler conjecture. 5. Background. 6. Inexhaustible. 7. Usual heuristic assumptions

Examples and Comments on Chapter XI, 1–23. [16. The general case. 19. No idea is really bad. 20. Some usual heuristic assumptions. 21. Optimism rewarded. 23. Numerical computation and the engineer.]

FINAL REMARK 210

SOLUTIONS TO PROBLEMS 213

BIBLIOGRAPHY 279

Volume I

Induction and Analogy in Mathematics

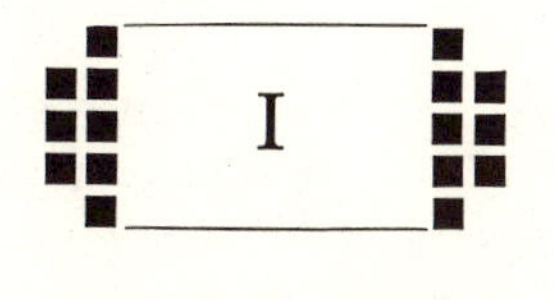

INDUCTION

It will seem not a little paradoxical to ascribe a great importance to observations even in that part of the mathematical sciences which is usually called Pure Mathematics, since the current opinion is that observations are restricted to physical objects that make impression on the senses. As we must refer the numbers to the pure intellect alone, we can hardly understand how observations and quasi-experiments can be of use in investigating the nature of the numbers. Yet, in fact, as I shall show here with very good reasons, the properties of the numbers known today have been mostly discovered by observation, and discovered long before their truth has been confirmed by rigid demonstrations. There are even many properties of the numbers with which we are well acquainted, but which we are not yet able to prove; only observations have led us to their knowledge. Hence we see that in the theory of numbers, which is still very imperfect, we can place our highest hopes in observations; they will lead us continually to new properties which we shall endeavor to prove afterwards. The kind of knowledge which is supported only by observations and is not yet proved must be carefully distinguished from the truth; it is gained by induction, as we usually say. Yet we have seen cases in which mere induction led to error. Therefore, we should take great care not to accept as true such properties of the numbers which we have discovered by observation and which are supported by induction alone. Indeed, we should use such a discovery as an opportunity to investigate more exactly the properties discovered and to prove or disprove them; in both cases we may learn something useful.—EULER[1]

1. Experience and belief. Experience modifies human beliefs. We learn from experience or, rather, we ought to learn from experience. To make the best possible use of experience is one of the great human tasks and to work for this task is the proper vocation of scientists.

A scientist deserving this name endeavors to extract the most correct belief from a given experience and to gather the most appropriate experience in order to establish the correct belief regarding a given question. The

[1] Euler, *Opera Omnia*, ser. 1, vol. 2, p. 459, Specimen de usu observationum in mathesi pura.

scientist's procedure to deal with experience is usually called *induction.* Particularly clear examples of the inductive procedure can be found in mathematical research. We start discussing a simple example in the next section.

2. Suggestive contacts. Induction often begins with observation. A naturalist may observe bird life, a crystallographer the shapes of crystals. A mathematician, interested in the Theory of Numbers, observes the properties of the integers 1, 2, 3, 4, 5, ...

If you wish to observe bird life with some chance of obtaining interesting results, you should be somewhat familiar with birds, interested in birds, perhaps you should even like birds. Similarly, if you wish to observe the numbers, you should be interested in, and somewhat familiar with, them. You should distinguish even and odd numbers, you should know the squares 1, 4, 9, 16, 25, . . . and the primes 2, 3, 5, 7, 11, 13, 17, 19, 23, 29, ... (It is better to keep 1 apart as "unity" and not to classify it as a prime.) Even with so modest a knowledge you may be able to observe something interesting.

By some chance, you come across the relations

$$3 + 7 = 10, \quad 3 + 17 = 20, \quad 13 + 17 = 30$$

and notice some resemblance between them. It strikes you that the numbers 3, 7, 13, and 17 are odd primes. The sum of two odd primes is necessarily an even number; in fact, 10, 20, and 30 are even. What about the *other* even numbers? Do they behave similarly? The first even number which is a sum of two odd primes is, of course,

$$6 = 3 + 3.$$

Looking beyond 6, we find that

$$8 = 3 + 5$$

$$10 = 3 + 7 = 5 + 5$$

$$12 = 5 + 7$$

$$14 = 3 + 11 = 7 + 7$$

$$16 = 3 + 13 = 5 + 11.$$

Will it go on like this forever? At any rate, the particular cases observed suggest a general statement: *Any even number greater than 4 is the sum of two odd primes.* Reflecting upon the exceptional cases, 2 and 4, which cannot be split into a sum of two odd primes, we may prefer the following more sophisticated statement: *Any even number that is neither a prime nor the square of a prime, is the sum of two odd primes.*

We arrived so at formulating a *conjecture.* We found this conjecture by *induction.* That is, it was suggested by observation, indicated by particular instances.

These indications are rather flimsy; we have only very weak grounds to believe in our conjecture. We may find, however, some consolation in the fact that the mathematician who discovered this conjecture a little more than two hundred years ago, Goldbach, did not possess much stronger grounds for it.

Is Goldbach's conjecture true? Nobody can answer this question today. In spite of the great effort spent on it by some great mathematicians, Goldbach's conjecture is today, as it was in the days of Euler, one of those "many properties of the numbers with which we are well acquainted, but which we are not yet able to prove" or disprove.

Now, let us look back and try to perceive such steps in the foregoing reasoning as might be typical of the inductive procedure.

First, we *noticed some similarity.* We recognized that 3, 7, 13, and 17 are primes, 10, 20, and 30 even numbers, and that the three equations $3 + 7 = 10$, $3 + 17 = 20$, $13 + 17 = 30$ are *analogous* to each other.

Then there was a step of *generalization.* From the examples 3, 7, 13, and 17 we passed to all odd primes, from 10, 20, and 30 to all even numbers, and then on to a possibly general relation

$$\text{even number} = \text{prime} + \text{prime}.$$

We arrived so at a clearly formulated general statement, which, however, is merely a conjecture, merely *tentative.* That is, the statement is by no means proved, it cannot have any pretension to be true, it is merely an attempt to get at the truth.

This conjecture has, however, some *suggestive points of contact* with experience, with "the facts," with "reality." It is true for the particular even numbers 10, 20, 30, also for 6, 8, 12, 14, 16.

With these remarks, we outlined roughly a first stage of the inductive process.

3. Supporting contacts. You should not put too much trust in any unproved conjecture, even if it has been propounded by a great authority, even if it has been propounded by yourself. You should try to prove it or to disprove it; you should *test* it.

We test Goldbach's conjecture if we examine some new even number and decide whether it is or is not a sum of two odd primes. Let us examine, for instance, the number 60. Let us perform a "quasi-experiment," as Euler expressed himself. The number 60 is even, but is it the sum of two primes? Is it true that

$$60 = 3 + \text{prime}?$$

No, 57 is not a prime. Is

$$60 = 5 + \text{prime}?$$

The answer is again "No": 55 is not a prime. If it goes on in this way, the conjecture will be exploded. Yet the next trial yields

$$60 = 7 + 53$$

and 53 is a prime. The conjecture has been verified in one more case.

The contrary outcome would have settled the fate of Goldbach's conjecture once and for all. If, trying all primes under a given even number, such as 60, you never arrive at a decomposition into a sum of two primes, you thereby explode the conjecture irrevocably. Having verified the conjecture in the case of the even number 60, you cannot reach such a definite conclusion. You certainly do not prove the theorem by a single verification. It is natural, however, to interpret such a verification as a *favorable sign*, speaking for the conjecture, rendering it *more credible*, although, of course, it is left to your personal judgement how much weight you attach to this favorable sign.

Let us return, for a moment, to the number 60. After having tried the primes 3, 5, and 7, we can try the remaining primes under 30. (Obviously, it is unnecessary to go further than 30 which equals 60/2, since one of the two primes, the sum of which should be 60, must be less than 30.) We obtain so all the decompositions of 60 into a sum of two primes:

$$60 = 7 + 53 = 13 + 47 = 17 + 43 = 19 + 41 = 23 + 37 = 29 + 31.$$

We can proceed systematically and examine the even numbers one after the other, as we have just examined the even number 60. We can *tabulate* the results as follows:

$$6 = 3 + 3$$

$$8 = 3 + 5$$

$$10 = 3 + 7 = 5 + 5$$

$$12 = 5 + 7$$

$$14 = 3 + 11 = 7 + 7$$

$$16 = 3 + 13 = 5 + 11$$

$$18 = 5 + 13 = 7 + 11$$

$$20 = 3 + 17 = 7 + 13$$

$$22 = 3 + 19 = 5 + 17 = 11 + 11$$

$$24 = 5 + 19 = 7 + 17 = 11 + 13$$

$$26 = 3 + 23 = 7 + 19 = 13 + 13$$

$$28 = 5 + 23 = 11 + 17$$

$$30 = 7 + 23 = 11 + 19 = 13 + 17.$$

The conjecture is verified in all cases that we have examined here. Each verification that lengthens the table strengthens the conjecture, renders it more credible, adds to its plausibility. Of course, no amount of such verifications could prove the conjecture.

We should examine our collected observations, we should compare and combine them, we should look for some clue that may be hidden behind them. In our case, it is very hard to discover some essential clue in the table. Still examining the table, we may realize more clearly the meaning of the conjecture. The table shows how often the even numbers listed in it can be represented as a sum of two primes (6 just once, 30 three times). The number of such representations of the even number $2n$ seems to "increase irregularly" with n. Goldbach's conjecture expresses the hope that the number of representations will never fall down to 0, however far we may extend the table.

Among the particular cases that we have examined we could distinguish two groups: those which preceded the formulation of the conjecture and those which came afterwards. The former suggested the conjecture, the latter supported it. Both kinds of cases provide some sort of contact between the conjecture and "the facts." The table does not distinguish between "suggestive" and "supporting" points of contact.

Now, let us look back at the foregoing reasoning and try to see in it traits typical of the inductive process.

Having conceived a conjecture, we tried to find out whether it is true or false. Our conjecture was a general statement suggested by certain particular instances in which we have found it true. We examined further particular instances. As it turned out that the conjecture is true in all instances examined, our confidence in it increased.

We did, it seems to me, only things that reasonable people usually do. In so doing, we seem to accept a principle: *A conjectural general statement becomes more credible if it is verified in a new particular case.*

Is this the principle underlying the process of induction?

4. The inductive attitude. In our personal life we often cling to illusions. That is, we do not dare to examine certain beliefs which could be easily contradicted by experience, because we are afraid of upsetting our emotional balance. There may be circumstances in which it is not unwise to cling to illusions, but in science we need a very different attitude, the *inductive attitude.* This attitude aims at adapting our beliefs to our experience as efficiently as possible. It requires a certain preference for what is matter of fact. It requires a ready ascent from observations to generalizations, and a ready descent from the highest generalizations to the most concrete observations. It requires saying "maybe" and "perhaps" in a thousand different shades. It requires many other things, especially the following three.

First, we should be ready to revise any one of our beliefs.

Second, we should change a belief when there is a compelling reason to change it.

Third, we should not change a belief wantonly, without some good reason.

These points sound pretty trivial. Yet one needs rather unusual qualities to live up to them.

The first point needs "intellectual courage." You need courage to revise your beliefs. Galileo, challenging the prejudice of his contemporaries and the authority of Aristotle, is a great example of intellectual courage.

The second point needs "intellectual honesty." To stick to my conjecture that has been clearly contradicted by experience just because it is *my* conjecture would be dishonest.

The third point needs "wise restraint." To change a belief without serious examination, just for the sake of fashion, for example, would be foolish. Yet we have neither the time nor the strength to examine seriously all our beliefs. Therefore it is wise to reserve the day's work, our questions, and our active doubts for such beliefs as we can reasonably expect to amend. "Do not believe anything, but question only what is worth questioning."

Intellectual courage, intellectual honesty, and wise restraint are the moral qualities of the scientist.

EXAMPLES AND COMMENTS ON CHAPTER I.

1. Guess the rule according to which the successive terms of the following sequence are chosen:

$$11, 31, 41, 61, 71, 101, 131, \ldots .$$

2. Consider the table:

$$\begin{aligned} 1 &= 0 + 1 \\ 2 + 3 + 4 &= 1 + 8 \\ 5 + 6 + 7 + 8 + 9 &= 8 + 27 \\ 10 + 11 + 12 + 13 + 14 + 15 + 16 &= 27 + 64 \end{aligned}$$

Guess the general law suggested by these examples, express it in suitable mathematical notation, and prove it.

3. Observe the values of the successive sums

$$1, \quad 1 + 3, \quad 1 + 3 + 5, \quad 1 + 3 + 5 + 7, \quad \ldots .$$

Is there a simple rule?

4. Observe the values of the consecutive sums

$$1, \quad 1+8, \quad 1+8+27, \quad 1+8+27+64, \quad \ldots .$$

Is there a simple rule?

5. The three sides of a triangle are of lengths l, m, and n, respectively. The numbers l, m, and n are positive integers, $l \leqq m \leqq n$. Find the number of different triangles of the described kind for a given n. [Take $n = 1, 2, 3, 4, 5, \ldots$.] Find a general law governing the dependence of the number of triangles on n.

6. The first three terms of the sequence 5, 15, 25, ... (numbers ending in 5) are divisible by 5. Are also the following terms divisible by 5?

The first three terms of the sequence 3, 13, 23, ... (numbers ending in 3) are prime numbers. Are also the following terms prime numbers?

7. By formal computation we find

$$(1 + 1!x + 2!x^2 + 3!x^3 + 4!x^4 + 5!x^5 + 6!x^6 + \ldots)^{-1}$$
$$= 1 - x - x^2 - 3x^3 - 13x^4 - 71x^5 - 461x^6 \ldots .$$

This suggests two conjectures about the following coefficients of the right hand power series: (1) they are all negative; (2) they are all primes. Are these two conjectures equally trustworthy?

8. Set

$$\left(1 - \frac{x}{1} + \frac{x^2}{2} - \frac{x^3}{3} + \ldots\right)^{-1} = A_0 + \frac{A_1 x}{1!} + \frac{A_2 x^2}{2!} + \ldots .$$

We find that for

$n =$	0	1	2	3	4	5	6	7	8	9
$A_n =$	1	1	1	2	4	14	38	216	600	6240.

State a conjecture.

9. The great French mathematician Fermat considered the sequence

$$5, 17, 257, 65537, \ldots,$$

the general term of which is $2^{2^n} + 1$. He observed that the first four terms (here given), corresponding to $n = 1, 2, 3$, and 4, are primes. He conjectured that the following terms are also primes. Although he did not prove it, he felt so sure of his conjecture that he challenged Wallis and other English mathematicians to prove it. Yet Euler found that the very next term, $2^{32} + 1$, corresponding to $n = 5$, is not a prime: it is divisible by 641.[2] See the passage of Euler at the head of this chapter: "Yet we have seen cases in which mere induction led to error."

10. In verifying Goldbach's conjecture for $2n = 60$ we tried successively the primes p under $n = 30$. We could have also tried, however, the primes p' between $n = 30$ and $2n = 60$. Which procedure is likely to be more advantageous for greater n?

11. In a dictionary, you will find among the explanations for the words "induction," "experiment," and "observation" sentences like the following.

"Induction is inferring a general law from particular instances, or a production of facts to prove a general statement."

"Experiment is a procedure for testing hypotheses."

"Observation is an accurate watching and noting of phenomena as they occur in nature with regard to cause and effect or mutual relations."

Do these descriptions apply to our example discussed in sect. 2 and 3?

12. *Yes and No.* The mathematician as the naturalist, in testing some consequence of a conjectural general law by a new observation, addresses a question to Nature: "I suspect that this law is true. Is it true?" If the consequence is clearly refuted, the law cannot be true. If the consequence is clearly verified, there is some indication that the law may be true. Nature may answer Yes or No, but it whispers one answer and thunders the other, its Yes is provisional, its No is definitive.

13. *Experience and behavior.* Experience modifies human behavior. And experience modifies human beliefs. These two things are not independent of each other. Behavior often results from beliefs, beliefs are potential behavior. Yet you can see the other fellow's behavior, you cannot see his beliefs. Behavior is more easily observed than belief. Everybody knows that "a burnt child dreads the fire," which expresses just what we said: experience modifies human behavior.

Yes, and it modifies animal behavior, too.

In my neighborhood there is a mean dog that barks and jumps at people without provocation. But I have found that I can protect myself rather easily. If I stoop and pretend to pick up a stone, the dog runs away howling. All dogs do not behave so, and it is easy to guess what kind of experience gave this dog this behavior.

The bear in the zoo "begs for food." That is, when there is an onlooker around, it strikes a ridiculous posture which quite frequently prompts the onlooker to throw a lump of sugar into the cage. Bears not in captivity probably never assume such a preposterous posture and it is easy to imagine what kind of experience led to the zoo bear's begging.

A thorough investigation of induction should include, perhaps, the study of animal behavior.

---

[2] Euler, *Opera Omnia*, ser. 1, vol. 2, p. 1–5. Hardy and Wright, *The Theory of Numbers*, p. 14–15.

14. *The logician, the mathematician, the physicist, and the engineer.* "Look at this mathematician," said the logician. "He observes that the first ninety-nine numbers are less than hundred and infers hence, by what he calls induction, that all numbers are less than a hundred."

"A physicist believes," said the mathematician, "that 60 is divisible by all numbers. He observes that 60 is divisible by 1, 2, 3, 4, 5, and 6. He examines a few more cases, as 10, 20, and 30, taken at random as he says. Since 60 is divisible also by these, he considers the experimental evidence sufficient."

"Yes, but look at the engineers," said the physicist. "An engineer suspected that all odd numbers are prime numbers. At any rate, 1 can be considered as a prime number, he argued. Then there come 3, 5, and 7, all indubitably primes. Then there comes 9; an awkward case, it does not seem to be a prime number. Yet 11 and 13 are certainly primes. 'Coming back to 9,' he said, 'I conclude that 9 must be an experimental error.' "

It is only too obvious that induction can lead to error. Yet it is remarkable that induction sometimes leads to truth, since the chances of error appear so overwhelming. Should we begin with the study of the obvious cases in which induction fails, or with the study of those remarkable cases in which induction succeeds? The study of precious stones is understandably more attractive than that of ordinary pebbles and, moreover, it was much more the precious stones than the pebbles that led the mineralogists to the wonderful science of crystallography.

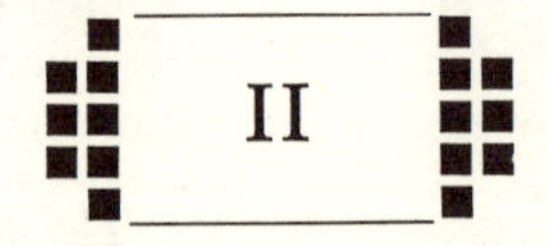

GENERALIZATION, SPECIALIZATION, ANALOGY

And I cherish more than anything else the Analogies, my most trustworthy masters. They know all the secrets of Nature, and they ought to be least neglected in Geometry.—KEPLER

1. Generalization, Specialization, Analogy, and Induction. Let us look again at the example of inductive reasoning that we have discussed in some detail (sect. 1.2, 1.3). We started from observing the *analogy* of the three relations

$$3 + 7 = 10, \quad 3 + 17 = 20, \quad 13 + 17 = 30,$$

we *generalized* in ascending from 3, 7, 13, and 17 to all primes, from 10, 20, and 30 to all even numbers, and then we *specialized* again, came down to test particular even numbers such as 6 or 8 or 60.

This first example is extremely simple. It illustrates quite correctly the role of generalization, specialization, and analogy in inductive reasoning. Yet we should examine less meager, more colorful illustrations and, before that, we should discuss generalization, specialization, and analogy, these great sources of discovery, for their own sake.

2. Generalization is passing from the consideration of a given set of objects to that of a larger set, containing the given one. For example, we generalize when we pass from the consideration of triangles to that of polygons with an arbitrary number of sides. We generalize also when we pass from the study of the trigonometric functions of an acute angle to the trigonometric functions of an unrestricted angle.

It may be observed that in these two examples the generalization was effected in two characteristically different ways. In the first example, in passing from triangles to polygons with n sides, we replace a constant by a variable, the fixed integer 3 by the arbitrary integer n (restricted only by the inequality $n \geqq 3$). In the second example, in passing from acute angles to arbitrary angles α, we remove a restriction, namely the restriction that $0° < \alpha < 90°$.

We often generalize in passing from just one object to a whole class containing that object.

3. Specialization is passing from the consideration of a given set of objects to that of a smaller set, contained in the given one. For example, we specialize when we pass from the consideration of polygons to that of regular polygons, and we specialize still further when we pass from regular polygons with n sides to the regular, that is, equilateral, triangle.

These two subsequent passages were effected in two characteristically different ways. In the first passage, from polygons to regular polygons, we introduced a restriction, namely that all sides and all angles of the polygon be equal. In the second passage we substituted a special object for a variable, we put 3 for the variable integer n.

Very often we specialize in passing from a whole class of objects to just one object contained in the class. For example, when we wish to check some general assertion about prime numbers we pick out some prime number, say 17, and we examine whether that general assertion is true or not for just this prime 17.

4. Analogy. There is nothing vague or questionable in the concepts of generalization and specialization. Yet as we start discussing analogy we tread on a less solid ground.

Analogy is a sort of similarity. It is, we could say, similarity on a more definite and more conceptual level. Yet we can express ourselves a little more accurately. The essential difference between analogy and other kinds of similarity lies, it seems to me, in the intentions of the thinker. Similar objects agree with each other in some aspect. If you intend to reduce the aspect in which they agree to definite concepts, you regard those similar objects as *analogous*. If you succeed in getting down to clear concepts, you have *clarified* the analogy.

Comparing a young woman to a flower, poets feel some similarity, I hope, but usually they do not contemplate analogy. In fact, they scarcely intend to leave the emotional level or reduce that comparison to something measurable or conceptually definable.

Looking in a natural history museum at the skeletons of various mammals, you may find them all frightening. If this is all the similarity you can find between them, you do not see much analogy. Yet you may perceive a wonderfully suggestive analogy if you consider the hand of a man, the paw of a cat, the foreleg of a horse, the fin of a whale, and the wing of a bat, these organs so differently used, as composed of similar parts similarly related to each other.

The last example illustrates the most typical case of clarified analogy; two *systems* are analogous, if they *agree in clearly definable relations of their respective parts*.

For instance, a triangle in a plane is analogous to a tetrahedron in space. In the plane, 2 straight lines cannot include a finite figure, but 3 may include a triangle. In space, 3 planes cannot include a finite figure but 4 may include a tetrahedron. The relation of the triangle to the plane is the same as that of the tetrahedron to space in so far as both the triangle and the tetrahedron are bounded by the minimum number of simple bounding elements. Hence the analogy.

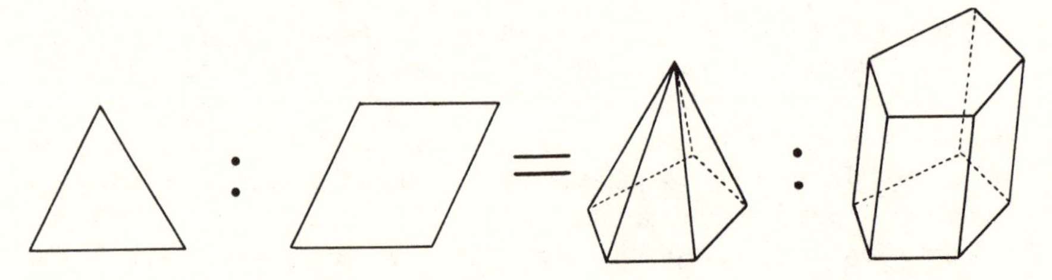

Fig. 2.1. Analogous relations in plane and space.

One of the meanings of the Greek word "analogia," from which the word "analogy" originates, is "proportion." In fact, the system of the two numbers 6 and 9 is "analogous" to the system of the two numbers 10 and 15 in so far as the two systems agree in the ratio of their corresponding terms,

$$6 : 9 = 10 : 15.$$

Proportionality, or agreement in the ratios of corresponding parts, which we may see intuitively in geometrically similar figures, is a very suggestive case of analogy.

Here is another example. We may regard a triangle and a pyramid as analogous figures. On the one hand take a segment of a straight line, and on the other hand a polygon. Connect all points of the segment with a point outside the line of the segment, and you obtain a triangle. Connect all points of the polygon with a point outside the plane of the polygon, and you obtain a pyramid. In the same manner, we may regard a parallelogram and a prism as analogous figures. In fact, move a segment or a polygon parallel to itself, across the direction of its line or plane, and the one will describe a parallelogram, the other a prism. We may be tempted to express these corresponding relations between plane and solid figures by a sort of proportion and if, for once, we do not resist temptation, we arrive at fig. 2.1. This figure modifies the usual meaning of certain symbols (: and =) in the same way as the meaning of the word "analogia" was modified in the course of linguistic history: from "proportion" to "analogy."

The last example is instructive in still another respect. Analogy, especially incompletely clarified analogy, may be ambiguous. Thus, comparing plane and solid geometry, we found first that a triangle in a plane is analogous to a tetrahedron in space and then that a triangle is analogous to a pyramid. Now, both analogies are reasonable, each is valuable at its place. There are several analogies between plane and solid geometry and not just one privileged analogy.

Fig. 2.2 exhibits how, starting from a triangle, we may ascend to a polygon by generalization, descend to an equilateral triangle by specialization, or pass to different solid figures by analogy—there are analogies on all sides.

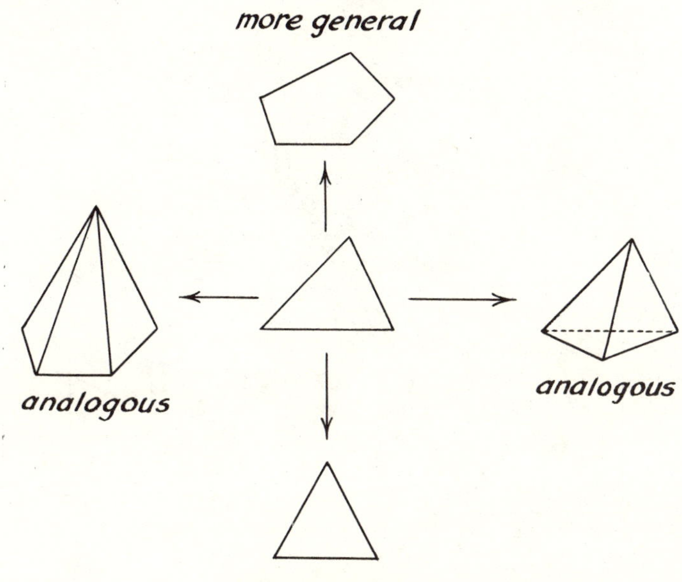

Fig. 2.2. Generalization, specialization, analogy.

And, remember, do not neglect vague analogies. Yet, if you wish them respectable, try to clarify them.

5. **Generalization, Specialization, and Analogy** often concur in solving mathematical problems.[1] Let us take as an example the proof of the best known theorem of elementary geometry, the theorem of Pythagoras. The proof that we shall discuss is not new; it is due to Euclid himself (Euclid VI, 31).

(1) We consider a right triangle with sides a, b, and c, of which the first, a, is the hypotenuse. We wish to show that

$$a^2 = b^2 + c^2. \tag{A}$$

---

[1] This section reproduces with slight changes a Note of the author in the *American Mathematical Monthly*, v. 55 (1948), p. 241–243.

This aim suggests that we describe squares on the three sides of our right triangle. And so we arrive at the not unfamiliar part I of our compound figure, fig. 2.3. (The reader should draw the parts of this figure as they arise, in order to see it in the making.)

(2) Discoveries, even very modest discoveries, need some remark, the recognition of some relation. We can discover the following proof by observing the *analogy* between the familiar part I of our compound figure

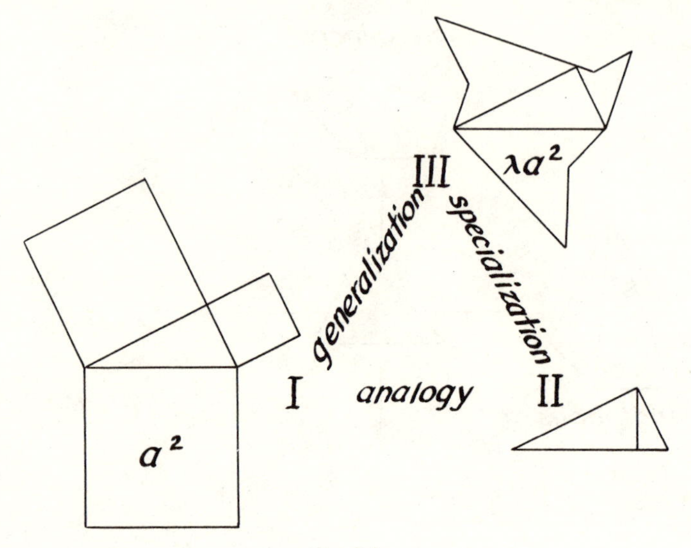

Fig. 2.3.

and the scarcely less familiar part II: the same right triangle that arises in I is divided in II into two parts by the altitude perpendicular to the hypotenuse.

(3) Perhaps, you fail to perceive the analogy between I and II. This analogy, however, can be made explicit by a common *generalization* of I and II which is expressed in III. There we find again the same right triangle, and on its three sides three polygons are described which are similar to each other but arbitrary otherwise.

(4) The area of the square described on the hypotenuse in I is a^2. The area of the irregular polygon described on the hypotenuse in III can be put equal to λa^2; the factor λ is determined as the ratio of two given areas. Yet then, it follows from the similarity of the three polygons described on the sides a, b, and c of the triangle in III that their areas are equal to λa^2, λb^2, and λc^2, respectively.

Now, if the equation (A) should be true (as stated by the theorem that we wish to prove), then also the following would be true:

(B) $$\lambda a^2 = \lambda b^2 + \lambda c^2.$$

In fact, very little algebra is needed to derive (B) from (A). Now, (B) represents a *generalization* of the original theorem of Pythagoras: *If three similar polygons are described on three sides of a right triangle, the one described on the hypotenuse is equal in area to the sum of the two others.*

It is instructive to observe that this generalization is *equivalent* to the special case from which we started. In fact, we can derive the equations (A) and (B) from each other, by multiplying or dividing by λ (which is, as the ratio of two areas, different from 0).

(5) The general theorem expressed by (B) is equivalent not only to the special case (A), but to any other special case. Therefore, if any such special case should turn out to be obvious, the general case would be demonstrated.

Now, trying to *specialize* usefully, we look around for a suitable special case. Indeed II represents such a case. In fact, the right triangle described on its own hypotenuse is similar to the two other triangles described on the two legs, as is well known and easy to see. And, obviously, the area of the whole triangle is equal to the sum of its two parts. And so, the theorem of Pythagoras has been proved.

The foregoing reasoning is eminently instructive. A case is instructive if we can learn from it something applicable to other cases, and the more instructive the wider the range of possible applications. Now, from the foregoing example we can learn the use of such fundamental mental operations as generalization, specialization, and the perception of analogies. There is perhaps no discovery either in elementary or in advanced mathematics or, for that matter, in any other subject that could do without these operations, especially without analogy.

The foregoing example shows how we can ascend by generalization from a special case, as from the one represented by I, to a more general situation as to that of III, and redescend hence by specialization to an analogous case, as to that of II. It shows also the fact, so usual in mathematics and still so surprising to the beginner, or to the philosopher who takes himself for advanced, that the general case can be logically equivalent to a special case. Our example shows, naïvely and suggestively, how generalization, specialization, and analogy are naturally combined in the effort to attain the desired solution. Observe that only a minimum of preliminary knowledge is needed to understand fully the foregoing reasoning.

6. Discovery by analogy. Analogy seems to have a share in all discoveries, but in some it has the lion's share. I wish to illustrate this by an example which is not quite elementary, but is of historic interest and far more impressive than any quite elementary example of which I can think.

Jacques Bernoulli, a Swiss mathematician (1654–1705), a contemporary of Newton and Leibnitz, discovered the sum of several infinite series, but did not succeed in finding the sum of the reciprocals of the squares,

$$1 + \frac{1}{4} + \frac{1}{9} + \frac{1}{16} + \frac{1}{25} + \frac{1}{36} + \frac{1}{49} + \dots .$$

"If somebody should succeed," wrote Bernoulli, "in finding what till now withstood our efforts and communicate it to us, we shall be much obliged to him."

The problem came to the attention of another Swiss mathematician, Leonhard Euler (1707–1783), who was born at Basle as was Jacques Bernoulli and was a pupil of Jacques' brother, Jean Bernoulli (1667–1748). He found various expressions for the desired sum (definite integrals, other series), none of which satisfied him. He used one of these expressions to compute the sum numerically to seven places (1.644934). Yet this is only an approximate value and his goal was to find the exact value. He discovered it, eventually. Analogy led him to an extremely daring conjecture.

(1) We begin by reviewing a few elementary algebraic facts essential to Euler's discovery. If the equation of degree n

$$a_0 + a_1x + a_2x^2 + \dots + a_nx^n = 0$$

has n different roots

$$\alpha_1, \alpha_2, \dots \alpha_n$$

the polynomial on its left hand side can be represented as a product of n linear factors,

$$a_0 + a_1x + a_2x^2 + \dots + a_nx^n =$$
$$a_n(x - \alpha_1)(x - \alpha_2) \dots (x - \alpha_n).$$

By comparing the terms with the same power of x on both sides of this identity, we derive the well known relations between the roots and the coefficients of an equation, the simplest of which is

$$a_{n-1} = -a_n(\alpha_1 + \alpha_2 + \dots + \alpha_n);$$

we find this by comparing the terms with x^{n-1}.

There is another way of presenting the decomposition in linear factors. If none of the roots $\alpha_1, \alpha_2, \dots \alpha_n$ is equal to 0, or (which is the same) if a_0 is different from 0, we have also

$$a_0 + a_1x + a_2x^2 + \dots + a_nx^n$$
$$= a_0\left(1 - \frac{x}{\alpha_1}\right)\left(1 - \frac{x}{\alpha_2}\right) \dots \left(1 - \frac{x}{\alpha_n}\right)$$

and

$$a_1 = -a_0\left(\frac{1}{\alpha_1} + \frac{1}{\alpha_2} + \dots + \frac{1}{\alpha_n}\right).$$

There is still another variant. Suppose that the equation is of degree $2n$, has the form

$$b_0 - b_1x^2 + b_2x^4 - \ldots + (-1)^n b_n x^{2n} = 0$$

and $2n$ different roots

$$\beta_1, \quad -\beta_1, \quad \beta_2, \quad -\beta_2, \quad \ldots \quad \beta_n, \quad -\beta_n.$$

Then

$$\begin{aligned} &b_0 - b_1x^2 + b_2x^4 - \ldots + (-1)^n b_n x^{2n} \\ &= b_0\left(1 - \frac{x^2}{\beta_1^2}\right)\left(1 - \frac{x^2}{\beta_2^2}\right)\ldots\left(1 - \frac{x^2}{\beta_n^2}\right) \end{aligned}$$

and

$$b_1 = b_0\left(\frac{1}{\beta_1^2} + \frac{1}{\beta_2^2} + \ldots + \frac{1}{\beta_n^2}\right).$$

(2) Euler considers the equation

$$\sin x = 0$$

or

$$\frac{x}{1} - \frac{x^3}{1 \cdot 2 \cdot 3} + \frac{x^5}{1 \cdot 2 \cdot 3 \cdot 4 \cdot 5} - \frac{x^7}{1 \cdot 2 \cdot 3 \cdots 7} + \ldots = 0.$$

The left hand side has an infinity of terms, is of "infinite degree." Therefore, it is no wonder, says Euler, that there is an infinity of roots

$$0, \quad \pi, \quad -\pi, \quad 2\pi, \quad -2\pi, \quad 3\pi, \quad -3\pi, \quad \ldots .$$

Euler discards the root 0. He divides the left hand side of the equation by x, the linear factor corresponding to the root 0, and obtains so the equation

$$1 - \frac{x^2}{2 \cdot 3} + \frac{x^4}{2 \cdot 3 \cdot 4 \cdot 5} - \frac{x^6}{2 \cdot 3 \cdot 4 \cdot 5 \cdot 6 \cdot 7} + \ldots = 0$$

with the roots

$$\pi, \quad -\pi, \quad 2\pi, \quad -2\pi, \quad 3\pi, \quad -3\pi, \quad \ldots .$$

We have seen an analogous situation before, under (1), as we discussed the last variant of the decomposition in linear factors. Euler concludes, by analogy, that

$$\begin{aligned} \frac{\sin x}{x} &= 1 - \frac{x^2}{2 \cdot 3} + \frac{x^4}{2 \cdot 3 \cdot 4 \cdot 5} - \frac{x^6}{2 \cdot 3 \cdots 7} + \ldots \\ &= \left(1 - \frac{x^2}{\pi^2}\right)\left(1 - \frac{x^2}{4\pi^2}\right)\left(1 - \frac{x^2}{9\pi^2}\right)\ldots, \end{aligned}$$

$$\frac{1}{2 \cdot 3} = \frac{1}{\pi^2} + \frac{1}{4\pi^2} + \frac{1}{9\pi^2} + \ldots,$$

$$1 + \frac{1}{4} + \frac{1}{9} + \frac{1}{16} + \ldots = \frac{\pi^2}{6}.$$

This is the series that withstood the efforts of Jacques Bernoulli—but it was a daring conclusion.

(3) Euler knew very well that his conclusion was daring. "The method was new and never used yet for such a purpose," he wrote ten years later. He saw some objections himself and many objections were raised by his mathematical friends when they recovered from their first admiring surprise.

Yet Euler had his reasons to trust his discovery. First of all, the numerical value for the sum of the series which he has computed before, agreed to the last place with $\pi^2/6$. Comparing further coefficients in his expression of $\sin x$ as a product, he found the sum of other remarkable series, as that of the reciprocals of the fourth powers,

$$1 + \frac{1}{16} + \frac{1}{81} + \frac{1}{256} + \frac{1}{625} + \dots = \frac{\pi^4}{90}.$$

Again, he examined the numerical value and again he found agreement.

(4) Euler also tested his method on other examples. Doing so he succeeded in rederiving the sum $\pi^2/6$ for Jacques Bernoulli's series by various modifications of his first approach. He succeeded also in rediscovering by his method the sum of an important series due to Leibnitz.

Let us discuss the last point. Let us consider, following Euler, the equation

$$1 - \sin x = 0.$$

It has the roots

$$\frac{\pi}{2}, \quad -\frac{3\pi}{2}, \quad \frac{5\pi}{2}, \quad -\frac{7\pi}{2}, \quad \frac{9\pi}{2}, \quad -\frac{11\pi}{2}, \quad \dots .$$

Each of these roots is, however, a double root. (The curve $y = \sin x$ does not intersect the line $y = 1$ at these abscissas, but is tangent to it. The derivative of the left hand side vanishes for the same values of x, but not the second derivative.) Therefore, the equation

$$1 - \frac{x}{1} + \frac{x^3}{1 \cdot 2 \cdot 3} - \frac{x^5}{1 \cdot 2 \cdot 3 \cdot 4 \cdot 5} + \dots = 0$$

has the roots

$$\frac{\pi}{2}, \quad \frac{\pi}{2}, \quad -\frac{3\pi}{2}, \quad -\frac{3\pi}{2}, \quad \frac{5\pi}{2}, \quad \frac{5\pi}{2}, \quad -\frac{7\pi}{2}, \quad -\frac{7\pi}{2}, \quad \dots$$

and Euler's analogical conclusion leads to the decomposition in linear factors

$$\begin{aligned} 1 - \sin x &= 1 - \frac{x}{1} + \frac{x^3}{1 \cdot 2 \cdot 3} - \frac{x^5}{1 \cdot 2 \cdot 3 \cdot 4 \cdot 5} + \dots \\ &= \left(1 - \frac{2x}{\pi}\right)^2 \left(1 + \frac{2x}{3\pi}\right)^2 \left(1 - \frac{2x}{5\pi}\right)^2 \left(1 + \frac{2x}{7\pi}\right)^2 \dots . \end{aligned}$$

Comparing the coefficient of x on both sides, we obtain

$$-1 = -\frac{4}{\pi} + \frac{4}{3\pi} - \frac{4}{5\pi} + \frac{4}{7\pi} - \dots ,$$

$$\frac{\pi}{4} = 1 - \frac{1}{3} + \frac{1}{5} - \frac{1}{7} + \frac{1}{9} - \frac{1}{11} + \dots .$$

This is Leibnitz's celebrated series; Euler's daring procedure led to a known result. "For our method," says Euler, "which may appear to some as not reliable enough, a great confirmation comes here to light. Therefore, we should not doubt at all of the other things which are derived by the same method."

(5) Yet Euler kept on doubting. He continued the numerical verifications described above under (3), examined more series and more decimal places, and found agreement in all cases examined. He tried other approaches, too, and, finally, he succeeded in verifying not only numerically, but exactly, the value $\pi^2/6$ for Jacques Bernoulli's series. He found a new proof. This proof, although hidden and ingenious, was based on more usual considerations and was accepted as completely rigorous. Thus, the most conspicuous consequence of Euler's discovery was satisfactorily verified.

These arguments, it seems, convinced Euler that his result was correct.[2]

7. Analogy and induction. We wish to learn something about the nature of inventive and inductive reasoning. What can we learn from the foregoing story?

(1) Euler's decisive step was daring. In strict logic, it was an outright fallacy: he applied a rule to a case for which the rule was not made, a rule about algebraic equations to an equation which is not algebraic. In strict logic, Euler's step was not justified. Yet it was justified by analogy, by the analogy of the most successful achievements of a rising science that he called himself a few years later the "Analysis of the Infinite." Other mathematicians, before Euler, passed from finite differences to infinitely small differences, from sums with a finite number of terms to sums with an infinity of terms, from finite products to infinite products. And so Euler passed from equations of finite degree (algebraic equations) to equations of infinite degree, applying the rules made for the finite to the infinite.

This analogy, this passage from the finite to the infinite, is beset with pitfalls. How did Euler avoid them? He was a genius, some people will answer, and of course that is no explanation at all. Euler had shrewd

[2] Much later, almost ten years after his first discovery, Euler returned to the subject, answered the objections, completed to some extent his original heuristic approach, and gave a new, essentially different proof. See L. Euler, *Opera Omnia*, ser. 1, vol. 14, p. 73–86, 138–155, 177–186, and also p. 156–176, containing a note by Paul Stäckel on the history of the problem.

reasons for trusting his discovery. We can understand his reasons with a little common sense, without any miraculous insight specific to genius.

(2) Euler's reasons for trusting his discovery, summarized in the foregoing,[3] are *not* demonstrative. Euler does not reexamine the grounds for his conjecture,[4] for his daring passage from the finite to the infinite; he examines only its consequences. He regards the verification of any such consequence as an argument in favor of his conjecture. He accepts both approximative and exact verifications, but seems to attach more weight to the latter. He examines also the consequences of closely related analogous conjectures[5] and he regards the verification of such a consequence as an argument for his conjecture.

Euler's reasons are, in fact, inductive. It is a typical inductive procedure to examine the consequences of a conjecture and to judge it on the basis of such an examination. In scientific research as in ordinary life, we believe, or ought to believe, a conjecture more or less according as its observable consequences agree more or less with the facts.

In short, Euler seems to think the same way as reasonable people, scientists or non-scientists, usually think. He seems to accept certain principles: *A conjecture becomes more credible by the verification of any new consequence.* And: *A conjecture becomes more credible if an analogous conjecture becomes more credible.*

Are the principles underlying the process of induction of this kind?

EXAMPLES AND COMMENTS ON CHAPTER II

First Part

1. *The right generalization.*

A. Find three numbers x, y, and z satisfying the following system of equations:

$$9x - 6y - 10z = 1,$$

$$-6x + 4y + 7z = 0,$$

$$x^2 + y^2 + z^2 = 9.$$

If you have to solve A, which one of the following three generalizations does give you a more helpful suggestion, B or C or D?

B. Find three unknowns from a system of three equations.

C. Find three unknowns from a system of three equations the first two of which are linear and the third quadratic.

---

[3] Under sect. 6 (3), (4), (5). For Euler's own summary see *Opera Omnia*, ser. 1, vol. 14, p. 140.

[4] The representation of $\sin x$ as an infinite product.

[5] Especially the product for $1 - \sin x$.

D. Find n unknowns from a system of n equations the first $n - 1$ of which are linear.

2. A point and a "regular" pyramid with hexagonal base are given in position. (A pyramid is termed "regular" if its base is a regular polygon the center of which is the foot of the altitude of the pyramid.) Find a plane that passes through the given point and bisects the volume of the given pyramid.

In order to help you, I ask you a question: What is the right generalization?

3. A. Three straight lines which are not in the same plane pass through the same point O. Pass a plane through O that is equally inclined to the three lines.

B. Three straight lines which are not in the same plane pass through the same point. The point P is on one of the lines; pass a plane through P that is equally inclined to the three lines.

Compare the problems A and B. Could you use the solution of one in solving the other? What is their logical connection?

4. A. Compute the integral

$$\int_{-\infty}^{\infty} (1 + x^2)^{-3}\, dx.$$

B. Compute the integral

$$\int_{-\infty}^{\infty} (p + x^2)^{-3}\, dx$$

where p is a given positive number.

Compare the problems A and B. Could you use the solution of one in solving the other? What is their logical connection?

5. *An extreme special case.* Two men are seated at a table of usual rectangular shape. One places a penny on the table, then the other does the same, and so on, alternately. It is understood that each penny lies flat on the table and not on any penny previously placed. The player who puts the last coin on the table takes the money. Which player should win, provided that each plays the best possible game?

This is a time-honored but excellent puzzle. I once had the opportunity to watch a really distinguished mathematician when the puzzle was proposed to him. He started by saying, "Suppose that the table is so small that it is covered by one penny. Then, obviously, the first player must win." That is, he started by picking out an *extreme special case* in which the solution is obvious.

From this special case, you can reach the full solution when you imagine the table gradually extending to leave place to more and more pennies. It may be still better to *generalize* the problem and to think of tables of various shapes and sizes. If you observe that the table has a center of symmetry and that the *right generalization* might be to consider tables with a center of symmetry, then you have got the solution, or you are at least very near to it.

6. Construct a common tangent to two given circles.

In order to help you, I ask you a question: Is there a more accessible extreme special case?

7. *A leading special case.* The area of a polygon is A, its plane includes with a second plane the angle α. The polygon is projected orthogonally onto the second plane. Find the area of the projection.

Observe that the shape of the polygon is not given. Yet there is an endless variety of possible shapes. Which shape should we discuss? Which shape should we discuss first?

There is a particular shape especially easy to handle: a rectangle, the base of which is parallel to the line l, intersection of the plane of the projected figure with the plane of the projection. If the base of such a rectangle is a, its height b, and therefore its area is ab, the corresponding quantities for the projection are a, $b \cos \alpha$, and $ab \cos \alpha$. If the area of such a rectangle is A, the area of its projection is $A \cos \alpha$.

This special case of the rectangle with base parallel to l is not only particularly accessible; it is a *leading special case.* The other cases follow; *the solution of the problem in the leading special case involves the solution in the general case.* In fact, starting from the rectangle with base parallel to l, we can extend the rule "area of the projection equals $A \cos \alpha$" successively to all other figures. First to right triangles with a leg parallel to l (by bisecting the rectangle we start from); then to any triangle with a side parallel to l (by combining two right triangles); finally to a general polygon (by dissecting it into triangles of the kind just mentioned). We could even pass to figures with curvilinear boundaries (by considering them as limits of polygons).

8. The angle at the center of a circle is double the angle at the circumference on the same base, that is, on the same arc. (Euclid III, 20.)

If the angle at the center is given, the angle at the circumference is not yet determined, but can have various positions. In the usual proof of the theorem (Euclid's proof), which is the "leading special position"?

9. Cauchy's theorem, fundamental in the theory of analytic functions, asserts that the integral of such a function vanishes along an arbitrary closed curve in the interior of which the function is regular. We may consider the special case of Cauchy's theorem in which the closed curve is a triangle as a leading special case: having proved the theorem for a triangle, we can easily extend it successively to polygons (by combining triangles) and to curves (by considering them as limits of polygons). Observe the analogy with ex. 7 and 8.

10. *A representative special case.* You have to solve some problem about polygons with n sides. You draw a pentagon, solve the problem for it, study your solution, and notice that it works just as well in the general case, for any n, as in the special case $n = 5$. Then you may call $n = 5$ a *representative* special case: it represents to you the general case. Of course, in order to be really representative, the case $n = 5$ should have no particular simplification that could mislead you. The representative special case should *not* be simpler than the general case.

Representative special cases are often convenient in teaching. We may prove a theorem on determinants with n rows in discussing carefully a determinant with just 3 rows.

11. *An analogous case.* The problem is to design airplanes so that the danger of skull fractures in case of accident is minimized. A medical doctor, studying this problem, experiments with eggs which he smashes under various conditions. What is he doing? He has *modified* the original problem, and is studying now an *auxiliary problem*, the smashing of eggs instead of the smashing of skulls. The link between the two problems, the original and the auxiliary, is *analogy*. From a mechanical viewpoint, a man's head and a hen's egg are roughly analogous: each consists of a rigid, fragile shell containing gelatinous material.

12. If two straight lines in space are cut by three parallel planes, the corresponding segments are proportional.

In order to help you to find a proof, I ask you a question: Is there a simpler analogous theorem?

13. The four diagonals of a parallelepiped have a common point which is the midpoint of each.

Is there a simpler analogous theorem?

14. The sum of any two face angles of a trihedral angle is greater than the third face angle.

Is there a simpler analogous theorem?

15. Consider a tetrahedron as the solid that is analogous to a triangle. List the concepts of solid geometry that are analogous to the following concepts of plane geometry: *parallelogram, rectangle, square, bisector of an angle.* State a theorem of solid geometry that is analogous to the following theorem of plane geometry: *The bisectors of the three angles of a triangle meet in one point which is the center of the circle inscribed in the triangle.*

16. Consider a pyramid as the solid that is analogous to a triangle. List the solids that are analogous to the following plane figures: *parallelogram, rectangle, circle.* State a theorem of solid geometry that is analogous to the following theorem of plane geometry: *The area of a circle is equal to the area of a triangle the base of which has the same length as the perimeter of the circle and the altitude of which is the radius.*

17. Invent a theorem of solid geometry that is analogous to the following theorem of plane geometry: *The altitude of an isosceles triangle passes through the midpoint of the base.*

What solid figure do you consider as analogous to an isosceles triangle?

18. *Great analogies.* (1) The foregoing ex. 12–17 insisted on the analogy between *plane geometry* and *solid geometry.* This analogy has many aspects and is therefore often ambiguous and not always clearcut, but it is an inexhaustible source of new suggestions and new discoveries.

(2) Numbers and figures are not the only objects of mathematics. Mathematics is basically inseparable from logic, and it deals with all objects which may be objects of an exact theory. Numbers and figures are, however, the most usual objects of mathematics, and the mathematician likes to illustrate facts about numbers by properties of figures and facts about figures by properties of numbers. Hence, there are countless aspects of the analogy between *numbers* and *figures.* Some of these aspects are very clear. Thus, in analytic geometry we study well-defined correspondences between algebraic and geometric objects and relations. Yet the variety of geometric figures is inexhaustible, and so is the variety of possible operations on numbers, and so are the possible correspondences between these varieties.

(3) The study of limits and limiting processes introduces another kind of analogy which we may call the analogy between the *infinite* and the *finite.* Thus, infinite series and integrals are in various ways analogous to the finite sums whose limits they are; the differential calculus is analogous to the calculus of finite differences; differential equations, especially linear and homogeneous differential equations, are somewhat analogous to algebraic equations, and so forth. An important, relatively recent, branch of mathematics is the theory of integral equations; it gives a surprising and beautiful answer to the question: What is the analogue, in the integral calculus, of a system of n linear equations with n unknowns? The analogy between the infinite and the finite is particularly challenging because it has characteristic difficulties and pitfalls. It may lead to discovery or error; see ex. 46.

(4) Galileo, who discovered the parabolic path of projectiles and the quantitative laws of their motion, was also a great discoverer in astronomy. With his newly invented telescope, he discovered the satellites of Jupiter. He noticed that these satellites circling the planet Jupiter are analogous to the moon circling the earth and also analogous to the planets circling the sun. He also discovered the phases of the planet Venus and noticed their similarity with the phases of the moon. These discoveries were received as a great confirmation of Copernicus's heliocentric theory, hotly debated at that time. It is strange that Galileo failed to consider the analogy between

the motion of heavenly bodies and the motion of projectiles, which can be seen quite intuitively. The path of a projectile turns its concave side towards the earth, and so does the path of the moon. Newton insisted on this analogy: " . . . a stone that is projected is by the pressure of its own weight forced out of the rectilinear path, which by the initial projection alone it should have pursued, and made to describe a curved line in the air, and . . . at last brought down to the ground; and the greater the

Fig. 2.4. From the path of the stone to the path of the moon. From Newton's *Principia*

velocity is with which it is projected, the farther it goes before it falls to the earth. We may therefore suppose the velocity to be so increased, that it would describe an arc of 1, 2, 5, 10, 100, 1000 miles before it arrived at the earth, till at last, exceeding the limits of the earth, it should pass into space without touching it."[6] See fig. 2.4.

Varying continuously, the path of the stone goes over into the path of the moon. And as the stone and the moon are to the earth, so are the satellites to Jupiter, or Venus and the other planets to the sun. Without visualizing this analogy, we can only very imperfectly understand Newton's discovery of universal gravitation, which we may still regard as the greatest scientific discovery ever made.

---

[6] Sir Isaac Newton's *Mathematical Principles of Natural Philosophy and his System of the World.* Translated by Motte, revised by Cajori. Berkeley, 1946; see p. 551.

19. *Clarified analogies.* Analogy is often vague. The answer to the question, what is analogous to what, is often ambiguous. The vagueness of analogy need not diminish its interest and usefulness; those cases, however, in which the concept of analogy attains the clarity of logical or mathematical concepts deserve special consideration.

(1) Analogy is similarity of relations. The similarity has a clear meaning if the *relations are governed by the same laws.* In this sense, the addition of numbers is analogous to the multiplication of numbers, in so far as addition and multiplication are subject to the same rules. Both addition and multiplication are commutative and associative,

$$a + b = b + a, \qquad ab = ba,$$

$$(a + b) + c = a + (b + c), \quad (ab)c = a(bc).$$

Both admit an inverse operation; the equations

$$a + x = b, \qquad ax = b$$

are similar, in so far as each admits a solution, and no more than one solution. (In order to be able to state the last rule without exceptions we must admit negative numbers when we consider addition, and we must exclude the case $a = 0$ when we consider multiplication.) In this connection subtraction is analogous to division; in fact, the solutions of the above equations are

$$x = b - a, \qquad x = \frac{b}{a},$$

respectively. Then, the number 0 is analogous to the number 1; in fact, the addition of 0 to any number, as the multiplication by 1 of any number, does not change that number,

$$a + 0 = a, \qquad a \cdot 1 = a.$$

These laws are the same for various classes of numbers; we may consider here rational numbers, or real numbers, or complex numbers. In general, *systems of objects subject to the same fundamental laws* (or axioms) may be considered as analogous to each other, and this kind of analogy has a completely clear meaning.

(2) The addition of *real* numbers is analogous to the multiplication of *positive* numbers in still another sense. Any real number r is the logarithm of some positive number p,

$$r = \log p.$$

(If we consider ordinary logarithms, $r = -2$ if $p = 0.01$.) By virtue of this relation, to each positive number corresponds a perfectly determined real number, and to each real number a perfectly determined positive number. In this correspondence the addition of real numbers corresponds to the multiplication of positive numbers. If

$$r = \log p, \quad r' = \log p', \quad r'' = \log p'',$$

then any of the following two relations implies the other:

$$r + r' = r'', \quad pp' = p''.$$

The formula on the left and that on the right tell the same story in two different languages. Let us call one of the coordinated numbers the translation of the other; for example, let us call the real number r (the logarithm of p) the *translation* of p, and p the *original* of r. (We could have interchanged the words "translation" and "original," but we had to choose, and having chosen, we stick to our choice.) In this terminology addition appears as the translation of multiplication, subtraction as the translation of division, 0 as the translation of 1, the commutative law and associative law for the addition of real numbers are conceived as translations of these laws for the multiplication of positive numbers. The translation is, of course, different from the original, but it is a correct translation in the following sense: from any relation between the original elements, we can conclude with certainty the corresponding relation between the corresponding elements of the translation, and *vice versa*. Such a correct translation, that is a *one-to-one correspondence that preserves the laws of certain relations*, is called *isomorphism* in the technical language of the mathematician. Isomorphism is a fully clarified sort of analogy.

(3) A third sort of fully clarified analogy is what the mathematicians call in technical language *homomorphism* (or *merohedral isomorphism*). It would take too much time to discuss an example sufficiently, or to give an exact description, but we may try to understand the following approximate description. Homomorphism is a kind of *systematically abridged translation*. The original is not only translated into another language, but also abridged so that what results finally from translation and abbreviation is uniformly, systematically condensed into one-half or one-third or some other fraction of the original extension. Subtleties may be lost by such abridgement but everything that is in the original is represented by something in the translation, and, on a reduced scale, the relations are preserved.

20. *Quotations.*

"Let us see whether we could, by chance, conceive some other general problem that contains the original problem and is easier to solve. Thus, when we are seeking the tangent at a given point, we conceive that we are just seeking a straight line which intersects the given curve in the given point and in another point that has a given distance from the given point. After having solved this problem, which is always easy to solve by algebra, we find the case of the tangent as a special case, namely, the special case in which the given distance is minimal, reduces to a point, vanishes." (LEIBNITZ)

"As it often happens, the general problem turns out to be easier than the special problem would be if we had attacked it directly." (P. G. LEJEUNE-DIRICHLET, R. DEDEKIND)

"[It may be useful] to reduce the genus to its several species, also to a few species. Yet the most useful is to reduce the genus to just one minimal species." (LEIBNITZ)

"It is proper in philosophy to consider the similar, even in things far distant from each other." (ARISTOTLE)

"Comparisons are of great value in so far as they reduce unknown relations to known relations.

"Proper understanding is, finally, a grasping of relations (un saisir de rapports). But we understand a relation more distinctly and more purely when we recognize it as the same in widely different cases and between completely heterogeneous objects." (ARTHUR SCHOPENHAUER)

You should not forget, however, that there are two kinds of generalizations. One is cheap and the other is valuable. It is easy to generalize by *diluting*; it is important to generalize by *condensing*. To dilute a little wine with a lot of water is cheap and easy. To prepare a refined and condensed extract from several good ingredients is much more difficult, but valuable. Generalization by condensing compresses into one concept of wide scope several ideas which appeared widely scattered before. Thus, the Theory of Groups reduces to a common expression ideas which were dispersed before in Algebra, Theory of Numbers, Analysis, Geometry, Crystallography, and other domains. The other sort of generalization is more fashionable nowadays than it was formerly. It dilutes a little idea with a big terminology. The author usually prefers to take even that little idea from somebody else, refrains from adding any original observation, and avoids solving any problem except a few problems arising from the difficulties of his own terminology. It would be very easy to quote examples, but I don't want to antagonize people.[7]

Second Part

The examples and comments of this second part are all connected with sect. 6 and each other. Many of them refer directly or indirectly to ex. 21, which should be read first.

21. *The conjecture E.* We regard the equation

$$\sin x = x\left(1 - \frac{x^2}{\pi^2}\right)\left(1 - \frac{x^2}{4\pi^2}\right)\left(1 - \frac{x^2}{9\pi^2}\right)\dots$$

as a conjecture; we call it the "conjecture E." Following Euler, we wish to investigate this conjecture inductively.

---

[7] Cf. G. Pólya and G. Szegö, *Aufgaben und Lehrsätze aus der Analysis*, vol. 1, p. VII.

Inductive investigation of a conjecture involves confronting its consequences with the facts. We shall often "predict from E and verify." "Predicting from E" means deriving under the assumption that E is true, "verifying" means deriving without this assumption. A fact "agrees with E" if it can be (easily) derived from the assumption that E is true.

In the following we take for granted the elements of the calculus (which, from the formal side, were completely known to Euler at the time of his discovery) including the rigorous concept of limits (about which Euler never attained full clarity). We shall use only limiting processes which can be justified (most of them quite easily) but we shall not enter into detailed justifications.

22. We know that $\sin(-x) = -\sin x$. Does this fact agree with E?

23. Predict from E and verify the value of the infinite product

$$\left(1-\frac{1}{4}\right)\left(1-\frac{1}{9}\right)\left(1-\frac{1}{16}\right)\cdots\left(1-\frac{1}{n^2}\right)\cdots .$$

24. Predict from E and verify the value of the infinite product

$$\left(1-\frac{4}{9}\right)\left(1-\frac{4}{16}\right)\left(1-\frac{4}{25}\right)\cdots\left(1-\frac{4}{n^2}\right)\cdots .$$

25. Compare ex. 23 and 24, and generalize.

26. Predict from E the value of the infinite product

$$\frac{2\cdot 4}{3\cdot 3}\cdot\frac{4\cdot 6}{5\cdot 5}\cdot\frac{6\cdot 8}{7\cdot 7}\cdot\frac{8\cdot 10}{9\cdot 9}\cdots .$$

27. Show that the conjecture E is equivalent to the statement

$$\frac{\sin \pi z}{\pi} = \lim_{n=\infty}\frac{(z+n)\ldots(z+1)z(z-1)\ldots(z-n)}{(-1)^n(n!)^2}.$$

28. We know that $\sin(x+\pi) = -\sin x$. Does this fact agree with E?

29. The method of sect. 6 (2) leads to the conjecture

$$\cos x = \left(1-\frac{4x^2}{\pi^2}\right)\left(1-\frac{4x^2}{9\pi^2}\right)\left(1-\frac{4x^2}{25\pi^2}\right)\cdots .$$

Show that this is not only analogous to, but a consequence of, the conjecture E.

30. We know that

$$\sin x = 2\sin(x/2)\cos(x/2).$$

Does this fact agree with E?

31. Predict from E and verify the value of the infinite product

$$\left(1-\frac{4}{1}\right)\left(1-\frac{4}{9}\right)\left(1-\frac{4}{25}\right)\left(1-\frac{4}{49}\right)\cdots .$$

32. Predict from E and verify the value of the infinite product

$$\left(1-\frac{16}{1}\right)\left(1-\frac{16}{9}\right)\left(1-\frac{16}{25}\right)\left(1-\frac{16}{49}\right)\cdots .$$

33. Compare ex. 31 and 32, and generalize.

34. We know that $\cos(-x) = \cos x$. Does this fact agree with E?

35. We know that $\cos(x+\pi) = -\cos x$. Does this fact agree with E?

36. Derive from E the product for $1 - \sin x$ conjectured in sect. 6 (4).

37. Derive from E that

$$\cot x = \cdots + \frac{1}{x+2\pi} + \frac{1}{x+\pi} + \frac{1}{x} + \frac{1}{x-\pi} + \frac{1}{x-2\pi} + \cdots .$$

38. Derive from E that

$$\begin{aligned}\cot x = \frac{1}{x} &- \frac{2x}{\pi^2}\left(1+\frac{1}{4}+\frac{1}{9}+\frac{1}{16}+\frac{1}{25}+\cdots\right)\\ &- \frac{2x^3}{\pi^4}\left(1+\frac{1}{16}+\frac{1}{81}+\frac{1}{256}+\frac{1}{625}+\cdots\right)\\ &- \frac{2x^5}{\pi^6}\left(1+\frac{1}{64}+\frac{1}{729}+\cdots\right)\\ &- \cdots\end{aligned}$$

and find the sum of the infinite series appearing as coefficients on the right hand side.

39. Derive from E that

$$\begin{aligned}\frac{\cos x}{1-\sin x} &= \cot\left(\frac{\pi}{4}-\frac{x}{2}\right)\\ &= -2\left(\frac{1}{x-\frac{\pi}{2}}+\frac{1}{x+\frac{3\pi}{2}}+\frac{1}{x-\frac{5\pi}{2}}+\frac{1}{x+\frac{7\pi}{2}}+\cdots\right)\\ &= \frac{4}{\pi}\left(1-\frac{1}{3}+\frac{1}{5}-\frac{1}{7}+\frac{1}{9}-\cdots\right)\\ &+ \frac{8x}{\pi^2}\left(1+\frac{1}{9}+\frac{1}{25}+\frac{1}{49}+\frac{1}{81}\cdots\right)\\ &+ \frac{16x^2}{\pi^3}\left(1-\frac{1}{27}+\frac{1}{125}-\frac{1}{343}+\cdots\right)\\ &+ \frac{32x^3}{\pi^4}\left(1+\frac{1}{81}+\frac{1}{625}+\cdots\right)\\ &+ \cdots\end{aligned}$$

and find the sum of the infinite series appearing as coefficients in the last expression.

40. Show that

$$1 + \frac{1}{4} + \frac{1}{9} + \frac{1}{16} + \frac{1}{25} + \cdots = \frac{4}{3}\left(1 + \frac{1}{9} + \frac{1}{25} + \frac{1}{49} + \cdots\right)$$

which yields a second derivation for the sum of the series on the left.

41 (continued). Try to find a third derivation, knowing that

$$\arcsin x = x + \frac{1}{2}\frac{x^3}{3} + \frac{1}{2}\frac{3}{4}\frac{x^5}{5} + \frac{1}{2}\frac{3}{4}\frac{5}{6}\frac{x^7}{7} + \cdots$$

and that, for $n = 0, 1, 2, \ldots$,

$$\int_0^1 (1 - x^2)^{-1/2} x^{2n+1}\, dx = \int_0^{\pi/2} (\sin t)^{2n+1}\, dt = \frac{2 \cdot 4 \cdots 2n}{3 \cdot 5 \cdots (2n+1)}.$$

42 (continued). Try to find a fourth derivation, knowing that

$$(\arcsin x)^2 = x^2 + \frac{2}{3}\frac{x^4}{2} + \frac{2}{3}\frac{4}{5}\frac{x^6}{3} + \frac{2}{3}\frac{4}{5}\frac{6}{7}\frac{x^8}{4} + \cdots$$

and that, for $n = 0, 1, 2, \ldots$

$$\int_0^1 (1 - x^2)^{-1/2} x^{2n}\, dx = \int_0^{\pi/2} (\sin t)^{2n}\, dt = \frac{1}{2}\frac{3}{4} \cdots \frac{2n-1}{2n}\frac{\pi}{2}.$$

43. Euler (*Opera Omnia*, ser. 1, vol. 14, p. 40–41) used the formula

$$1 + \frac{1}{4} + \frac{1}{9} + \frac{1}{16} + \cdots$$

$$= \log x \cdot \log (1 - x) + \frac{x + (1 - x)}{1} + \frac{x^2 + (1 - x)^2}{4} + \frac{x^3 + (1 - x)^3}{9} + \cdots,$$

valid for $0 < x < 1$, to compute numerically the sum of the series on the left hand side.

(a) Prove the formula.

(b) Which value of x is the most advantageous in computing the sum on the left?

44. *An objection and a first approach to a proof.* There is no reason to admit *a priori* that $\sin x$ can be decomposed into linear factors corresponding to the roots of the equation

$$\sin x = 0.$$

Yet even if we should admit this, there remains an objection: Euler did *not* prove that

$$0, \quad \pi, \quad -\pi, \quad 2\pi, \quad -2\pi, \quad 3\pi, \quad -3\pi, \quad \ldots$$

are *all* the roots of this equation. We can satisfy ourselves (by discussing the curve $y = \sin x$) that there are no other real roots, yet Euler did by no means exclude the existence of complex roots.

This objection was raised by Daniel Bernoulli (a son of Jean, 1700–1788). Euler answered it by considering

$$\sin x = (e^{ix} - e^{-ix})/(2i) = \lim_{n \to \infty} P_n(x)$$

where

$$P_n(x) = \frac{1}{2i}\left[\left(1 + \frac{ix}{n}\right)^n - \left(1 - \frac{ix}{n}\right)^n\right]$$

is a polynomial (of degree n if n is odd).

Show that $P_n(x)$ has no complex roots.

45. *A second approach to a proof.* Assuming that n is odd in ex. 44, factorize $P_n(x)/x$ so that its k-th factor approaches

$$1 - \frac{x^2}{k^2\pi^2}$$

as n tends to ∞, for any fixed k ($k = 1, 2, 3, \ldots$).

46. *Dangers of analogy.* In short, the analogy between the finite and the infinite led Euler to a great discovery. Yet he skirted a fallacy. Here is an example showing the danger on a smaller scale.

The series

$$1 - \frac{1}{2} + \frac{1}{3} - \frac{1}{4} + \frac{1}{5} - \frac{1}{6} + \frac{1}{7} - \frac{1}{8} + \ldots = l$$

converges. Its sum l can be roughly estimated by the first two terms:

$$1/2 < l < 1.$$

Now

$$2l = \frac{2}{1} - \frac{1}{1} + \frac{2}{3} - \frac{1}{2} + \frac{2}{5} - \frac{1}{3} + \frac{2}{7} - \frac{1}{4} + \ldots .$$

In this series, there is just one term with a given even denominator (it is negative), but two terms with a given odd denominator (one positive, and the other negative). Let us bring together the terms with the same odd denominator:

$$\begin{aligned}
&\frac{2}{1} - \frac{1}{2} + \frac{2}{3} - \frac{1}{4} + \frac{2}{5} - \ldots \\
&-\frac{1}{1} \qquad\; -\frac{1}{3} \qquad\; -\frac{1}{5} \\
&= 1 - \frac{1}{2} + \frac{1}{3} - \frac{1}{4} + \frac{1}{5} - \ldots \\
&= l.
\end{aligned}$$

Yet $2l \neq l$, since $l \neq 0$. Where is the mistake and how can you protect yourself from repeating it?

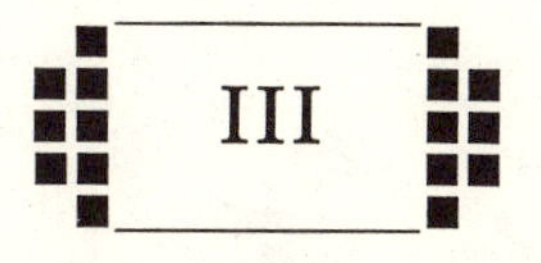

INDUCTION IN SOLID GEOMETRY

Even in the mathematical sciences, our principal instruments to discover the truth are induction and analogy.—LAPLACE[1]

1. Polyhedra. "A complicated polyhedron has many faces, corners, and edges." Some vague remark of this sort comes easily to almost anybody who has had some contact with solid geometry. Not so many people will, however, make a serious effort to deepen this remark and seek some more precise information behind it. The right thing to do is to distinguish clearly the quantities involved and to ask some definite question. Let us denote, therefore, the number of faces, the number of vertices and the number of edges of the polyhedron by F, V, and E, respectively (corresponding initials), and let us ask a clear question as: "Is it generally true that the number of faces increases when the number of vertices increases? Does F necessarily increase with V?"

To begin with, we can scarcely do anything better than examine examples, particular polyhedra. Thus, for a cube (the solid I in fig. 3.1)

$$F = 6, \quad V = 8, \quad E = 12.$$

Or, for a prism with triangular base (the solid II in fig. 3.1)

$$F = 5, \quad V = 6, \quad E = 9.$$

Once launched in this direction, we naturally examine and compare various solids, for example, those exhibited in fig. 3.1 which are, besides No. I and No. II already mentioned, the following: a prism with pentagonal base (No. III), pyramids with square, triangular, or pentagonal base (Nos. IV, V, VI), an octahedron (No. VII), a "tower with roof" (No. VIII; a pyramid is placed upon the upper face of a cube as base), and a "truncated cube" (No. IX). Let us make a little effort of imagination and represent

[1] Essai philosophique sur les probabilités; see *Oeuvres complètes de Laplace*, vol. 7, p. V.

these solids, one after the other, clearly enough to count faces, vertices, and edges. The numbers found are listed in the following table.

	Polyhedron	*F*	*V*	*E*
I.	cube	6	8	12
II.	triangular prism	5	6	9
III.	pentagonal prism	7	10	15
IV.	square pyramid	5	5	8
V.	triangular pyramid	4	4	6
VI.	pentagonal pyramid	6	6	10
VII.	octahedron	8	6	12
VIII.	"tower"	9	9	16
IX.	"truncated cube"	7	10	15

Our fig. 3.1 has some superficial similarity with a mineralogical display, and the above table is somewhat similar to the notebook in which the physicist enters the results of his experiments. We examine and compare our figures and the numbers in our table as the mineralogist or the physicist would examine and compare their more laboriously collected specimens and data. We now have something in hand that could answer our original question: "Does *V* increase with *F*?" In fact, the answer is "No"; comparing the cube and the octahedron (Nos. I and VII) we see, that one has more vertices and the other more faces. Thus, our first attempt at establishing a thoroughgoing regularity failed.

We can, however, try something else. Does *E* increase with *F*? Or with *V*? To answer these questions systematically, we rearrange our table. We dispose our polyhedra so that *E* increases when we read down the successive items:

Polyhedron	*F*	*V*	*E*
triangular pyramid	4	4	6
square pyramid	5	5	8
triangular prism	5	6	9
pentagonal pyramid	6	6	10
cube	6	8	12
octahedron	8	6	12
pentagonal prism	7	10	15
"truncated cube"	7	10	15
"tower"	9	9	16

Looking at our more conveniently arranged data, we can easily observe that no regularity of the suspected kind exists. As *E* increases from 15 to 16, *V* drops from 10 to 9. Again, as we pass from the octahedron to the pentagonal prism, *E* increases from 12 to 15 but *F* drops from 8 to 7. Neither *F* nor *V* increases steadily with *E*.

We again failed in finding a generally valid regularity. Yet we do not like to admit that our original idea was completely wrong. Some modification of our idea may still be right. Neither F nor V increases with E, it is true, but they appear to increase "on the whole." Examining our

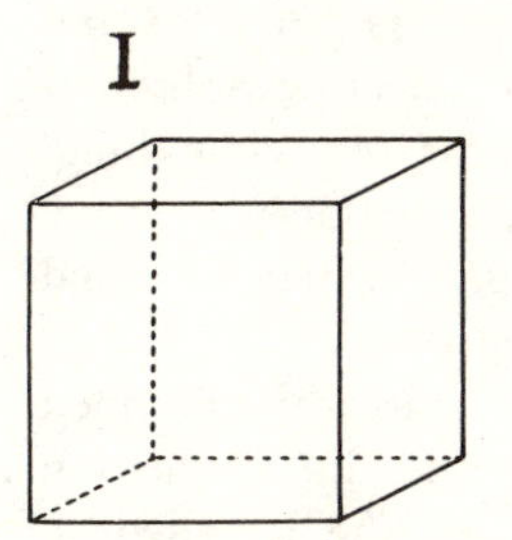

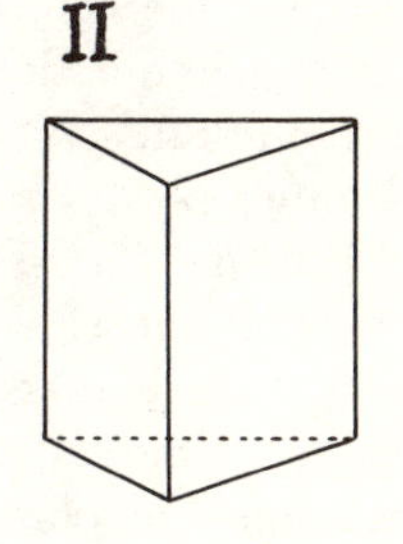

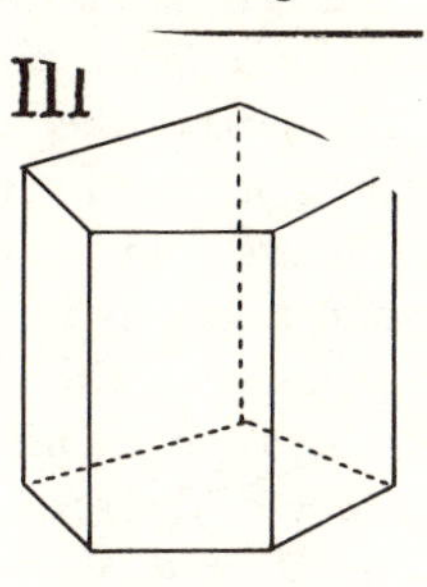

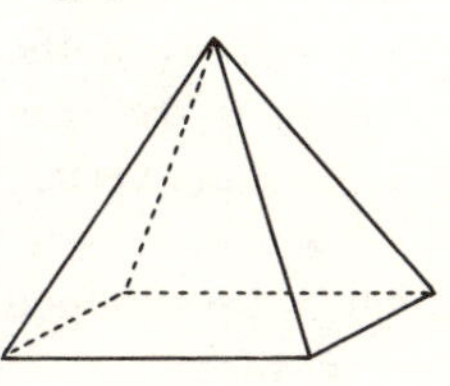

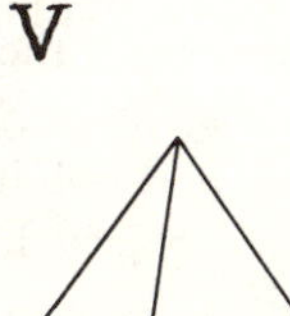

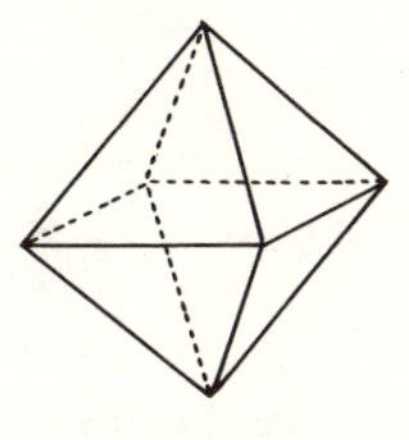

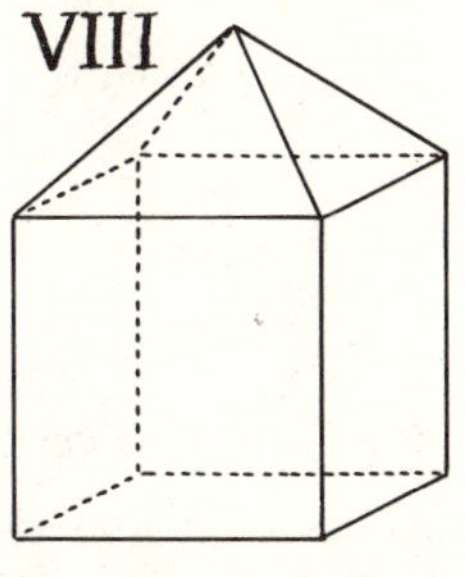

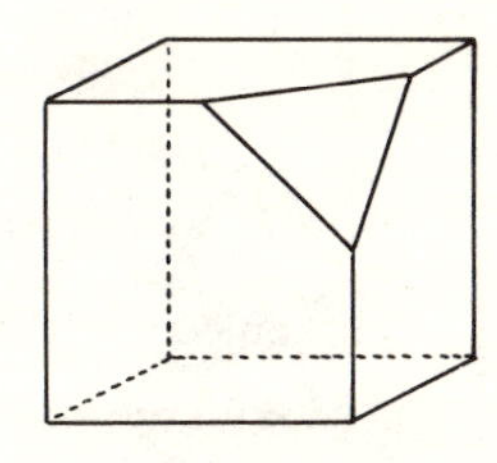

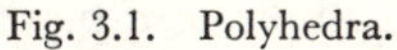

Fig. 3.1. Polyhedra.

well-arranged data, we may observe that F and V increase "jointly": $F + V$ *increases* as we read down the lines. And then a more precise regularity may strike us: throughout the table

$$F + V = E + 2.$$

This relation is verified in all nine cases listed in our table. It seems unlikely that such a persistent regularity should be mere coincidence.

Thus, we are led to the *conjecture* that, not only in the observed cases, but in any polyhedron the *number of faces increased by the number of vertices is equal to the number of edges increased by two.*

2. First supporting contacts. A well-trained naturalist does not easily admit a conjecture. Even if the conjecture appears plausible and has been verified in some cases, he will question it and collect new observations or design new experiments to test it. We may do the very same thing. We are going to examine still other polyhedra, count their faces, vertices, and edges, and compare $F + V$ to $E + 2$. These numbers may be equal or not. It will be interesting to find out which is the case.

Looking at fig. 3.1, we may observe that we have already examined three of the regular solids, the cube, the tetrahedron, and the octahedron (I, V, and VII). Let us examine the remaining two, the icosahedron and the dodecahedron.

The icosahedron has 20 faces, all triangles, and so $F = 20$. The 20 triangles have together $3 \times 20 = 60$ sides of which 2 coincide in the same edge of the icosahedron. Therefore, the number of edges is $60/2 = 30 = E$. We can find V analogously. We know that 5 faces of the icosahedron are grouped around each of its vertices. The 20 triangles have together $3 \times 20 = 60$ angles, of which 5 belong to the same vertex. Therefore, the number of vertices is $60/5 = 12 = V$.

The dodecahedron has 12 faces, all pentagons, of which 3 are grouped around each vertex. We conclude hence, similarly as before, that

$$F = 12, \quad V = \frac{12 \times 5}{3} = 20, \quad E = \frac{12 \times 5}{2} = 30.$$

We can now add to our list on p. 36 two more lines:

Polyhedron	F	V	E
icosahedron . .	20	12	30
dodecahedron . .	12	20	30

Our conjecture, that $F + V = E + 2$, is verified in both cases.

3. More supporting contacts. Thanks to the foregoing verifications, our conjecture became perceptibly more plausible; but is it proved now? By no means. In a similar situation, a conscientious naturalist would feel satisfaction over the success of his experiments, but would go on devising further experiments. Which polyhedron should we test now?

The point is that our conjecture is so well verified by now that verification in just one more instance would add only little to our confidence, so little perhaps that it would be scarcely worth the trouble of choosing a polyhedron and counting its parts. Could we find some more worthwhile way of testing our conjecture?

Looking at fig. 3.1, we may observe that all solids in the first line are of the same nature; they are prisms. Again, all solids in the second line are pyramids. Our conjecture is certainly true of the three prisms and the three pyramids shown in fig. 3.1; but is it true of *all* prisms and pyramids?

If a prism has n lateral faces, it has $n + 2$ faces in all, $2n$ vertices and $3n$ edges. A pyramid with n lateral faces has $n + 1$ faces in all, $n + 1$ vertices and $2n$ edges. Thus, we may add two more lines to our list on p. 36:

Polyhedron	F	V	E
Prism with n lateral faces . .	$n + 2$	$2n$	$3n$
Pyramid with n lateral faces .	$n + 1$	$n + 1$	$2n$

Our conjecture asserting that $F + V = E + 2$ turned out to be true not only for one or two more polyhedra but for two unlimited series of polyhedra.

4. A severe test. The last remark adds considerably to our confidence in our conjecture, but does not prove it, of course. What should we do? Should we go on testing further particular cases? Our conjecture seems to withstand simple tests fairly well. Therefore we should submit it to some severe, searching test that stands a good chance to refute it.

Let us look again at our collection of polyhedra (fig. 3.1). There are prisms (I, II, III), pyramids (IV, V, VI), regular solids (I, V, VII); yet we have already considered all these kinds of solids exhaustively. What else is there? Fig. 3.1 contains also the "tower" (No. VIII) which is obtained by placing a "roof" on the top of a cube. Here we may perceive the possibility of a generalization. We take any polyhedron instead of the cube, choose any face of the polyhedron, and place a "roof" on it. Let the original polyhedron have F faces, V vertices, and E edges, and let its face chosen have n sides. We place on this face a pyramid with n lateral faces and so obtain a new polyhedron. How many faces, vertices, and edges has the new "roofed" polyhedron? One face (the chosen one) is lost in the process, and n new ones are won (the n lateral faces of the pyramid) so that the new polyhedron has $F - 1 + n$ faces. All vertices of the polyhedron belong also to the new one, but one vertex is added (the summit of the pyramid) and so the new polyhedron has $V + 1$ vertices. Again, all edges of the old polyhedron belong also to the new one, but n edges are added (the lateral edges of the pyramid) and so the new polyhedron has $E + n$ edges.

Let us summarize. The original polyhedron had F, V, and E faces, vertices, and edges, respectively, whereas the new "roofed" polyhedron has

$$F + n - 1, \quad V + 1, \quad \text{and} \quad E + n$$

parts of the corresponding kind. Are these facts consistent with our conjecture?

If the relation $F + V = E + 2$ holds, then, obviously,

$$(F + n - 1) + (V + 1) = (E + n) + 2$$

holds *also*. That is, if our conjecture happens to be verified in the case of the original polyhedron, it must be verified also in the case of the new "roofed" polyhedron. Our conjecture survives the "roofing," and so it has passed a very severe test, indeed. There is such an inexhaustible variety of polyhedra which we can derive from those already examined by repeated "roofing," and we have proved that our conjecture is true for all of them.

By the way, the last solid of our fig. 3.1, the "truncated cube" (IX), opens the way to a similar consideration. Instead of the cube, let us "truncate" any polyhedron, cutting off an arbitrarily chosen vertex. Let the original polyhedron have

$$F,\ V, \text{ and } E$$

faces, vertices, and edges, respectively, and let n be the number of the edges radiating from the vertex we have chosen. Cutting off this vertex we introduce 1 new face (which has n sides), n new edges, and also n new vertices, but we lose 1 vertex. To sum up, the new "truncated" polyhedron has

$$F+1, \quad V+n-1, \quad \text{and} \quad E+n$$

faces, vertices, and edges, respectively. Now, from

$$F+V=E+2$$

follows

$$(F+1)+(V+n-1)=(E+n)+2.$$

That is, our conjecture is tenacious enough to survive the "truncating." It has passed another severe test.

It is natural to regard the foregoing remarks as a very strong argument for our conjecture. We can perceive in them even something else: the first hint of a proof. Starting from some simple polyhedron, as the tetrahedron or the cube, for which the conjecture holds, we can derive by roofing and truncating a vast variety of other polyhedra for which the conjecture also holds. Could we derive *all* polyhedra? Then we would have a proof! Besides, there may be other operations which, like truncating and roofing, preserve the conjectural relation.

5. Verifications and verifications. The mental procedures of the trained naturalist are not essentially different from those of the common man, but they are more thorough. Both the common man and the scientist are led to conjectures by a few observations and they are both paying attention to later cases which could be in agreement or not with the conjecture. A case in agreement makes the conjecture more likely, a conflicting case disproves it, and here the difference begins: Ordinary people are usually more apt to look for the first kind of cases, but the scientist looks for the second kind. The reason is that everybody has a little vanity, the common man as the scientist, but different people take pride in different things. Mr. Anybody does not like to confess, even to himself, that he was mistaken and so he does not like conflicting cases, he avoids them, he is even inclined to explain them away when they present themselves. The scientist, on the contrary, is ready enough to recognize a mistaken conjecture, but he does not like to leave questions undecided. Now, a case in agreement does not settle the question definitively, but a conflicting case does. The scientist, seeking a definitive decision, looks for cases which have a chance to upset the conjecture, and the more chance they have, the more they are welcome. There is an important point to observe. If a case which threatens to upset the conjecture turns out, after all, to be in agreement with it, the conjecture emerges greatly strengthened from the test. The more danger, the more honor; passing the most threatening examination grants the highest recognition, the strongest experimental evidence to the conjecture. There are instances and instances, verifications and verifications. An instance which is *more likely to be conflicting* brings the conjecture in any case nearer to decision than an instance which is less so, and this explains the preference of the scientist.

Now, we may get down to our own particular problem and see how the foregoing remarks apply to the "experimental research on polyhedra" that we have undertaken. Each new case in which the relation $F + V = E + 2$ is verified adds to the confidence that this relation is generally true. Yet we soon get tired of a monotonous sequence of verifications. A case little different from the previously examined cases, if it agrees with the conjecture, adds to our confidence, of course, but it adds little. In fact we easily believe, before the test, that the case at hand will behave as the previous cases from which it differs but little. We desire not only another verification, but a *verification of another kind*. In fact, looking back at the various phases of our research (sect. 2, 3, and 4), we may observe that each one yielded a kind of verification that surpassed essentially those obtained in the foregoing. Each phase verified the conjecture for a *more extensive variety of cases than the foregoing*.

6. A very different case. Variety being important, let us look for some polyhedron very different from those heretofore examined. Thus, we may hit upon the idea of regarding a picture frame as a polyhedron. We take a very long triangular rod, we cut four pieces of it, we adjust these pieces at the ends, and fit them together to a framelike polyhedron. Fig. 3.2 suggests that the frame is placed on a table so that the edges which have already been on the uncut rod all lie horizontally. There are 4 times 3, that is, 12, horizontal edges, and also 4 times 3 non-horizontal edges, so that the total number of edges is $E = 12 + 12 = 24$. Counting the faces and vertices, we find that $F = 4 \times 3 = 12$, and $V = 4 \times 3 = 12$. Now, $F + V = 24$ is different from $E + 2 = 26$. Our conjecture, taken in full generality, turned out to be false!

We can say, of course, that we have never intended to state the proposition in such generality, that we meant all the time polyhedra which are convex or, so to say, "sphere-shaped" and not polyhedra which are "doughnut-shaped" as the picture frame. But these are excuses. In fact, we have to shift our position and modify our original statement. It is quite possible that the blow that we received will be beneficial in the end and lead us eventually to an amended and more precise statement of our conjecture. Yet it was a blow to our confidence, anyway.

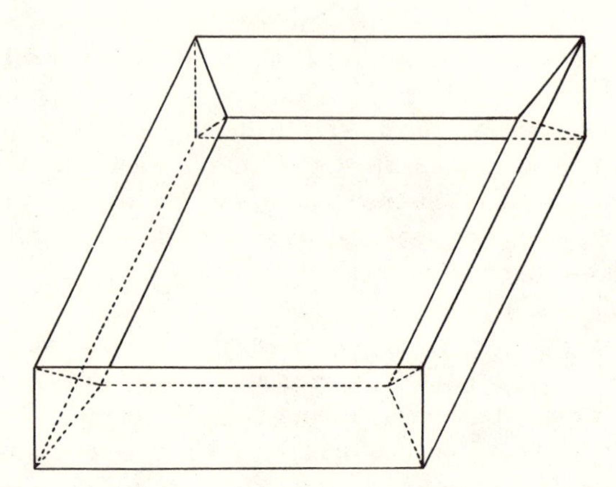

Fig. 3.2. A doughnut-shaped polyhedron.

7. Analogy. The example of the "picture frame" killed our conjecture in its original form but it can be promptly revived in a revised (and, let us hope, improved) form, with an important restriction.

The tetrahedron is convex, and so is the cube, and so are the other polyhedra in our collection (fig. 3.1), and so are all the polyhedra that we can derive from them by truncating and by "moderate" roofing (by placing *sufficiently flat* roofs on their various faces). At any rate, there is no danger that these operations could lead from a convex or "sphere-shaped" polyhedron to a "doughnut-shaped" solid.

Observing this, we introduce some much-needed precision. We conjecture that in any *convex* polyhedron the relation $F + V = E + 2$ holds between the numbers of faces, vertices, and edges. (The restriction to "sphere-shaped" polyhedra may be even preferable, but we do not wish to stop to define here the meaning of the term.)

This conjecture has some chance to be true. Nevertheless, our confidence was shaken and we look around for some new support for our conjecture. We cannot hope for much help from further verifications. It seems that we have exhausted the most obvious sources. Yet we may still hope for some help from analogy. Is there any simpler analogous case that could be instructive?

Polygons are analogous to polyhedra. A polygon is a part of a plane as a polyhedron is a part of space. A polygon has a certain number, V, of vertices (the vertices of its angles) and a certain number, E, of edges (or sides). Obviously

$$V = E.$$

Yet this relation, valid for convex polygons, appears too simple and throws little light on the more intricate relation

$$F + V = E + 2$$

which we suspect to be valid for all convex polyhedra.

If we are genuinely concerned in the question, we naturally try to bring these two relations nearer to each other. There is an ingenious way of doing so. We have to bring first the various numbers into a natural order. The polyhedron is 3-dimensional; its faces (polygons) are 2-dimensional, its edges 1-dimensional, and its vertices (points) 0-dimensional, of course. We may now rewrite our equations, arranging the quantities in the order of increasing dimensions. The relation for polygons, written in the form

$$V - E + 1 = 1,$$

becomes comparable to the relation for polyhedra, written in the form

$$V - E + F - 1 = 1.$$

The 1, on the left hand side of the equation for polygons, stands for the only two-dimensional element concerned, the interior of the polygon. The 1, on the left hand side of the equation for polyhedra, stands for the only three-dimensional element concerned, the interior of the polyhedron. The numbers, on the left hand side, counting elements of 0, 1, 2, and 3 dimensions, respectively, are disposed in this natural order, and have alternating signs. The right hand side is the same in both cases; the analogy seems complete. As the first equation, for polygons, is obviously true, the analogy adds to our confidence in the second equation, for polyhedra, which we have conjectured.

8. The partition of space. We pass now to another example of inductive research in solid geometry. In our foregoing example, we started from a general, somewhat vague remark. Our point of departure now is a particular clear-cut problem. We consider a simple but not too familiar problem of solid geometry: *Into how many parts is space divided by 5 planes?*

This question is easily answered if the five given planes are all parallel to each other, in which case space is visibly divided into 6 parts. This case, however, is too particular. If our planes are in a "general position," no two among them will be parallel and there will be considerably more parts than 6. We have to restate our problem more precisely, adding an essential clause: *Into how many parts is space divided by 5 planes, provided that these planes are in a general position?*

The idea of a "general position" is quite intuitive; planes are in such a position when they are not linked by any particular relation, when they are given independently, chosen at random. It would not be difficult to clear up the term completely by a technical definition but we shall not do so, for two reasons. First, this presentation should not be too technical. Second, leaving the notion somewhat hazy, we come nearer to the mental attitude of the naturalist who is often obliged to start with somewhat hazy notions, but clears them up as he goes ahead.

9. Modifying the problem. Let us concentrate upon our problem. We are given 5 planes in general position. They cut space into a certain number of partitions. (We may think of a cheese sliced into pieces by 5 straight cuts with a sharp knife.) We have to find the number of these partitions. (Into how many pieces is the cheese cut?)

It seems difficult to see at once all the partitions effected by the 5 planes. (It may be impossible to "see" them. At any rate, do not overstrain your geometric imagination; rather, try to think. Your reason may carry you farther than your imagination.) But why just 5 planes? Why not any number of planes? In how many parts is space divided by 4 planes? By 3 planes? Or by 2 planes? Or by just 1 plane?

We reach here cases which are accessible to our geometric intuition. One plane divides space obviously into 2 parts. Two planes divide space into 3 parts if they are parallel. We have to discard, however, this particular position; 2 planes in a general position intersect, and divide space into 4 parts. Three planes in a general position divide space into 8 parts. In order to realize this last, more difficult, case, we may think of 2 vertical walls inside a building, crossing each other, and of a horizontal layer, supported by beams, crossing both walls and forming around the point where it crosses both the ceiling of 4 rooms and the floor of 4 other rooms.

10. Generalization, specialization, analogy. Our problem is concerned with 5 planes but, instead of considering 5 planes, we first played with 1, 2, and 3 planes. Have we squandered our time? Not at all. We have prepared ourselves for our problem by examining *simpler analogous cases*. We have tried our hand at these simpler cases; we have clarified the intervening concepts and familiarized ourselves with the kind of problem we have to face.

Even the way that led us to those simpler analogous problems is typical and deserves our attention. First, we passed from the case of 5 planes to the case of any number of planes, let us say, to n planes: we *generalized*. Then, from n planes, we passed back to 4 planes, to 3 planes, to 2 planes, to just 1 plane, that is, we put $n = 4, 3, 2, 1$ in the general problem: we *specialized*. But the problem about dividing space by, let us say, 3 planes is *analogous* to our original question involving 5 planes. Thus, we have reached analogy in a typical manner, by *introductory generalization and subsequent specialization*.

11. An analogous problem. What about the next case of 4 planes? Four planes, in general position, determine various portions of space, one of which is limited, contained by four triangular faces, and is called a tetrahedron (see fig. 3.3). This configuration reminds us of three straight lines in a plane, in general position, which determine various portions of the plane, one of which is limited, contained by three line-segments, and is a triangle (see fig. 3.4). We have to ascertain the number of portions of space determined by the four planes. Let us try our hand at the simpler analogous problem: Into how many portions is the plane divided by three lines?

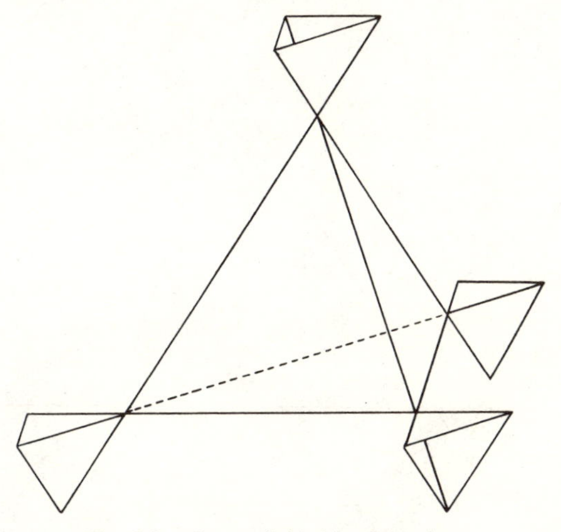

Fig. 3.3. Space divided by four planes.

Many of us will see the answer immediately, even without drawing a figure, and anybody may see it, by using a rough sketch (see fig. 3.4). The required number of parts is 7.

We have found the solution of the simpler analogous problem; but can we use this solution for our original problem? Yes, we can, if we handle the analogy of the two configurations intelligently. We ought to consider the dissection of the plane by 3 straight lines so that we may apply afterwards the same consideration to the dissection of space by 4 planes.

Thus, let us look again at the dissection of the plane by 3 lines, bounding a triangle. One division is finite, it is the interior of the triangle. And the infinite divisions have either a common side with the triangle (there are 3 such divisions), or a common vertex (there are also 3 of this kind). Thus the number of all divisions is $1 + 3 + 3 = 7$.

Now, we consider the dissection of space by 4 planes bounding a tetrahedron. One division is finite, it is the interior of the tetrahedron. An infinite division may have a common face (a 2-dimensional part of the boundary) with the tetrahedron (there are 4 such divisions), or a common edge (1-dimensional part of the boundary; there are 6 divisions of this kind) or a common vertex (0-dimensional part of the boundary; there are 4 divisions of this kind, emphasized in fig. 3.3). Thus the number of all divisions is $1 + 4 + 6 + 4 = 15$.

We have reached this result by analogy, and we have used analogy in a typical, important way. First, we devised an easier analogous problem and we solved it. Then, in order to solve the original, more difficult problem (about the tetrahedron), we used the new easier analogous problem

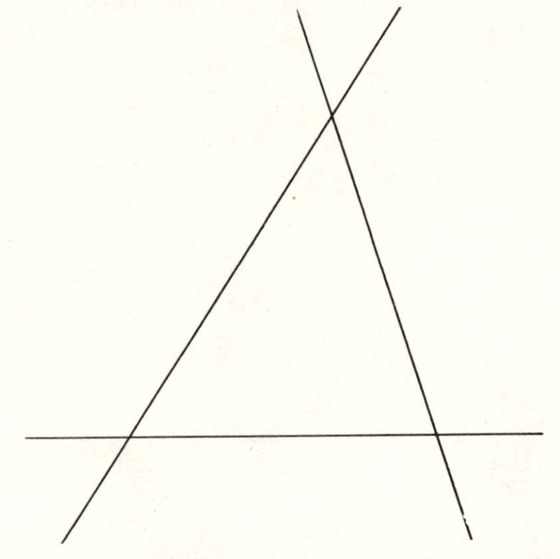

Fig. 3.4. Plane divided by three lines.

(about the triangle) as a *model*; in solving the more difficult problem we followed the pattern of the solution of the easier problem. But before doing this, we had to reconsider the solution of the easier problem. We rearranged it, remade it into a new pattern fit for imitation.

To single out an analogous easier problem, to solve it, to remake its solution so that it may serve as a model and, at last, to reach the solution of the original problem by following the model just created—this method may seem roundabout to the uninitiated, but is frequently used in mathematical and non-mathematical scientific research.

12. An array of analogous problems. Yet our original problem is still unsolved. It is concerned with the dissection of space by 5 planes. What is the analogous problem in two dimensions? Dissection by 5 straight lines? Or by 4 straight lines? It may be better for us to consider these problems in full generality, the dissection of space by n planes, and the dissection of a plane by n straight lines. These dissecting straight lines must be, of course, in general position (no 2 are parallel and no 3 meet in the same point).

If we are accustomed to use geometrical analogy, we may go one step further and consider also the division of the straight line by n different points. Although this problem seems to be rather trivial, it may be instructive. We easily see that a straight line is divided by 1 point into 2 parts, by 2 points into 3, by 3 points into 4, and, generally, by n points into $n + 1$ different parts.

Again, if we are accustomed to pay attention to extreme cases, we may consider the undivided space, plane or line, and regard it as a "division effected by 0 dividing elements."

Let us set up the following table that exhibits all our results obtained hitherto.

Number of dividing elements	Number of divisions of		
	space by planes	plane by straight lines	line by points
0	1	1	1
1	2	2	2
2	4	4	3
3	8	7	4
4	15		5
...	...	...	...
n			$n + 1$

13. Many problems may be easier than just one. We started out to solve a problem, that about the dissection of space by 5 planes. We have not yet solved it, but we set up many new problems. Each unfilled case of our table corresponds to an open question.

This procedure of heaping up new problems may seem foolish to the uninitiated. But some experience in solving problems may teach us that many problems together may be easier to solve than just one of them—if the many problems are well coordinated, and the one problem by itself is isolated. Our original problem appears now as one in an array of unsolved problems. But the point is that all these unsolved problems form an array: they are well disposed, grouped together, in close analogy with each other and with a few problems solved already. If we compare the present position of our question, well inserted into an array of analogous questions, with its original position, as it was still completely isolated, we are naturally inclined to believe that some progress has been made.

14. A conjecture. We look at the results displayed in our table as a naturalist looks at the collection of his specimens. This table is a challenge to our inventive ability, to our faculties of observation. Can we discover any connection, any regularity?

Looking at the second column (division of space by planes) we may notice the sequence 1, 2, 4, 8—there is a clear regularity; we see here the successive powers of 2. Yet, what a disappointment! The next term in the column is 15, and not 16 as we have expected. Our first guess was not so good; we must look for something else.

Eventually, we may chance upon adding two juxtaposed numbers and observe that their sum is in the table. We recognize a peculiar connection; we obtain a number of the table by adding two others, the number above it and the number to the right of the latter. For example,

$$\begin{matrix} 8 & 7 \\ 15 & \end{matrix}$$

are linked by the relation

$$8 + 7 = 15.$$

This is a remarkable connection, a striking clue. It seems unlikely that this connection which we can observe throughout the whole table so far calculated should result from mere chance.

Thus, the situation suggests that the regularity observed extends beyond the limits of our observation, that the numbers of the table not yet calculated are connected in the same way as those already calculated, and so we are led to conjecture that the law we chanced upon is generally valid.

If this is so, however, we can solve our original problem. By adding juxtaposed numbers we can extend our table till we reach the number we wished to obtain:

0	1	1	1
1	2	2	2
2	4	4	3
3	8	7	4
4	15	**11**	5
5	**26**		

In the table as it is reprinted here two new numbers appear in heavy type, computed by addition, $11 = 7 + 4$, $26 = 15 + 11$. If our guess is correct, 26 should be the number of portions into which space is dissected by 5 planes in general position. We have solved the proposed problem, it seems. Or, at least, we have succeeded in hitting on a plausible conjecture supported by all the evidence heretofore collected.

15. Prediction and verification. In the foregoing we have followed exactly the typical procedure of the naturalist. If a naturalist observes a striking regularity, which cannot be reasonably attributed to mere chance, he conjectures that the regularity extends beyond the limits of his actual observations. Making such a conjecture is often the decisive step in inductive research.

The following step may be prediction. On the basis of his former observations and their concordance with conjectural law, the naturalist predicts the result of his next observation. Much depends on the outcome of that next observation. Will the prediction turn out to be true or not? We are very much in the same position. We have found, or, rather, guessed or

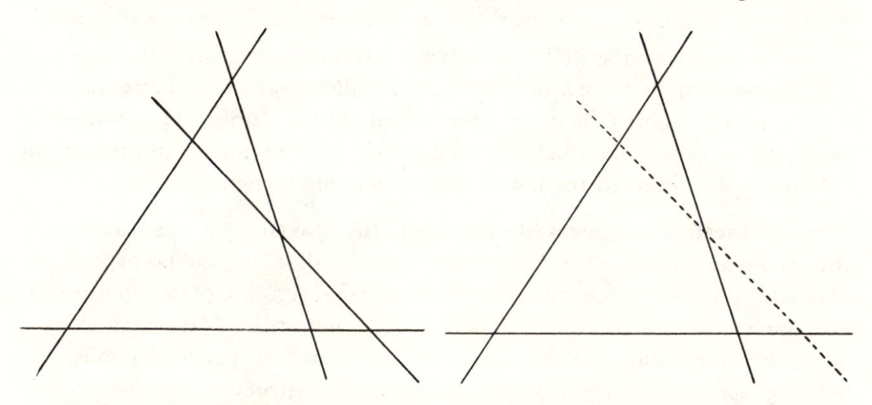

Fig. 3.5. Plane divided by four lines. Fig. 3.6. Transition from three lines to four.

predicted that 11 should be the number of regions into which a plane is dissected by 4 straight lines in general position. Is that so? Is our prediction correct?

Examining a rough sketch (see fig. 3.5) we can convince ourselves that our guess was good, that 11 *is* actually the correct number. This confirmation of our prediction yields inductive evidence in favor of the rule on the basis of which we made our prediction. Having passed the test successfully, our conjecture comes out strengthened.

16. Again and better. We have verified that number 11 by looking at the figure and counting. Yes, 4 lines in general position seem to divide the plane into 11 portions. But let us do it again and do it better. We have counted those portions in some way. Let us count them again and count them so that we should be certain of avoiding confusion and miscounts and traps set by special positions.

Let us start from the fact that 3 lines determine exactly 7 portions of the plane. We have some reasons to believe that 4 lines determine 11 portions. Why just 4 portions more? Why does the number 4 intervene in this connection? Why does the introduction of a new line increase the number of portions just by 4?

We emphasize one line in fig. 3.5, we redraw it in short strokes (see fig. 3.6). The new figure does not look very different but it expresses a very different conception. We regard the emphasized line as *new* and the three other lines as *old*. The old lines cut the plane into 7 portions. What happens when a new line is added?

The new line, drawn at random, must intersect each old line and each one in a different point. That makes 3 points. These 3 points divide the new line into 4 segments. Now each segment bisects an old division of the plane, makes two new divisions out of an old one. Taken together, the 4 segments of the new line create 8 new divisions and abolish 4 old divisions —the *number* of divisions increases by just 4. *This* is the reason that the number of divisions now is just 4 more than it was before: $7 + 4 = 11$.

This way of arriving at the number 11 is convincing and illuminating. We may begin now to see a reason for the regularity which we have observed and on which we have based our prediction of that number 11. We begin to suspect an explanation behind the facts and our confidence in the general validity of the observed regularity is greatly strengthened.

17. Induction suggests deduction; the particular case suggests the general proof. We have been careful all along to point out the parallelism between our reasoning and the procedures of the naturalist. We started from a special problem as the naturalist may start from a puzzling observation. We advanced by tentative generalizations, by noticing accessible special cases, by observing instructive analogies. We tried to guess some regularity and failed, we tried again and did better. We succeeded in conjecturing a general law which is supported by all experimental evidence at our disposal. We put one more special case to the test and found concordance with the conjectured law, the authority of which gained by such verification. At last, we began to see a reason for that general law, a sort of explanation, and our confidence was greatly strengthened. A naturalist's research may pass through exactly the same phases.

There is, however, a parting of the ways at which the mathematician turns sharply away from the naturalist. Observation is the highest authority for the naturalist, but not for the mathematician. Verification in many well-chosen instances is the only way of confirming a conjectural law in the natural sciences. Verification in many well-chosen instances may be very helpful as an encouragement, but can never prove a conjectural law in the mathematical sciences. Let us consider our own concrete case. By examining various special cases and comparing them, we have been led to conjecturing a general rule from which it would follow that the solution of our originally proposed problem is 26. Are all our observations and verifications sufficient to prove the general rule? Or can they prove the special result that the solution of our problem is actually 26? Not in the least. For a mathematician with rigid standards, the number 26 is just a clever guess and no amount of experimental verification could demonstrate the suspected general rule. Induction renders its results probable, it never proves them.

It may be observed, however, that inductive research may be useful in mathematics in another respect that we have not yet mentioned. The careful observation of special cases that leads us to a general mathematical result may also *suggest its proof*. From the intent examination of a particular case a general insight may emerge.

In fact, this actually happened to us already in the foregoing section. The general rule that we have discovered by induction is concerned with two juxtaposed numbers in our table, such as 7 and 4, and with their sum, which is 11 in the case at hand. Now, in the foregoing section we have visualized the geometrical significance of 7, 4, and 11 in our problem and, in doing so, we have understood *why* the relation $7 + 4 = 11$ arises there. We dealt, in fact, with the passage from 3 lines dividing the plane to 4 such lines. Yet there is no particular virtue in the numbers 3 and 4; we could pass just as well from any number to the following, from n to $n + 1$. The special case discussed may represent to us the general situation (ex. 2.10). I leave to the reader the pleasure of fully extracting the general idea from the particular observation of the foregoing section. In doing so, he may give a formal proof for the rule discovered inductively, at least as far as the last two columns are concerned.

Yet, in order to complete the proof, we have to consider not only the dissection of a plane by straight lines, but also the dissection of space by planes. We may hope, however, that if we are able to clear up the dissection of a plane, analogy will help us to clear up the dissection of space. Again, I leave to the reader the pleasure of profiting from the advice of analogy.

18. More conjectures. We have not yet exhausted the subject of plane and space partitions. There are a few more little discoveries to make and they are well accessible to inductive reasoning. We may be easily led to them by careful observation and understanding combination of particular instances.

We may wish to find a *formula* for the number of divisions of a plane by n lines in general position. In fact, we have already a formula in a simpler analogous case: n different points divide a straight line into $n + 1$ segments. This analogous formula, the particular cases listed in our table, our inductively discovered general rule (which we have almost proved), all our results hitherto obtained may help us to solve this new problem. I do not enter into details. I just note the solution which we may find, following the foregoing hints, in various manners.

$1 + n$ is the number of portions into which a straight line is divided by n different points.

$1 + n + \frac{n(n-1)}{2}$ is the number of portions into which a plane is divided by n straight lines in general position.

The reader may derive the latter formula or at least he can check it in the simplest cases, for $n = 0, 1, 2, 3, 4$. I leave also to the reader the pleasure of discovering a third formula of the same kind, for the number of space partitions. In making this little discovery, the reader may broaden his experience of inductive reasoning in mathematical matters and enjoy the help that analogy lends us in the solution of problems little or great.

EXAMPLES AND COMMENTS ON CHAPTER III

The formula $F + V = E + 2$, conjectured in sect. 1, is due to Leonhard Euler. We call it "Euler's formula," regard it as a conjecture, and examine it in various ways, sometimes inductively and sometimes with a view to finding a proof, in ex. 1–10. We return to it in ex. 21–30 and ex. 31–41. Before attempting any example in these two divisions, read ex. 21 and ex. 31, respectively.

1. Two pyramids, standing on opposite sides of their common base, form jointly a "double pyramid." An octahedron is a particular double pyramid; the common base is a square. Does Euler's formula hold for the general double pyramid?

2. Take a convex polyhedron with F faces, V vertices, and E edges, choose a point P in its interior (its centroid, for example), describe a sphere with center P and project the polyhedron from the center P onto the surface of the sphere. This projection transforms the F faces into F regions or "countries" on the surface of the sphere, it transforms any of the E edges into a boundary line separating two neighboring countries and any of the V vertices into a "corner" or a common boundary point of three or more countries (a "three-country corner" or a "four-country corner," etc.). This projection yields boundary lines of particularly simple nature (arcs of great circles) but, obviously, the validity of Euler's formula for such a *subdivision of the surface of the sphere into countries* is independent of the precise form of the boundary lines; the numbers F, V, and E are not influenced by continuous deformation of these lines.

(1) A *meridian* is one half of a great circle connecting the two poles, South and North. A *parallel circle* is the intersection of the globe's surface with a plane parallel to the equator. The earth's surface is divided by m meridians and p parallel circles into F countries. Compute F, V, and E. Does Euler's formula hold?

(2) The projection of the octahedron from its center P onto the surface of the sphere is a special case of the situation described in (1). For which values of m and p?

3. Chance plays a rôle in discovery. Inductive discovery obviously depends on the observational material. In sect. 1 we came across certain polyhedra, but we could have chanced upon others. Probably we would not have missed the regular solids, but our list could have come out thus:

Polyhedron	F	V	E
tetrahedron	4	4	6
cube	6	8	12
octahedron	8	6	12
pentagonal prism	7	10	15
pentagonal double pyramid	10	7	15
dodecahedron	12	20	30
icosahedron	20	12	30

Do you observe some regularity? Can you explain it? What is the connection with Euler's formula?

4. Try to generalize the relation between two polyhedra observed in the table of ex. 3. [The relation described in the solution of ex. 3 under (2) is too "narrow," too "detailed." Take, however, the cube and the octahedron in the situation there described, color the edges of one in red, those of the other in blue, and project them from their common center P onto a sphere as described in ex. 2. Then generalize.]

5. It would be sufficient to prove Euler's formula in a particular case: for convex polyhedra that have only triangular faces. Why? [Sect. 4.]

6. It would be sufficient to prove Euler's formula in a particular case: for convex polyhedra that have only three-edged vertices. Why? [Sect. 4.]

7. In proving Euler's formula we can restrict ourselves to figures in a plane. In fact, imagine that $F - 1$ faces of the polyhedron are made of cardboard, but one face is made of glass; we call this face the "window." You look through the window into the interior of the polyhedron, holding your eyes so close to the window that you see the whole interior. (This may be impossible if the polyhedron is not convex.) You can interpret what you see as a plane figure drawn on the window pane: you see a *subdivision* of the window into smaller polygons.

In this subdivision there are N_2 polygons, N_1 straight boundary lines (some outer, some inner) and N_0 vertices.

(1) Express N_0, N_1, N_2 in terms of F, V, E.

(2) If Euler's formula holds for F, V, and E, which formula holds for N_0, N_1, and N_2?

8. A rectangle is l inches long and m inches wide; l and m are integers. The rectangle is subdivided into lm equal squares by straight lines parallel to its sides.

(1) Express N_0, N_1, and N_2 (defined in ex. 7) in terms of l and m.

(2) Is the relation ex. 7 (2) valid in the present case?

9. Ex. 5 and 7 suggest that we should examine the subdivision of a triangle into N_2 triangles with $N_0 - 3$ vertices in the interior of the subdivided triangle. In computing the sum of all the angles in those N_2 triangles in two different ways, you may prove Euler's formula.

10. Sect. 7 suggests the extension of Euler's formula to four and more dimensions. How can we make such an extension tangible? How can we visualize it?

Ex. 7 shows that the case of polyhedra can be reduced to the subdivision of a plane polygon. Analogy suggests that the case of four dimensions may be reduced to the subdivision of a polyhedron in our visible three-dimensional space. If we wish to proceed inductively, we have to examine some example of such a subdivision. By analogy, ex. 8 suggests the following.

A box (that is, a rectangular parallelepiped) has the dimensions l, m, and n; these three numbers are integers. The box is subdivided into lmn equal cubes by planes parallel to its faces. Let N_0, N_1, N_2, and N_3 denote the number of vertices, edges, faces, and polyhedra (cubes) forming the subdivision, respectively.

(1) Express N_0, N_1, N_2, and N_3 in terms of l, m, and n.

(2) Is there a relation analogous to equation (2) in the solution of ex. 7?

11. Let P_n denote the number of parts into which the plane is divided by n straight lines in general position. Prove that $P_{n+1} = P_n + (n + 1)$.

12. Let S_n denote the number of parts into which space is divided by n planes in general position. Prove that $S_{n+1} = S_n + P_n$.

13. Verify the conjectural formula

$$P_n = 1 + n + \frac{n(n-1)}{2}$$

for $n = 0, 1, 2, 3, 4$.

14. Guess a formula for S_n and verify it for $n = 0, 1, 2, 3, 4, 5$.

15. How many parts out of the 11 into which the plane is divided by 4 straight lines in general position are finite? [How many are infinite?]

16. Generalize the foregoing problem.

17. How many parts out of the 26 into which space is divided by 5 planes in general position are infinite?

18. Five planes pass through the center of a sphere, but in other respects their position is general. Find the number of the parts into which the surface of the sphere is divided by the five planes.

19. Into how many parts is the plane divided by 5 mutually intersecting circles in general position?

20. Generalize the foregoing problems.

21. *Induction: adaptation of the mind, adaptation of the language.* Induction results in adapting our mind to the facts. When we compare our ideas with the observations, there may be agreement or disagreement. If there is agreement, we feel more confident of our ideas; if there is disagreement, we modify our ideas. After repeated modification our ideas may fit the facts somewhat better. Our first ideas about any new subject are almost bound to be wrong, at least in part; the inductive process gives us a chance to correct them, to adapt them to reality. Our examples show this process on a small scale, but pretty clearly. In sect. 1, after two or three wrong conjectures, we arrived eventually at the right conjecture. We arrived at it by accident, you may say. "Yet such accidents happen only to people who deserve them," as Lagrange said once when an incomparably greater discovery, by Newton, was discussed.

Adaptation of the mind may be more or less the same thing as adaptation of the language; at any rate, one goes hand in hand with the other. The progress of science is marked by the progress of terminology. When the physicists started to talk about "electricity," or the physicians about "contagion," these terms were vague, obscure, muddled. The terms that the scientists use today, such as "electric charge," "electric current," "fungus infection," "virus infection," are incomparably clearer and more definite. Yet what a tremendous amount of observation, how many ingenious experiments lie between the two terminologies, and some great discoveries too. Induction changed the terminology, clarified the concepts. We can illustrate also this aspect of the process, the inductive clarification of concepts, by a suitable small-scale mathematical example. The situation, not infrequent in mathematical research, is this: A theorem has been already formulated, but we have to give a more precise meaning to the terms in which it is formulated in order to render it strictly correct. This can be done conveniently by an inductive process, as we shall see.

Let us look back at ex. 2 and its solution. We talked about the "subdivision of the sphere into countries" without proposing a formal definition of this term. We hoped that Euler's formula remains valid if F, V, and E denote the number of countries, boundary lines, and corners in such a subdivision. Yet again, we relied on examples and a rough description and did not give formal definitions for F, V, and E. In what exact sense should we take these terms to render Euler's formula strictly correct? This is our question.

Let us say that a subdivision of the sphere (that is, of the spherical *surface*) with a corresponding interpretation of F, V, and E is "right" if Euler's formula holds, and is "wrong" if this formula does not hold. Propose examples of subdivisions which could help us to discover some clear and simple distinction between "right" and "wrong" cases.

22. The whole surface of the globe consists of just one country. Is that right? (We mean "right" from the viewpoint of Euler's formula.)

23. The globe's surface is divided into just two countries, the western hemisphere and the eastern hemisphere, separated by a great circle. Is that wrong?

24. Two parallel circles divide the sphere into three countries. Is it right or wrong?

25. Three meridians divide the sphere into three countries. Is it right or wrong?

26. Call the division of the sphere by m meridians and p parallel circles the "division (m,p)"; cf. ex. 2 (1). Is the extreme case $(0,p)$ right or wrong?

27. Is the extreme case $(m,0)$ right or wrong? (Cf. ex. 26.)

28. Which subdivisions (m,p) (cf. ex. 26) can be generated by the process described in ex. 2? (Projection of a convex polyhedron onto the sphere, followed by continuous shifting of the boundaries which leaves the number of countries and the number of the boundary lines around each country unaltered.) Which conditions concerning m and p characterize such subdivisions?

29. What is wrong with the examples in which Euler's formula fails? Which geometrical conditions, rendering more precise the meaning of F, V, and E, would ensure the validity of Euler's formula?

30. Propose more examples to illustrate the answer to ex. 29.

31. *Descartes' work on polyhedra.* Among the manuscripts left by Descartes there were brief notes on the general theory of polyhedra. A copy of these notes (by the hand of Leibnitz) was discovered and published in 1860, more than two hundred years after Descartes' death; cf. Descartes' *Oeuvres*, vol. 10, pp. 257–276. These notes treat of subjects closely related to Euler's theorem: although the notes do not state the theorem explicitly, they contain results from which it immediately follows.

We consider, with Descartes, a convex polyhedron. Let us call any angle of any face of the polyhedron a *surface angle*, and let $\Sigma\alpha$ stand for the sum of all surface angles. Descartes computes $\Sigma\alpha$ in two different manners, and Euler's theorem results immediately from the comparison of the two expressions.

The following examples give the reader an opportunity to reconstruct some of Descartes' conclusions. We shall use the following notation:

F_n denotes the number of faces with n edges,

V_n the number of vertices in which n edges end, so that

$$F_3 + F_4 + F_5 + \dots = F,$$
$$V_3 + V_4 + V_5 + \dots = V.$$

We continue to call E the number of all edges of the polyhedron.

32. Express the number of all surface angles in three different ways: in terms of $F_3, F_4, F_5, \ldots$, of $V_3, V_4, V_5, \ldots$, and of E, respectively.

33. Compute $\Sigma\alpha$ for the five regular solids: the tetrahedron, the cube, the octahedron, the dodecahedron, and the icosahedron.

34. Express $\Sigma\alpha$ in terms of $F_3, F_4, F_5, \ldots$.

35. Express $\Sigma\alpha$ in terms of E and F.

36. *Supplementary solid angles, supplementary spherical polygons.* We call *solid angle* what is more usually called a polyhedral angle.

Two convex solid angles have the same number of faces and the same vertex, but no other point in common. To each face of one solid angle corresponds an edge of the other, and the face is perpendicular to the corresponding edge. (This relation between the two solid angles is reciprocal: the edge e, intersection of two contiguous faces of the first solid angle, corresponds to the face f', of the second solid angle, if f' is bounded by the two edges corresponding to the two above mentioned faces.) Two solid angles in this mutual relation are called *supplementary* solid angles. (This name is not usual, but two ordinary supplementary angles can be brought into an analogous mutual position.) Each of two supplementary solid angles is called the supplement of the other.

The sphere with radius 1, described about the common vertex of two supplementary solid angles as center, is intersected by these in two spherical polygons: also these polygons are called supplementary.

We consider two supplementary spherical polygons. Let $a_1, a_2, \ldots a_n$ denote the sides of the first polygon, $\alpha_1, \alpha_2, \ldots \alpha_n$ its angles, A its area, P its perimeter, and let $a_1', a_2', \ldots a_n', \alpha_1', \alpha_2', \ldots \alpha_n', A', P'$ stand for the analogous parts of the other polygon. Then, if the notation is appropriately chosen,

$$a_1 + \alpha_1' = a_2 + \alpha_2' = \ldots = a_n + \alpha_n' = \pi,$$

$$a_1' + \alpha_1 = a_2' + \alpha_2 = \ldots = a_n' + \alpha_n = \pi;$$

this is well known and easily seen.

Prove that

$$P + A' = P' + A = 2\pi.$$

[Assume as known that the area of a spherical triangle with angles α, β, and γ is the "spherical excess" $\alpha + \beta + \gamma - \pi$.]

37. "As in a plane figure all exterior angles jointly equal 4 right angles, so in a solid figure all exterior solid angles jointly equal 8 right angles." Try to interpret this sentence found in Descartes' notes as a theorem which you can prove. [See fig. 3.7.]

38. Express $\Sigma\alpha$ in terms of V.

39. Prove Euler's theorem.

40. The initial remark of sect. 1 is vague, but can suggest several precise statements. Here is one that we have not considered in sect. 1: "If any one of the three quantities F, V, and E tends to ∞, also the other two must tend to ∞." Prove the following inequalities which hold generally for convex polyhedra and give still more precise information:

$$2E \geqq 3F, \quad 2V \geqq F + 4, \quad 3V \geqq E + 6,$$
$$2E \geqq 3V, \quad 2F \geqq V + 4, \quad 3F \geqq E + 6,$$

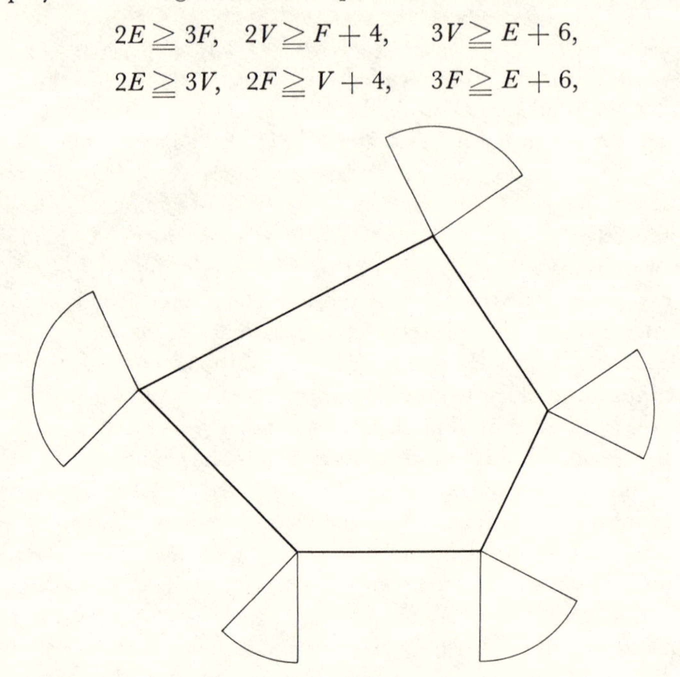

Fig. 3.7. Exterior angles of a polygon.

Can the case of equality be attained in these inequalities? For which kind of polyhedra can it be attained?

41. There are convex polyhedra all faces of which are polygons of the same kind, that is, polygons with the same number of sides. For example, all faces of a tetrahedron are triangles, all faces of a parallelepiped quadrilaterals, all faces of a regular dodecahedron pentagons. "And so on," you may be tempted to say. Yet such simple induction may be misguiding: there exists *no* convex polyhedron with faces which are all hexagons. Try to prove this. [Ex. 31.]

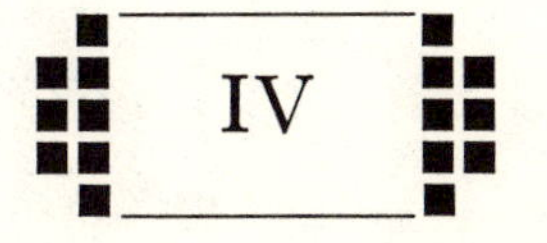

INDUCTION IN THE THEORY OF NUMBERS

In the Theory of Numbers it happens rather frequently that, by some unexpected luck, the most elegant new truths spring up by induction. —GAUSS[1]

1. Right triangles in integers.[2] The triangle with sides 3, 4, and 5 is a right triangle since

$$3^2 + 4^2 = 5^2.$$

This is the simplest example of a right triangle of which the sides are measured by integers. Such "right triangles in integers" have played a rôle in the history of the Theory of Numbers; even the ancient Babylonians discovered some of their properties.

One of the more obvious problems about such triangles is the following: *Is there a right triangle in integers, the hypotenuse of which is a given integer n?*

We concentrate upon this question. We seek a triangle the hypotenuse of which is measured by the given integer n and the legs by some integers x and y. We may assume that x denotes the longer of the two legs. Therefore, being given n, we seek two integers x and y such that

$$n^2 = x^2 + y^2, \quad 0 < y \leqq x < n.$$

We may attack the problem inductively and, unless we have some quite specific knowledge, we cannot attack it any other way. Let us take an example. We choose $n = 12$. Therefore, we seek two positive integers x and y, such that $x \geqq y$ and

$$144 = x^2 + y^2.$$

[1] *Werke*, vol. 2, p. 3.

[2] Parts of this chapter appeared already under the title "Let us teach guessing" in the volume *Études de philosophie des sciences en hommage à Ferdinand Gonseth*. Éditions du Griffon, 1950; see pp. 147–154.

Which values are available for x^2? The following:

$$1, 4, 9, 16, 25, 36, 49, 64, 81, 100, 121.$$

Is $x^2 = 121$? That is, is

$$144 - x^2 = 144 - 121 = y^2$$

a square? No, 23 is not a square. We should now try other squares but, in fact, we need not try too many of them. Since $y \leqq x$,

$$144 = x^2 + y^2 \leqq 2x^2$$
$$x^2 \geqq 72.$$

Therefore, $x^2 = 100$ and $x^2 = 81$ are the only remaining possibilities. Now, neither of the numbers

$$144 - 100 = 44, \quad 144 - 81 = 63$$

is a square and hence the answer: there is no right triangle in integers with hypotenuse 12.

We treat similarly the hypotenuse 13. Of the three numbers

$$169 - 144 = 25, \quad 169 - 121 = 48, \quad 169 - 100 = 69$$

only one is a square and so there is just one right triangle in integers with hypotenuse 13:

$$169 = 144 + 25.$$

Proceeding similarly, we can examine with a little patience all the numbers under a given not too high limit, such as 20. We find only five "hypotenuses" less than 20, the numbers 5, 10, 13, 15, and 17:

$$\begin{aligned} 25 &= 16 + 9 \\ 100 &= 64 + 36 \\ 169 &= 144 + 25 \\ 225 &= 144 + 81 \\ 289 &= 225 + 64. \end{aligned}$$

By the way, the cases 10 and 15 are not very interesting. The triangle with sides 10, 8, and 6 is similar to the simpler triangle with sides 5, 4, and 3, and the same is true of the triangle with sides 15, 12, and 9. The remaining three right triangles, with hypotenuse 5, 13, and 17, respectively, are essentially different, none is similar to another among them.

We may notice that all three numbers 5, 13, and 17 are *odd primes*. They are, however, not all the odd primes under 20; none of the other odd primes, 3, 7, 11, and 19 is a hypotenuse. Why that? What is the difference between the two sets? *When, under which circumstances, is an odd prime the hypotenuse of some right triangle in integers, and when is it not?*

This is a modification of our original question. It may appear more hopeful; at any rate, it is new. Let us investigate it—again, inductively. With a little patience, we construct the following table (the dash indicates that there is no right triangle with hypotenuse p).

Odd prime p	*Right triangles with hypotenuse p*
3	—
5	$25 = 16 + 9$
7	—
11	—
13	$169 = 144 + 25$
17	$289 = 225 + 64$
19	—
23	—
29	$841 = 441 + 400$
31	—

When is a prime a hypotenuse; when is it not? What is the difference between the two cases? A physicist could easily ask himself some very similar questions. For instance, he investigates the double refraction of crystals. Some crystals do show double refraction; others do not. Which crystals are doubly refracting, which are not? What is the difference between the two cases?

The physicist looks at his crystals and we look at our two sets of primes

$$5, 13, 17, 29, \ldots \quad \text{and} \quad 3, 7, 11, 19, 23, 31, \ldots .$$

We are looking for some characteristic difference between the two sets. The primes in both sets increase by irregular jumps. Let us look at the lengths of these jumps, at the successive differences:

$$\begin{array}{ccccccc} 5 & & 13 & & 17 & & 29 \\ & 8 & & 4 & & 12 & \end{array} \qquad \begin{array}{ccccccccccc} 3 & & 7 & & 11 & & 19 & & 23 & & 31 \\ & 4 & & 4 & & 8 & & 4 & & 8 & \end{array}$$

Many of these differences are equal to 4, and, as it is easy to notice, *all are divisible by 4.* The primes in the first set, led by 5, leave the remainder 1 when divided by 4, are of the form $4n + 1$ with integral n. The primes in the second set, led by 3, are of the form $4n + 3$. Could this be the characteristic difference we are looking for? If we do not discard this possibility from the outset, we are led to the following conjecture: *A prime of the form $4n + 1$ is the hypotenuse of just one right triangle in integers; a prime of the form $4n + 3$ is the hypotenuse of no such triangle.*

2. Sums of squares. The problem of the right triangles in integers, one aspect of which we have just discussed (in sect. 1), played, as we have said, an important rôle in the history of the Theory of Numbers. It leads on, in fact, to many further questions. Which numbers, squares or not, can be decomposed into two squares? What about the numbers which cannot be decomposed into two squares? Perhaps, they are decomposable into three squares; but what about the numbers which are not decomposable into three squares?

We could go on indefinitely, but, and this is highly remarkable, we need not. Bachet de Méziriac (author of the first printed book on mathematical recreations) remarked that *any number* (that is, positive integer) *is either a square, or the sum of two, three, or four squares.* He did not pretend to possess a proof. He found indications pointing to his statement in certain problems of Diophantus and verified it up to 325.

In short, Bachet's statement was just a conjecture, found inductively. It seems to me that his main achievement was to put the question: *HOW MANY squares are needed to represent all integers?* Once this question is clearly put, there is not much difficulty in discovering the answer inductively. We construct a table beginning with

$$\begin{aligned}
1 &= 1\\
2 &= 1 + 1\\
3 &= 1 + 1 + 1\\
4 &= 4\\
5 &= 4 + 1\\
6 &= 4 + 1 + 1\\
7 &= 4 + 1 + 1 + 1\\
8 &= 4 + 4\\
9 &= 9\\
10 &= 9 + 1.
\end{aligned}$$

This verifies the statement up to 10. Only the number 7 requires four squares; the others are representable by one or two or three. Bachet went on tabulating up to 325 and found many numbers requiring four squares and none requiring more. Such inductive evidence satisfied him, it seems, at least to a certain degree, and he published his statement. He was lucky. His conjecture turned out to be true and so he became the discoverer of the "four-square theorem" which we can state also in the form: The equation

$$n = x^2 + y^2 + z^2 + w^2$$

where n is any given positive integer has a solution in which x, y, z, and w are non-negative integers.

The decomposition of a number into a sum of squares has still other aspects. Thus, we may investigate the number of solutions of the equation

$$n = x^2 + y^2$$

in integers x and y. We may admit only positive integers, or all integers, positive, negative, and 0. If we choose the latter conception of the problem and take as example $n = 25$, we find 12 solutions of the equation

$$25 = x^2 + y^2,$$

namely the following

$$\begin{aligned} 25 &= 5^2 + 0^2 = (-5)^2 + 0^2 = 0^2 + 5^2 = 0^2 + (-5)^2 \\ &= 4^2 + 3^2 = (-4)^2 + 3^2 = 4^2 + (-3)^2 = (-4)^2 + (-3)^2 \\ &= 3^2 + 4^2 = (-3)^2 + 4^2 = 3^2 + (-4)^2 = (-3)^2 + (-4)^2. \end{aligned}$$

By the way, these solutions have an interesting geometric interpretation, but we need not discuss it now. See ex. 2.

3. On the sum of four odd squares. Of the many problems concerned with sums of squares I choose one that looks somewhat far-fetched, but will turn out to be exceptionally instructive.

Let u denote a positive odd integer. Investigate inductively the number of the solutions of the equation

$$4u = x^2 + y^2 + z^2 + w^2$$

in positive odd integers x, y, z, and w.

For example, if $u = 1$ we have the equation

$$4 = x^2 + y^2 + z^2 + w^2$$

and there is obviously just one solution, $x = y = z = w = 1$. In fact, we do not regard

$$x = -1, \quad y = 1, \quad z = 1, \quad w = 1$$

or

$$x = 2, \quad y = 0, \quad z = 0, \quad w = 0$$

as a solution, since we admit only positive odd numbers for x, y, z, and w. If $u = 3$, the equation is

$$12 = x^2 + y^2 + z^2 + w^2$$

and the following two solutions:

$$\begin{aligned} x &= 3, \quad y = 1, \quad z = 1, \quad w = 1 \\ x &= 1, \quad y = 3, \quad z = 1, \quad w = 1 \end{aligned}$$

are different.

In order to emphasize the restriction laid upon the values of x, y, z, and w, we shall avoid the term "solution" and use instead the more specific description: "representation of $4u$ as a sum of four odd squares." As this description is long, we shall abbreviate it in various ways, sometimes even to the one word "representation."

4. Examining an example. In order to familiarize ourselves with the meaning of our problem, let us consider an example. We choose $u = 25$. Then $4u = 100$, and we have to find all representations of 100 as a sum of four odd squares. Which odd squares are available for this purpose? The following:

$$1, \quad 9, \quad 25, \quad 49, \quad 81.$$

If 81 is one of the four squares the sum of which is 100, then the sum of the three others must be

$$100 - 81 = 19.$$

The only odd squares less than 19 are 1 and 9, and $19 = 9 + 9 + 1$ is evidently the only possibility to represent 19 as a sum of 3 odd squares if the terms are arranged in order of magnitude. We obtain

$$100 = 81 + 9 + 9 + 1.$$

We find similarly

$$100 = 49 + 49 + 1 + 1,$$
$$100 = 49 + 25 + 25 + 1,$$
$$100 = 25 + 25 + 25 + 25.$$

Proceeding systematically, by splitting off the largest square first, we may convince ourselves that we have exhausted all possibilities, provided that the 4 squares are arranged in descending order (or rather in non-ascending order). But there are more possibilities if we take into account, as we should, all arrangements of the terms. For example,

$$\begin{aligned} 100 &= 49 + 49 + 1 + 1 \\ &= 49 + 1 + 49 + 1 \\ &= 49 + 1 + 1 + 49 \\ &= 1 + 49 + 49 + 1 \\ &= 1 + 49 + 1 + 49 \\ &= 1 + 1 + 49 + 49. \end{aligned}$$

These 6 sums have the same terms, but the order of the terms is different; they are to be considered, according to the statement of our problem, as 6 different representations; the one representation

$$100 = 49 + 49 + 1 + 1$$

with non-increasing terms is a source of 5 other representations, of 6 representations in all. We have similarly

Non-increasing terms	*Number of arrangements*
$81 + 9 + 9 + 1$	12
$49 + 49 + 1 + 1$	6
$49 + 25 + 25 + 1$	12
$25 + 25 + 25 + 25$	1

To sum up, we found in our case where $u = 25$ and $4u = 100$

$$12 + 6 + 12 + 1 = 31$$

representations of $4u = 100$ as a sum of 4 odd squares.

5. Tabulating the observations. The special case $u = 25$ where $4u = 100$ and the number of representations is 31 has shown us clearly the meaning of the problem. We may now explore systematically the simplest cases, $u = 1, 3, 5, \ldots$ up to $u = 25$. We construct Table I. (See below; the reader should construct the table by himself, or at least check a few items.)

Table I

u	$4u$	*Non-increasing*	*Arrangements*	*Representations*
1	4	1 + 1 + 1 + 1	1	1
3	12	9 + 1 + 1 + 1	4	4
5	20	9 + 9 + 1 + 1	6	6
7	28	25 + 1 + 1 + 1	4	8
		9 + 9 + 9 + 1	4	
9	36	25 + 9 + 1 + 1	12	13
		9 + 9 + 9 + 9	1	
11	44	25 + 9 + 9 + 1	12	12
13	52	49 + 1 + 1 + 1	4	14
		25 + 25 + 1 + 1	6	
		25 + 9 + 9 + 9	4	
15	60	49 + 9 + 1 + 1	12	24
		25 + 25 + 9 + 1	12	
17	68	49 + 9 + 9 + 1	12	18
		25 + 25 + 9 + 9	6	
19	76	49 + 25 + 1 + 1	12	20
		49 + 9 + 9 + 9	4	
		25 + 25 + 25 + 1	4	
21	84	81 + 1 + 1 + 1	4	32
		49 + 25 + 9 + 1	24	
		25 + 25 + 25 + 9	4	
23	92	81 + 9 + 1 + 1	12	24
		49 + 25 + 9 + 9	12	
25	100	81 + 9 + 9 + 1	12	31
		49 + 49 + 1 + 1	6	
		49 + 25 + 25 + 1	12	
		25 + 25 + 25 + 25	1	

6. What is the rule? Is there any recognizable law, any simple connection between the odd number u and the number of different representations of $4u$ as a sum of four odd squares?

This question is the kernel of our problem. We have to answer it on the basis of the observations collected and tabulated in the foregoing section. We are in the position of the naturalist trying to extract some rule, some general formula from his experimental data. Our experimental material available at this moment consists of two parallel series of numbers

1	3	5	7	9	11	13	15	17	19	21	23	25
1	4	6	8	13	12	14	24	18	20	32	24	31.

The first series consists of the successive odd numbers, but what is the rule governing the second series?

As we try to answer this question, our first feeling may be close to despair. That second series looks quite irregular, we are puzzled by its complex origin, we can scarcely hope to find any rule. Yet, if we forget about the complex origin and concentrate upon what is before us, there is a point easy enough to notice. It happens rather often that a term of the second series exceeds the corresponding term of the first series by just one unit. Emphasizing these cases by heavy print in the first series, we may present our experimental material as follows:

1	**3**	**5**	**7**	9	**11**	**13**	15	**17**	**19**	21	**23**	25
1	4	6	8	13	12	14	24	18	20	32	24	31.

The numbers in heavy print attract our attention. It is not difficult to recognize them: they are *primes*. In fact, they are *all* the primes in the first row as far as our table goes. This remark may appear very surprising if we remember the origin of our series. We considered squares, we made no reference whatever to primes. Is it not strange that the prime numbers play a rôle in our problem? It is difficult to avoid the impression that our observation is significant, that there is something remarkable behind it.

What about those numbers of the first series which are not in heavy print? They are odd numbers, but not primes. The first, 1, is unity, the others are composite

$$9 = 3 \times 3, \quad 15 = 3 \times 5, \quad 21 = 3 \times 7, \quad 25 = 5 \times 5.$$

What is the nature of the corresponding numbers in the second series?

If the odd number u is a prime, the corresponding number is $u + 1$; if u is not a prime, the corresponding number is not $u + 1$. This we have observed already. We may add one little remark. If $u = 1$, the corresponding number is also 1, and so *less* than $u + 1$, but in all other cases in which u is not a prime the corresponding number is *greater* than $u + 1$. That is, the number corresponding to u is less than, equal to, or greater than $u + 1$ accordingly as u is unity, a prime, or a composite number. There is some regularity.

Let us concentrate upon the composite numbers in the upper line and the corresponding numbers in the lower line:

$$\begin{array}{cccc} 3 \times 3 & 3 \times 5 & 3 \times 7 & 5 \times 5 \\ 13 & 24 & 32 & 31 \end{array} .$$

There is something strange. Squares in the first line correspond to primes in the second line. Yet we have too few observations; probably we should not attach too much weight to this remark. Still, it is true that, conversely, under the composite numbers in the first line which are not squares, we find numbers in the second line which are not primes:

$$\begin{array}{cc} 3 \times 5 & 3 \times 7 \\ 4 \times 6 & 4 \times 8. \end{array}$$

Again, there is something strange. Each factor in the second line exceeds the corresponding factor in the first line by just one unit. Yet we have too few observations; we had better not attach too much weight to this remark. Still, our remark shows some parallelism with a former remark. We noticed before

$$\begin{array}{c} p \\ p+1 \end{array}$$

and we notice now

$$\begin{array}{c} pq \\ (p+1)(q+1) \end{array}$$

where p and q are primes. There is some regularity.

Perhaps we shall see more clearly if we write the entry corresponding to pq differently:

$$(p+1)(q+1) = pq + p + q + 1.$$

What can we see there? What are these numbers pq, p, q, 1? At any rate, the cases

$$\begin{array}{cc} 9 & 25 \\ 13 & 31 \end{array}$$

remain unexplained. In fact, the entries corresponding to 9 and 25 are greater than $9 + 1$ and $25 + 1$, respectively, as we have already observed:

$$13 = 9 + 1 + 3 \qquad 31 = 25 + 1 + 5 .$$

What are these numbers?

If one more little spark comes from somewhere, we may succeed in combining our fragmentary remarks into a coherent whole, our scattered indications into an illuminating view of the full correspondence:

$$\begin{array}{ccccc} p & pq & 9 & 25 & 1 \\ p+1 & pq+p+q+1 & 9+3+1 & 25+5+1 & 1 . \end{array}$$

DIVISORS! The second line shows the divisors of the numbers in the first line. This may be the desired rule, and a discovery, a real discovery: *To each number in the first line corresponds the sum of its divisors.*

And so we have been led to a conjecture, perhaps to one of those "most elegant new truths" of Gauss: *If u is an odd number, the number of representations of $4u$ as a sum of four odd squares is equal to the sum of the divisors of u.*

7. On the nature of inductive discovery. Looking back at the foregoing sections (3 to 6) we may find many questions to ask.

What have we obtained? Not a proof, not even the shadow of a proof, just a conjecture: a simple description of the facts within the limits of our experimental material, and a certain hope that this description may apply beyond the limits of our experimental material.

How have we obtained our conjecture? In very much the same manner that ordinary people, or scientists working in some non-mathematical field, obtain theirs. We collected relevant observations, examined and compared them, noticed fragmentary regularities, hesitated, blundered, and eventually succeeded in *combining the scattered details into an apparently meaningful whole.* Quite similarly, an archaeologist may reconstitute a whole inscription from a few scattered letters on a worn-out stone, or a palaeontologist may reconstruct the essential features of an extinct animal from a few of its petrified bones. In our case the meaningful whole appeared at the same moment when we recognized the appropriate unifying concept (the divisors).

8. On the nature of inductive evidence. There remain a few more questions.

How strong is the evidence? Your question is incomplete. You mean, of course, the inductive evidence for our conjecture stated in sect. 6 that we can derive from Table I of sect. 5; this is understood. Yet what do you mean by "strong"? The evidence is strong if it is convincing; it is convincing if it convinces somebody. Yet you did not say whom it should convince—me, or you, or Euler, or a beginner, or whom?

Personally, I find the evidence pretty convincing. I feel sure that Euler would have thought very highly of it. (I mention Euler because he came very near to discovering our conjecture; see ex. 6.24.) I think that a beginner who knows a little about the divisibility of numbers ought to find the evidence pretty convincing, too. A colleague of mine, an excellent mathematician who however was not familiar with this corner of the Theory of Numbers, found the evidence "hundred per cent convincing."

I am not concerned with subjective impressions. What is the precise, objectively evaluated degree of rational belief, justified by the inductive evidence? You give me one thing (A), you fail to give me another thing (B), and you ask me a third thing (C).

(A) You give me exactly the inductive evidence: the conjecture has been verified in the first thirteen cases, for the numbers 4, 12, 20, . . . , 100. This is perfectly clear.

(B) You wish me to evaluate the degree of rational belief justified by this evidence. Yet such belief must depend, if not on the whims and the temperament, certainly on the *knowledge* of the person receiving the evidence. He may know a proof of the conjectural theorem or a counter-example exploding it. In these two cases the degree of his belief, already firmly established, will remain unchanged by the inductive evidence. Yet if he knows something that comes very close to a complete proof, or to a complete refutation, of the theorem, his belief is still capable of modification and will be affected by the inductive evidence here produced, although different degrees of belief will result from it according to the kind of knowledge he has. Therefore, if you wish a definite answer, you should specify a definite level of knowledge on which the proposed inductive evidence (A) should be judged. You should give me a definite set of relevant known facts (an explicit list of known elementary propositions in the Theory of Numbers, perhaps).

(C) You wish me to evaluate the degree of rational belief justified by the inductive evidence exactly. Should I give it to you perhaps expressed in percentages of "full credence"? (We may agree to call "full credence" the degree of belief justified by a complete mathematical proof of the theorem in question.) Do you expect me to say that the given evidence justifies a belief amounting to 99% or to 2.875% or to .000001% of "full credence"?

In short, you wish me to solve a problem: Given (A) the inductive evidence and (B) a definite set of known facts or propositions, compute the percentage of full credence rationally resulting from both (C).

To solve this problem is much more than I can do. I do not know anybody who could do it, or anybody who would dare to do it. I know of some philosophers who promise to do something of this sort in great generality. Yet, faced with the concrete problem, they shrink and hedge and find a thousand excuses why not to do just this problem.

Perhaps the problem is one of those typical philosophical problems about which you can talk a lot in general, and even worry genuinely, but which fade into nothingness when you bring them down to concrete terms.

Could you compare the present case of inductive inference with some standard case and so arrive at a reasonable estimate of the strength of the evidence? Let us compare the inductive evidence for our conjecture with Bachet's evidence for his conjecture.

Bachet's conjecture was: For $n = 1, 2, 3, \ldots$ the equation

$$n = x^2 + y^2 + z^2 + w^2$$

has at least one solution in non-negative integers x, y, z, and w. He verified this conjecture for $n = 1, 2, 3, \ldots, 325$. (See sect. 2, especially the short table.)

Our conjecture is: For a given odd u, the number of solutions of the equation

$$4u = x^2 + y^2 + z^2 + w^2$$

in positive odd integers x, y, z, and w is equal to the sum of the divisors of u. We verified this conjecture for $u = 1, 3, 5, 7, \ldots, 25$ (13 cases). (See sect. 3 to 6.)

I shall compare these two conjectures and the inductive evidence yielded by their respective verifications in three respects.

Number of verifications. Bachet's conjecture was verified in 325 cases, ours in 13 cases only. The advantage in this respect is clearly on Bachet's side.

Precision of prediction. Bachet's conjecture predicts that the number of solutions is $\geqq 1$; ours predicts that the number of solutions is exactly equal to such and such a quantity. It is obviously reasonable to assume, I think, that *the verification of a more precise prediction carries more weight* than that of a less precise prediction. The advantage in this respect is clearly on our side.

Rival conjectures. Bachet's conjecture is concerned with the maximum number of squares, say M, needed in representing an arbitrary positive integer as sum of squares. In fact, Bachet's conjecture asserts that $M = 4$. I do not think that Bachet had any *a priori* reason to prefer $M = 4$ to, say, $M = 5$, or to any other value, as $M = 6$ or $M = 7$; even $M = \infty$ is not excluded *a priori*. (Naturally, $M = \infty$ would mean that there are larger and larger integers demanding more and more squares. On the face, $M = \infty$ could appear as the most likely conjecture.) In short, Bachet's conjecture has many obvious rivals. Yet ours has none. Looking at the irregular sequence of the numbers of representations (sect. 6) we had the impression that we might not be able to find any rule. Now we did find an admirably clear rule. We hardly expect to find any other rule.

It may be difficult to choose a bride if there are many desirable young ladies to choose from; if there is just one eligible girl around, the decision may come much quicker. It seems to me that our attitude toward conjectures is somewhat similar. Other things being equal, a conjecture that has many obvious rivals is more difficult to accept than one that is unrivalled. If you think as I do, you should find that in this respect the advantage is on the side of our conjecture, not on Bachet's side.

Please observe that the evidence for Bachet's conjecture is stronger in one respect and the evidence for our conjecture is stronger in other respects, and do not ask unanswerable questions.

EXAMPLES AND COMMENTS ON CHAPTER IV

1. *Notation.* We assume that n and k are positive integers and consider the Diophantine equation

$$n = x_1^2 + x_2^2 + \ldots + x_k^2.$$

We say that two solutions $x_1, x_2, \ldots x_k$ and $x_1', x_2', \ldots x_k'$ are equal if, and only if, $x_1 = x_1', x_2 = x_2', \ldots x_k = x_k'$. If we admit for $x_1, x_2, \ldots x_k$ *all* integers, positive, negative, or null, we call the number of solutions $R_k(n)$. If we admit only *positive odd* integers, we call the number of solutions $S_k(n)$. This notation is important in the majority of the following problems.

Bachet's conjecture (sect. 2) is expressed in this notation by the inequality

$$R_4(n) > 0 \text{ for } n = 1, 2, 3, \ldots .$$

The conjecture that we discovered in sect. 6 affirms that $S_4(4(2n - 1))$ equals the sum of the divisors of $2n - 1$, for $n = 1, 2, 3, \ldots$.

Find $R_2(25)$ and $S_3(11)$.

2. Let x and y be rectangular coordinates in a plane. The points for which both x and y are integers are called the "lattice points" of the plane. Lattice points in space are similarly defined.

Interpret $R_2(n)$ and $R_3(n)$ geometrically, in terms of lattice points.

3. Express the conjecture encountered in sect. 1 in using the symbol $R_2(n)$.

4. When is an odd prime the sum of two squares? Try to answer this question inductively, by examining the table

$$\begin{array}{rl}
3 & \quad — \\
5 = & 4 + 1 \\
7 & \quad — \\
11 & \quad — \\
13 = & 9 + 4 \\
17 = & 16 + 1 \\
19 & \quad — \\
23 & \quad — \\
29 = & 25 + 4 \\
31 & \quad —
\end{array}$$

Extend this table if necessary and compare it with the table in sect. 1.

5. Could you verify by mathematical deduction some part of your answer to ex. 4 obtained by induction? After such a verification, would it be reasonable to change your confidence in the conjecture?

6. Verify Bachet's conjecture (sect. 2) up to 30 inclusively. Which numbers require actually four squares?

7. In order to understand better Table I in sect. 5, let a^2, b^2, c^2, and d^2 denote four different odd squares and consider the sums

$$(1)\ a^2 + b^2 + c^2 + d^2$$
$$(2)\ a^2 + a^2 + b^2 + c^2$$
$$(3)\ a^2 + a^2 + b^2 + b^2$$
$$(4)\ a^2 + a^2 + a^2 + b^2$$
$$(5)\ a^2 + a^2 + a^2 + a^2\ .$$

How many different representations (in the sense of sect. 3) can you derive from each by permuting the terms?

8. The number of representations of $4u$, as a sum of four odd squares is odd if, and only if, u is a square. (Following the notation of sect. 3, we assume that u is odd.) Prove this statement and show that it agrees with the conjecture of sect. 6. How does this remark influence your confidence in the conjecture?

9. Now, let a, b, c, and d denote different positive integers (odd or even). Consider the five sums mentioned in ex. 7 and also the following:

$$(6)\ a^2 + b^2 + c^2 \qquad (9)\ a^2 + b^2$$
$$(7)\ a^2 + a^2 + b^2 \qquad (10)\ a^2 + a^2$$
$$(8)\ a^2 + a^2 + a^2 \qquad (11)\ a^2\ .$$

Find in each of these eleven cases the contribution to $R_4(n)$. You derive from each sum all possible representations by the following obvious operations: you add 0^2 as many times as necessary to bring the number of terms to 4, you change the arrangement, and you replace some (or none, or all) of the numbers a, b, c, d by $-a$, $-b$, $-c$, $-d$, respectively. (Check examples in Table II.)

10. Investigate inductively the number of solutions of the equation $n = x^2 + y^2 + z^2 + w^2$ in integers x, y, z, and w, positive, negative, or 0. Start by constructing a table analogous to Table I.

11 (continued). Try to use the method, or the result, of sect. 6.

12 (continued). Led by the analogy of sect. 6 or by your observation of Table II, distinguish appropriate classes of integers and investigate each class by itself.

13 (continued). Concentrate upon the most stubborn class.

14 (continued). Try to summarize all fragmentary regularities and express the law in one sentence.

15 (continued). Check the rule found in the first three cases not contained in Table II.

16. Find $R_8(5)$ and $S_8(40)$.

17. Check at least two entries of Table III, p. 75, not yet given in Tables I and II.

18. Using Table III, investigate inductively $R_8(n)$ and $S_8(8n)$.

19 (continued). Try to use the method, or the result, of sect. 6 and ex. 10–15.

20 (continued). Led by analogy or observation, distinguish appropriate classes of integers and investigate each class by itself.

21 (continued). Try to discover a cue in the most accessible case.

22 (continued). Try to find some unifying concept that could summarize the fragmentary regularities.

23 (continued). Try to express the law in one sentence.

24. Which numbers can and which numbers cannot be expressed in the form $3x + 5y$, where x and y are non-negative integers?

25. Try to guess the law of the following table:

a	b	Last integer not expressible in form $ax + by$
2	3	1
2	5	3
2	7	5
2	9	7
3	4	5
3	5	7
3	7	11
3	8	13
4	5	11
5	6	19

It is understood that x and y are non-negative integers. Check a few items and extend the table, if necessary. [Observe the change in the last column when just one of the two numbers a and b changes.]

26. *Dangers of induction.* Examine inductively the following assertions:

(1) $(n - 1)! + 1$ is divisible by n when n is a prime, but not divisible by n when n is composite.

(2) $2^{n-1} - 1$ is divisible by n when n is an odd prime, but not divisible by n when n is composite.

Table II

n	Non-increasing	Repres.	$R_4(n)/8$
1	1	4×2	1
2	$1 + 1$	6×4	3
3	$1 + 1 + 1$	4×8	4
4	4	4×2	3
	$1 + 1 + 1 + 1$	1×16	
5	$4 + 1$	12×4	6
6	$4 + 1 + 1$	12×8	12
7	$4 + 1 + 1 + 1$	4×16	8
8	$4 + 4$	6×4	3
9	9	4×2	13
	$4 + 4 + 1$	12×8	
10	$9 + 1$	12×4	18
	$4 + 4 + 1 + 1$	6×16	
11	$9 + 1 + 1$	12×8	12
12	$9 + 1 + 1 + 1$	4×16	12
	$4 + 4 + 4$	4×8	
13	$9 + 4$	12×4	14
	$4 + 4 + 4 + 1$	4×16	
14	$9 + 4 + 1$	24×8	24
15	$9 + 4 + 1 + 1$	12×16	24
16	16	4×2	3
	$4 + 4 + 4 + 4$	1×16	
17	$16 + 1$	12×4	18
	$9 + 4 + 4$	12×8	
18	$16 + 1 + 1$	12×8	39
	$9 + 9$	6×4	
	$9 + 4 + 4 + 1$	12×16	
19	$16 + 1 + 1 + 1$	4×16	20
	$9 + 9 + 1$	12×8	
20	$16 + 4$	12×4	18
	$9 + 9 + 1 + 1$	6×16	
21	$16 + 4 + 1$	24×8	32
	$9 + 4 + 4 + 4$	4×16	
22	$16 + 4 + 1 + 1$	12×16	36
	$9 + 9 + 4$	12×8	
23	$9 + 9 + 4 + 1$	12×16	24
24	$16 + 4 + 4$	12×8	12
25	25	4×2	31
	$16 + 9$	12×4	
	$16 + 4 + 4 + 1$	12×16	

Table II (*continued*)

n	Non-increasing	Repres.	$R_4(n)/8$
26	25 + 1	12 × 4	42
	16 + 9 + 1	24 × 8	
	9 + 9 + 4 + 4	6 × 16	
27	25 + 1 + 1	12 × 8	40
	16 + 9 + 1 + 1	12 × 16	
	9 + 9 + 9	4 × 8	
28	25 + 1 + 1 + 1	4 × 16	24
	16 + 4 + 4 + 4	4 × 16	
	9 + 9 + 9 + 1	4 × 16	
29	25 + 4	12 × 4	30
	16 + 9 + 4	24 × 8	
30	25 + 4 + 1	24 × 8	72
	16 + 9 + 4 + 1	24 × 16	

Table III

n	$R_4(n)/8$	$R_8(n)/16$	$S_8(8n)$	$S_4(4(2n-1))$	$2n-1$
1	1	1	1	1	1
2	3	7	8	4	3
3	4	28	28	6	5
4	3	71	64	8	7
5	6	126	126	13	9
6	12	196	224	12	11
7	8	344	344	14	13
8	3	583	512	24	15
9	13	757	757	18	17
10	18	882	1008	20	19
11	12	1332	1332	32	21
12	12	1988	1792	24	23
13	14	2198	2198	31	25
14	24	2408	2752	40	27
15	24	3528	3528	30	29
16	3	4679	4096	32	31
17	18	4914	4914	48	33
18	39	5299	6056	48	35
19	20	6860	6860	38	37
20	18	8946	8064	56	39

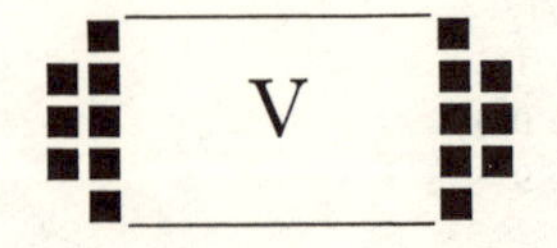

MISCELLANEOUS EXAMPLES OF INDUCTION

When you have satisfied yourself that the theorem is true, you start proving it.—THE TRADITIONAL MATHEMATICS PROFESSOR[1]

1. Expansions. In dealing with problems of any kind, we need inductive reasoning of some kind. In various branches of mathematics there are certain problems which call for inductive reasoning in a typical manner. The present chapter illustrates this point by a few examples. We begin with a relatively simple example.

Expand into powers of x the function $1/(1 - x + x^2)$.

This problem can be solved in many ways. The following solution is somewhat clumsy, but is based on a sound principle and may occur naturally to an intelligent beginner who knows little, yet knows at least the sum of the geometric series:

$$1 + r + r^2 + r^3 + \dots = \frac{1}{1 - r}.$$

There is an opportunity to use this formula in our problem:

$$\frac{1}{1 - x + x^2} = \frac{1}{1 - x(1 - x)}$$

$$\begin{aligned}
&= 1 + x(1 - x) + x^2(1 - x)^2 + x^3(1 - x)^3 + \dots \\
&= 1 + x - x^2 \\
&\qquad + x^2 - 2x^3 + x^4 \\
&\qquad\qquad + x^3 - 3x^4 + 3x^5 - x^6 \\
&\qquad\qquad\qquad + x^4 - 4x^5 + 6x^6 - 4x^7 + x^8 \\
&\qquad\qquad\qquad\qquad + x^5 - 5x^6 + 10x^7 - 10x^8 + \dots \\
&\qquad\qquad\qquad\qquad\qquad + x^6 - 6x^7 + 15x^8 - \dots \\
&\qquad\qquad\qquad\qquad\qquad\qquad + x^7 - 7x^8 + \dots \\
&\qquad\qquad\qquad\qquad\qquad\qquad\qquad + x^8 - \dots \\
&\qquad\qquad\qquad\qquad\qquad\qquad\qquad\qquad \dots \\
&= 1 + x \qquad - x^3 - x^4 \qquad + x^6 + x^7 \qquad \dots .
\end{aligned}$$

[1] This dictum of the well-known pedagogue (*How to Solve It*, p. 181) is sometimes preceded by the following exhortation: "If you have to prove a theorem, do not rush. First of all, understand fully what the theorem says, try to see clearly what it means. Then check the theorem; it could be false. Examine its consequences, verify as many particular instances as are needed to convince yourself of its truth. When . . ."

The result is striking. Any non-vanishing coefficient has either the value 1 or the value -1. The succession of the coefficients seems to show some regularity which becomes more apparent if we compute more terms:

$$\frac{1}{1-x+x^2} = 1 + x - x^3 - x^4 + x^6 + x^7 - x^9 - x^{10} + x^{12} + x^{13} - \dots .$$

Periodic! The sequence of coefficients appears to be periodical with the period 6:

$$1, 1, 0, -1, -1, 0 \mid 1, 1, 0, -1, -1, 0 \mid 1, 1, \dots .$$

We naturally expect that the periodicity observed extends beyond the limit of our observations. Yet this is an inductive conclusion, or a mere guess, which we should regard with due skepticism. The guess, however, is based on facts, and so it deserves serious examination. Examining it means, among other things, restating it. There is an interesting way of restating our conjecture:

$$\frac{1}{1-x+x^2} = 1 - x^3 + x^6 - x^9 + x^{12} - \dots$$
$$+ x - x^4 + x^7 - x^{10} + x^{13} - \dots .$$

Now, we may easily notice two geometric series on the right hand side, both with the same ratio $-x^3$, which we can sum. And so our conjecture boils down to

$$\frac{1}{1-x+x^2} = \frac{1}{1+x^3} + \frac{x}{1+x^3} = \frac{1+x}{1+x^3},$$

which is, of course, true. We have proved our conjecture.

Our example, simple as it is, is typical in many respects. If we have to expand a given function, we can often obtain the first few coefficients without much trouble. Looking at these coefficients, we should try, as we did here, to guess the law governing the expansion. Having guessed the law, we should try, as we did here, to prove it. It may be a great advantage, as it was here, to work out the proof backwards, starting from an appropriate, clear statement of the conjecture.

By the way, our example is quite rewarding (which is also typical). It leads to a curious relation between binomial coefficients.

It is not superfluous to add that the problem of expanding a given function in a series frequently arises in various branches of mathematics. See the next section, and the Examples and Comments on Chapter VI.

2. Approximations.[2] Let E denote the length of the perimeter of an ellipse with semiaxes a and b. There is no simple expression for E in terms of a and b, but several approximate expressions have been proposed among which the following two are perhaps the most obvious:

$$P = \pi(a+b), \quad P' = 2\pi(ab)^{1/2};$$

[2] Cf. Putnam, 1949.

P is a proximate, P' another proximate, E an exact, expression for the same quantity, the length of the perimeter of the ellipse. When a coincides with b, the ellipse becomes a circle, and both P and P' coincide with E.

How closely do P and P' approximate E when a is different from b? Which one comes nearer to the truth, P or P'? Questions of this kind frequently arise in all branches of applied mathematics and there is a widely accepted procedure to deal with them which we may describe roughly as follows. *Expand $(P - E)/E$, the relative error of the approximation, in powers of a suitable small quantity and base your judgement upon the initial term* (the first non-vanishing term) *of the expansion.*

Let us see what this means and how the procedure works when applied to our case. First, we should choose a "suitable small quantity." We try ε, the numerical eccentricity of the ellipse, defined by the formula

$$\varepsilon = \frac{(a^2 - b^2)^{1/2}}{a};$$

we take a as the major, and b as the minor, semiaxis. When a becomes b and the ellipse a circle, ε vanishes. When the ellipse is not very different from a circle, ε is small. Therefore, let us expand the relative error into powers of ε. We obtain (let us skip the details here)

$$\frac{P - E}{E} = -\frac{1}{64}\varepsilon^4 + \ldots, \qquad \frac{P' - E}{E} = -\frac{3}{64}\varepsilon^4 + \ldots.$$

We computed only the initial term which, in both cases, is of order 4, contains ε^4. We omitted in both expansions the terms of higher order, containing $\varepsilon^5, \varepsilon^6, \ldots$. The terms omitted are negligible in comparison with the initial terms when ε is very small (infinitely small), that is, when the ellipse is almost circular. Therefore, for almost circular ellipses, P comes nearer to the true value E than P'. (In fact, the ratio of the errors becomes $1 : 3$ as ε tends to 0.) Both P and P' approximate E from below:

$$E > P > P'.$$

All this holds for very small ε, for almost circular ellipses. We do not know yet how much of these results remains valid when ε is not so small. In fact, at this moment we know only limit relations, valid for $\varepsilon \to 0$. We do not yet know anything definitive about the error of our approximations when $\varepsilon = 0.5$ or $\varepsilon = 0.1$. Of course, what we need in practice is information about such concrete cases.

In such circumstances, practical people test their formulas numerically. We may follow them, but which case should we test first? It is advisable not to forget the extreme cases. The numerical eccentricity ε varies between the extreme values 0 and 1. When $\varepsilon = 0$, $b = a$ and the ellipse becomes a circle. Yet we know this case fairly well by now and we turn rather to the other extreme case. When $\varepsilon = 1$, $b = 0$, the ellipse becomes a line segment of length $2a$, and the length of the perimeter is $4a$. We have

$$E = 4a, \quad P = \pi a, \quad P' = 0 \quad \text{when } \varepsilon = 1.$$

It may be worth noticing that in both extreme cases, for $\varepsilon = 1$ just as for very small ε, $E > P > P'$. Are these inequalities generally valid?

For the second inequality, the answer is easy. In fact, we have, for $a > b$,

$$P = \pi(a + b) > 2\pi(ab)^{1/2} = P'$$

since this is equivalent to

$$(a + b)^2 > 4ab$$

or to

$$(a - b)^2 > 0.$$

We focus our attention upon the remaining question. Is the inequality $E > P$ generally valid? It is natural to conjecture that what we found true in the extreme cases (ε small, and $\varepsilon = 1$) remains true in the intermediate cases (for all values of ε between 0 and 1). Our conjecture is not supported by many observations, that is true, but it is supported by analogy. A similar question (concerning $P > P'$) which we asked in the same breath and based on similar grounds has been answered in the affirmative.

Let us test a case numerically. We know a little more about the case where ε is nearly 0 than about the case where it is nearly 1. We choose a simple value for ε, nearer to 1 than to 0: $a = 5$, $b = 3$, $\varepsilon = 4/5$. We find for this ε (using appropriate tables)

$$E = 2\pi \times 4.06275, \quad P = 2\pi \times 4.00000.$$

The inequality $E > P$ is verified. This numerical verification of our conjecture comes from a new side, from a different source, and therefore carries some weight. Let us note also that

$$(P - E)/E = -0.0155, \qquad -\varepsilon^4/64 = -0.0064.$$

The relative error is about 1.5%. It is considerably larger than the initial term of its expansion, but has the same sign. As $\varepsilon = 4/5 = 0.8$ is not too small, our remark fits into the whole picture and tends to increase our confidence in the conjecture.

Approximate formulas play an important role in applied mathematics. Trying to judge such a formula, we often adopt in practice the procedure followed in this section. We compute the initial term in the expansion of the relative error and supplement the information so gained by numerical tests, considerations of analogy, etc., in short, by inductive, non-demonstrative reasoning.

3. Limits. In order to see inductive reasoning at work in still another domain, we consider the following problem.[3]

---

[3] See Putnam, 1948.

Let $a_1, a_2, \ldots, a_n, \ldots$ be an arbitrary sequence of positive numbers. Show that

$$\limsup_{n\to\infty} \left(\frac{a_1 + a_{n+1}}{a_n}\right)^n \geq e.$$

This problem requires some preliminary knowledge, especially familiarity with the concept of "lim sup" or "upper limit of indetermination."[4] Yet even if you are thoroughly familiar with this concept, you may experience some difficulty in finding a proof. My congratulations to any undergraduate who can do the problem by his own means in a few hours.

If you have struggled with the problem yourself a little while, you may follow with more sympathy the struggle described in the following sections.

4. Trying to disprove it. We begin with the usual questions.

What is the hypothesis? Just $a_n > 0$, nothing else.

What is the conclusion? That inequality with e on the right and that complicated limit on the left.

Do you know a related theorem? No, indeed. It is very different from anything I know.

Is it likely that the theorem is true? Or is it more likely that it is false? False, of course. In fact I cannot believe that such a precise consequence can be derived from such a broad hypothesis, just $a_n > 0$.

What are you required to do? To prove the theorem. Or to disprove it. I am very much for disproving it.

Can you test any particular case of the theorem? Yes, that is what I am about to do.

[In order to simplify the formulas, we set

$$\left(\frac{a_1 + a_{n+1}}{a_n}\right)^n = b_n$$

and write $b_n \to b$ for $\lim_{n\to\infty} b_n = b$.]

I try $a_n = 1$, for $n = 1, 2, 3, \ldots$. Then

$$b_n = \left(\frac{1+1}{1}\right)^n = 2^n \to \infty.$$

In this case, the assertion of the theorem is verified.

Yet I could set $a_1 = 0$, $a_n = 1$ for $n = 2, 3, 4, \ldots$. Then

$$b_n = \left(\frac{0+1}{1}\right)^n = 1^n \to 1 < e.$$

The theorem is exploded! No, it is not. The hypothesis allows $a_1 = 0.00001$, but it prohibits $a_1 = 0$. What a pity!

[4] See e.g., G. H. Hardy, *Pure Mathematics*, sect. 82.

Let me try something else. Let $a_n = n$. Then

$$b_n = \left(\frac{1 + (n+1)}{n}\right)^n = \left(1 + \frac{2}{n}\right)^n \to e^2.$$

Again verified.

Now, let $a_n = n^2$. Then

$$b_n = \left(\frac{1 + (n+1)^2}{n^2}\right)^n = \left(1 + \frac{2}{n}\frac{n+1}{n}\right)^n \to e^2.$$

Again verified. And again e^2. Should e^2 stand on the right hand side in the conclusion instead of e? That would improve the theorem.

Let me *introduce a parameter.* Let me take .... Yes, let me take $a_1 = c$, where I can dispose of c, but $a_n = n$ for $n = 2, 3, 4, \ldots$. Then

$$b_n = \left(\frac{c + (n+1)}{n}\right)^n = \left(1 + \frac{1+c}{n}\right)^n \to e^{1+c}.$$

This is always $> e$, since $c = a_1 > 0$. Yet it can come as close to e as we please, since c can be arbitrarily small. I cannot disprove it, I cannot prove it.

Just one more trial. Let me take $a_n = n^c$. Then [we skip some computations]

$$b_n = \left[\frac{1 + (n+1)^c}{n^c}\right]^n \to \begin{cases} \infty \text{ if } 0 < c < 1, \\ e^2 \text{ if } c = 1, \\ e^c \text{ if } c > 1. \end{cases}$$

Again, the limit can come as close to e as we please, but remains always superior to e. I shall never succeed in bringing down this ... limit below e. It is time to turn round.

5. Trying to prove it. In fact, the indications for a *volte-face* are quite strong. In the light of the accumulated inductive evidence the prospects of disproving the theorem appear so dim that the prospects of proving it look relatively bright.

Therefore, nothing remains but to start reexamining the theorem, its statement, its hypothesis, its conclusion, the concepts involved, etc.

Can you relax the hypothesis? No, I cannot. If I admit $a_n = 0$, the conclusion is no more valid, the theorem becomes false ($a_1 = 0, a_2 = a_3 = a_4 = \ldots = 1$).

Can you improve the conclusion? I certainly cannot improve it by substituting some greater number for e, since then the conclusion is no more valid, the theorem becomes false (examples in the foregoing sect. 4).

Have you taken into account all essential notions involved in the problem? No, I have not. That may be the trouble.

What have you failed to take into account? The definition of lim sup. The definition of the number e.

What is $\limsup b_n$? It is the upper limit of indetermination of b_n as n tends to infinity.

What is e? I could define e in various ways. The above examples suggest that the most familiar definition of e may be the best:

$$e = \lim_{n\to\infty}\left(1 + \frac{1}{n}\right)^n.$$

Could you restate the theorem? . . . .

Could you restate the theorem in some more accessible form? . . . .

Could you restate the conclusion? What is the conclusion? The conclusion contains e. What is e? (I failed to ask this before.) Oh, yes—the conclusion is

$$\limsup_{n\to\infty}\left(\frac{a_1 + a_{n+1}}{a_n}\right)^n \geqq \lim_{n\to\infty}\left(1 + \frac{1}{n}\right)^n$$

or, which is the same,

$$\limsup_{n\to\infty}\left[\frac{n(a_1 + a_{n+1})}{(n+1)a_n}\right]^n \geqq 1.$$

This looks much better!

Can the conclusion be false, when the hypothesis is fulfilled? Yes, that is the question. Let me see it. Let me look squarely at the negation of the assertion, at the exactly opposite assertion. Let me write it down:

$$(?) \qquad \limsup_{n\to\infty}\left[\frac{n(a_1 + a_{n+1})}{(n+1)a_n}\right]^n < 1.$$

I put a query in front of it, because just this point is in doubt. Let me call it the "formula (?)." What does (?) mean? It certainly implies that there is an N such that

$$\left[\frac{n(a_1 + a_{n+1})}{(n+1)a_n}\right]^n < 1 \text{ for } n \geqq N.$$

It follows hence that

$$\frac{n(a_1 + a_{n+1})}{(n+1)a_n} < 1 \text{ for } n \geqq N.$$

It follows further . . . . Let me try something. Yes, I can write it neatly! It follows further from (?) that

$$\frac{a_1}{n+1} + \frac{a_{n+1}}{n+1} < \frac{a_n}{n}$$

or

$$\frac{a_{n+1}}{n+1} - \frac{a_n}{n} < -\frac{a_1}{n+1} \text{ for } n \geqq N.$$

Let me write this out broadly. It follows that

$$\frac{a_n}{n} - \frac{a_{n-1}}{n-1} < -\frac{a_1}{n}$$

$$\frac{a_{n-1}}{n-1} - \frac{a_{n-2}}{n-2} < -\frac{a_1}{n-1}$$

$$\cdots\cdots\cdots\cdots$$

$$\frac{a_{N+1}}{N+1} - \frac{a_N}{N} < -\frac{a_1}{N+1}$$

and so

$$\frac{a_n}{n} < \frac{a_N}{N} - a_1\left(\frac{1}{N+1} + \frac{1}{N+2} + \ldots + \frac{1}{n-1} + \frac{1}{n}\right)$$

$$= C - a_1\left(1 + \frac{1}{2} + \frac{1}{3} + \ldots + \frac{1}{n}\right)$$

where C is a constant, independent of n provided that $n \geqq N$. It does not really matter but, in fact,

$$C = \frac{a_N}{N} + a_1\left(1 + \frac{1}{2} + \ldots + \frac{1}{N}\right).$$

It matters, however, that n can be arbitrarily large, and that the harmonic series diverges. It follows, therefore, that

$$\lim_{n\to\infty} \frac{a_n}{n} = -\infty.$$

Now, this contradicts flatly the hypothesis that $a_n > 0$ for $n = 1, 2, 3, \ldots$. Yet this contradiction follows faultlessly from the formula (?). Therefore, in fact, (?) must be responsible for the contradiction; (?) is incompatible with the hypothesis $a_n > 0$; the opposite to (?) must be true—the theorem is proved!

6. The role of the inductive phase. Looking back at the foregoing solution superficially, we could think that the first, inductive, phase of the solution (sect. 4) is not used at all in the second, demonstrative, phase (sect. 5). Yet this is not so. The inductive phase was useful in several respects.

First, examining concrete particular cases of the theorem, we understood it thoroughly, and realized its full meaning. We satisfied ourselves that its hypothesis is essential, its conclusion sharp. This information was helpful in the second phase: we knew that we must use the whole hypothesis and that we must take into account the precise value of the constant e.

Second, having verified the theorem in several particular cases, we gathered strong inductive evidence for it. The inductive phase overcame our initial suspicion and gave us a strong confidence in the theorem. Without such confidence we would have scarcely found the courage to undertake the proof which did not look at all a routine job. "When you have satisfied yourself that the theorem is true, you start proving it"—the traditional mathematics professor is quite right.

Third, the examples in which the familiar limit formula for e popped up again and again, gave us reasonable ground for introducing that limit formula into the statement of the theorem. And introducing it turned out the crucial step toward the solution.

On the whole, it seems natural and reasonable that the inductive phase precedes the demonstrative phase. First guess, then prove.

EXAMPLES AND COMMENTS ON CHAPTER V.

1. By multiplying the series

$$(1 - x^2)^{-1/2} = 1 + \frac{1}{2}x^2 + \frac{1\;3}{2\;4}x^4 + \ldots$$

$$\arcsin x = \frac{x}{1} + \frac{1}{2}\frac{x^3}{3} + \frac{1\;3}{2\;4}\frac{x^5}{5} + \ldots$$

you find the first terms of the expansion

$$y = (1 - x^2)^{-1/2} \arcsin x = x + \frac{2}{3}x^3 + \ldots .$$

(a) Compute a few more terms and try to guess the general term.

(b) Show that y satisfies the differential equation

$$(1 - x^2)y' - xy = 1$$

and use this equation to prove your guess.

2. By multiplying the series

$$e^{x^2/2} = 1 + \frac{x^2}{2} + \frac{x^4}{2 \cdot 4} + \ldots$$

$$\int_0^x e^{-t^2/2}\,dt = \frac{x}{1} - \frac{1}{2}\frac{x^3}{3} + \frac{1}{2 \cdot 4}\frac{x^5}{5} - \ldots$$

you find the first terms of the expansion

$$y = e^{x^2/2} \int_0^x e^{-t^2/2}\,dt = x + \frac{1}{3}x^3 + \ldots .$$

(a) Compute a few more terms and try to guess the general term.

(b) Your guess, if correct, suggests that y satisfies a simple differential equation. By establishing this equation, prove your guess.

3. The functional equation

$$f(x) = \frac{1}{1+x} f\left(\frac{2\sqrt{x}}{1+x}\right)$$

is satisfied by the power series

$$f(x) = 1 + \frac{1}{4}x^2 + \frac{9}{64}x^4 + \frac{25}{256}x^6 + \frac{1225}{16384}x^8 + \dots .$$

Verify these coefficients, derive a few more, if necessary, and try to guess the general term.

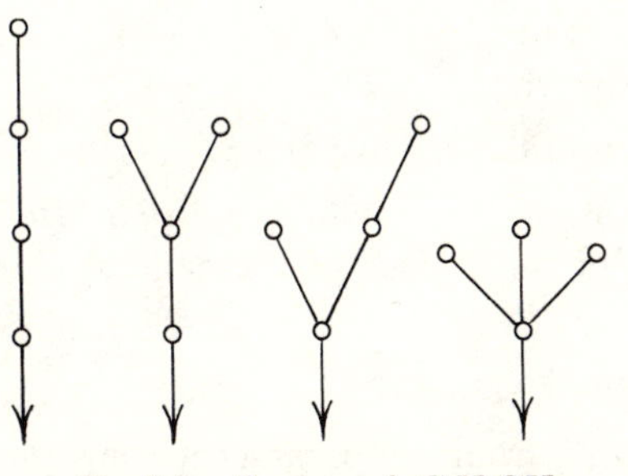

Fig. 5.1. Compounds C_4H_9OH.

4. The functional equation

$$f(x) = 1 + \frac{x}{6}[f(x)^3 + 3f(x)f(x^2) + 2f(x^3)]$$

is satisfied by the power series

$$f(x) = 1 + x + x^2 + 2x^3 + 4x^4 + \dots + a_n x^n + \dots .$$

It is asserted that a_n is the number of the structurally different chemical compounds (aliphatic alcohols) having the same chemical formula $C_nH_{2n+1}OH$. In the case $n = 4$, the answer is true. There are $a_4 = 4$ alcohols C_4H_9OH; they are represented in fig. 5.1, each compound as a "tree," each C as a little circle or "knot," and the radical —OH as an arrow; the H's are dropped. Test other values of n.

5.
$$\sum_{k=0}^{n/2} (-1)^k \binom{n-k}{k} = 1, 1, 0, -1, -1, 0$$
according as $n \equiv 0, 1, 2, 3, 4, 5 \pmod{6}$.

6. An ellipse describes a prolate, or an oblate, spheroid according as it rotates about its major, or minor, axis.

For the area of the surface of the prolate spheriod

$$E = 2\pi ab[(1 - \varepsilon^2)^{1/2} + (\arcsin \varepsilon)/\varepsilon], \quad P = 4\pi(a^2 + 2b^2)/3$$

are the exact, and a proximate expression, respectively (a, b, and ε as in sect. 2). Find

(a) the initial term of the relative error
(b) the relative error when $b = 0$.

What about the sign of the relative error?

7. For the area of the surface of the oblate ellipsoid

$$E = 2\pi a^2 \left[1 + \frac{1-\varepsilon^2}{2\varepsilon} \log \frac{1+\varepsilon}{1-\varepsilon}\right], \qquad P = \frac{4\pi(2a^2+b^2)}{3}$$

are the exact, and a proximate, expression, respectively. Find

(a) the initial term of the relative error
(b) the relative error when $b = 0$.

What about the sign of the relative error?

8. Comparing ex. 6 and ex. 7, which approximate formula would you propose for the area of the surface of the general ellipsoid with semiaxes a, b, and c?

What about the sign of the error?

9. [Sect. 2.] Starting from the parametric representation of the ellipse, $x = a \sin t$, $y = b \cos t$, show that

$$E = 4a \int_0^{\pi/2} (1 - \varepsilon^2 \sin^2 t)^{1/2}\, dt$$

$$= 2\pi a \left[1 - \sum_1^{\infty} \left(\frac{1}{2}\frac{3}{4} \cdots \frac{2n-1}{2n}\right)^2 \frac{\varepsilon^{2n}}{2n-1}\right]$$

and derive hence the initial terms given without proof in sect. 2.

10 (continued). Using the expansions in powers of ε, prove that $E > P$ for $0 < \varepsilon \leqq 1$.

11. [Sect. 2.] Determine the number α so that the expression

$$P'' = \alpha P + (1 - \alpha)P'$$

should yield the best possible approximation to E for small ε. (That is, the order of the initial term of $(P'' - E)/E$ should be as high as possible.)

12 (continued). Investigate the approximation by P'' following the method of sect. 2. (Inductively!)

13. Given a positive integer p and a sequence of positive numbers $a_1, a_2, a_3, \ldots, a_n, \ldots$. Show that

$$\limsup_{n\to\infty} \left(\frac{a_1 + a_{n+p}}{a_n}\right)^n \geqq e^p.$$

14 (continued). Point out a sequence $a_1, a_2, a_3, \ldots$ for which equality is attained.

15. *Explain the observed regularities.* A discovery in physics is often attained in two steps. First a certain regularity is noticed in the data of observation. Then this regularity is explained as a consequence of some general law. Different persons may take the two steps which may be separated by a long interval of time. A great example is that of Kepler and Newton: the regularities in the motion of the planets observed by Kepler have been explained by the law of gravitation discovered by Newton. Something similar may happen in mathematical research, and here is a neat example which requires little preliminary knowledge.

The usual table of four-place common logarithms lists 900 mantissas, those of the logarithms of the integers from 100 to 999. We may be inclined to think, before observation, that the ten figures, 0, 1, . . . , 9 are about equally frequent in these tables, but this is not so: they certainly do not turn up equally often as the *first* figure of the mantissa. By counting the mantissas that have the same first figure, we obtain Table I. (Check it!)

Table I. Mantissas with the same first figure in four-place logarithms

First figure	Nr. of mantissas	Ratios
0	26	
		1.269
1	33	
		1.242
2	41	
		1.268
3	52	
		1.250
4	65	
		1.262
5	82	
		1.256
6	103	
		1.252
7	129	
		1.271
8	164	
		1.250
9	205	
	Total 900	

Inspecting the second column of Table I we may notice that any two consecutive numbers in it have approximately the same ratio. This induces us to compute these ratios to a few decimals: they are listed in the last column of Table I.

Why are these ratios approximately equal? Try to perceive some precise regularity behind the observed approximate regularity. The numbers in the second column of Table I are approximately the terms of a geometric progression. Could you discover an exact geometric progression to which the terms of the approximate progression are simply related? [The ratio of the exact progression should be, perhaps, some kind of average of the ratios listed in the last column of Table I.]

16. *Classify the observed facts.* A great part of the naturalist's work is aimed at describing and classifying the objects that he observes. Such work was

Fig. 5.2. Symmetries of friezes.

predominant for a long time after Linnaeus when the main activity of the naturalists consisted in describing new species and genera of plants and animals, and in reclassifying the known species and genera. Not only plants and animals are described and classified by the naturalists, but also other objects, especially minerals; the classification of crystals is based on their symmetry. A good classification is important; it reduces the observable variety to relatively few clearly characterized and well ordered types. The mathematician has not often opportunity to indulge in description and classification, but it may happen.

If you are acquainted with a few simple notions of plane geometry (line of symmetry, center of symmetry) you may have fun with ornaments. Fig. 5.2

exhibits fourteen ornamental bands each of which is generated by a simple figure, repeated periodically along a (horizontal) straight line. Let us call such a band a "frieze." Match each frieze on the left-hand side of fig. 5.2 (marked with a numeral) with a frieze on the right-hand side (marked with a letter) so that the two friezes matched have the *same type of symmetry*. Moreover, examine ornamental bands which you can find on all sorts of objects, or in older architectural works, and try to match each with a frieze in fig. 5.2. Finally, give a complete list of the various types of symmetry that a frieze may have and an exhaustive description of each type of symmetry. [Consider a frieze as infinitely long, in both directions, and the generating figure as periodically repeated an infinity of times. Observe that the term "type of symmetry" was not formally defined: to arrive at an appropriate interpretation of this term is an important part of your task.]

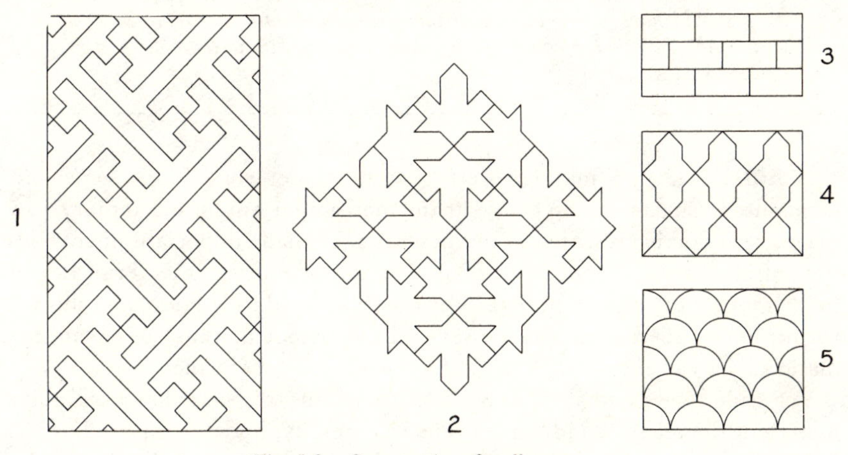

Fig. 5.3. Symmetries of wallpapers.

17. Find two ornaments in fig. 5.3 that have the same type of symmetry. Each ornament must be conceived as covering the whole plane with its repeated patterns.

18. *What is the difference?* The twenty-six capital letters are divided into five batches as follows:

A M T U V W Y
B C D E K
N S Z
H I O X
F G J L P Q R.

What is the difference? What could be a simple basis for the exhibited classification? [Look at the five equations:

$$y = x^2, \quad y^2 = x, \quad y = x^3, \quad x^2 + 2y^2 = 1, \quad y = x + x^4.$$

What is the difference?]

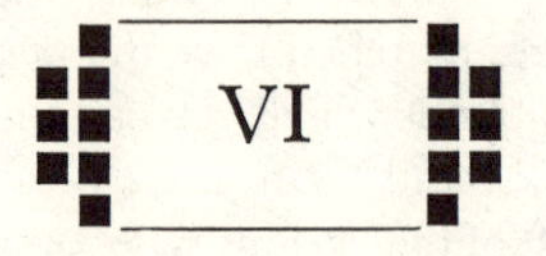

A MORE GENERAL STATEMENT

He [Euler] preferred instructing his pupils to the little satisfaction of amazing them. He would have thought not to have done enough for science if he should have failed to add to the discoveries, with which he enriched science, the candid exposition of the ideas that led him to those discoveries.
—CONDORCET

1. Euler. Of all mathematicians with whose work I am somewhat acquainted, Euler seems to be by far the most important for our inquiry. A master of inductive research in mathematics, he made important discoveries (on infinite series, in the Theory of Numbers, and in other branches of mathematics) by induction, that is, by observation, daring guess, and shrewd verification. In this respect, however, Euler is not unique; other mathematicians, great and small, used induction extensively in their work.

Yet Euler seems to me almost unique in one respect: he takes pains to present the relevant inductive evidence carefully, in detail, in good order. He presents it convincingly but honestly, as a genuine scientist should do. His presentation is "the candid exposition of the ideas that led him to those discoveries" and has a distinctive charm. Naturally enough, as any other author, he tries to impress his readers, but, as a really good author, he tries to impress his readers only by such things as have genuinely impressed himself.

The next section brings a sample of Euler's writing. The memoir chosen can be read with very little previous knowledge and is entirely devoted to the exposition of an inductive argument.

2. Euler's memoir is given here, in English translation, *in extenso*, except for a few unessential alterations which should make it more accessible to a modern reader.[1]

[1] The original is in French; see Euler's *Opera Omnia*, ser. 1, vol. 2, p. 241–253. The alterations consist in a different notation (footnote 2), in the arrangement of a table (explained in footnote 3), in slight changes affecting a few formulas, and in dropping a repetition of former arguments in the last No. 13 of the memoir. The reader may consult the easily available original.

DISCOVERY OF A MOST EXTRAORDINARY LAW OF THE NUMBERS CONCERNING THE SUM OF THEIR DIVISORS

1. Till now the mathematicians tried in vain to discover some order in the sequence of the prime numbers and we have every reason to believe that there is some mystery which the human mind shall never penetrate. To convince oneself, one has only to glance at the tables of the primes, which some people took the trouble to compute beyond a hundred thousand, and one perceives that there is no order and no rule. This is so much more surprising as the arithmetic gives us definite rules with the help of which we can continue the sequence of the primes as far as we please, without noticing, however, the least trace of order. I am myself certainly far from this goal, but I just happened to discover an extremely strange law governing the sums of the divisors of the integers which, at the first glance, appear just as irregular as the sequence of the primes, and which, in a certain sense, comprise even the latter. This law, which I shall explain in a moment, is, in my opinion, so much more remarkable as it is of such a nature that we can be assured of its truth without giving it a perfect demonstration. Nevertheless, I shall present such evidence for it as might be regarded as almost equivalent to a rigorous demonstration.

2. A prime number has no divisors except unity and itself, and this distinguishes the primes from the other numbers. Thus 7 is a prime, for it is divisible only by 1 and itself. Any other number which has, besides unity and itself, further divisors, is called composite, as for instance, the number 15, which has, besides 1 and 15, the divisors 3 and 5. Therefore, generally, if the number p is prime, it will be divisible only by 1 and p; but if p was composite, it would have, besides 1 and p, further divisors. Therefore, in the first case, the sum of its divisors will be $1 + p$, but in the latter it would exceed $1 + p$. As I shall have to consider the sum of divisors of various numbers, I shall use[2] the sign $\sigma(n)$ to denote the sum of the divisors of the number n. Thus, $\sigma(12)$ means the sum of all the divisors of 12, which are 1, 2, 3, 4, 6, and 12; therefore, $\sigma(12) = 28$. In the same way, one can see that $\sigma(60) = 168$ and $\sigma(100) = 217$. Yet, since unity is only divisible by itself, $\sigma(1) = 1$. Now, 0 (zero) is divisible by all numbers. Therefore, $\sigma(0)$ should be properly infinite. (However, I shall assign to it later a finite value, different in different cases, and this will turn out serviceable.)

3. Having defined the meaning of the symbol $\sigma(n)$, as above, we see clearly that if p is a prime $\sigma(p) = 1 + p$. Yet $\sigma(1) = 1$ (and not $1 + 1$); hence we see that 1 should be excluded from the sequence of the primes; 1 is the beginning of the integers, neither prime nor composite. If, however, n is composite, $\sigma(n)$ is greater than $1 + n$.

---

[2] Euler was the first to introduce a symbol for the sum of the divisors; he used $\int n$, not the modern $\sigma(n)$ of the text.

In this case we can easily find $\sigma(n)$ from the factors of n. If $a, b, c, d, \ldots$ are different primes, we see easily that

$$\sigma(ab) = 1 + a + b + ab = (1 + a)(1 + b) = \sigma(a)\sigma(b),$$
$$\sigma(abc) = (1 + a)(1 + b)(1 + c) = \sigma(a)\sigma(b)\sigma(c),$$
$$\sigma(abcd) = \sigma(a)\sigma(b)\sigma(c)\sigma(d)$$

and so on. We need particular rules for the powers of primes, as

$$\sigma(a^2) = 1 + a + a^2 = \frac{a^3 - 1}{a - 1}$$

$$\sigma(a^3) = 1 + a + a^2 + a^3 = \frac{a^4 - 1}{a - 1}$$

and, generally,

$$\sigma(a^n) = \frac{a^{n+1} - 1}{a - 1}.$$

Using this, we can find the sum of the divisors of any number, composite in any way whatever. This we see from the formulas

$$\sigma(a^2b) = \sigma(a^2)\sigma(b)$$
$$\sigma(a^3b^2) = \sigma(a^3)\sigma(b^2)$$
$$\sigma(a^3b^4c) = \sigma(a^3)\sigma(b^4)\sigma(c)$$

and, generally,

$$\sigma(a^\alpha b^\beta c^\gamma d^\delta e^\varepsilon) = \sigma(a^\alpha)\sigma(b^\beta)\sigma(c^\gamma)\sigma(d^\delta)\sigma(e^\varepsilon).$$

For instance, to find $\sigma(360)$ we set, since 360 factorized is $2^3 \cdot 3^2 \cdot 5$,

$$\sigma(360) = \sigma(2^3)\sigma(3^2)\sigma(5) = 15 \cdot 13 \cdot 6 = 1170.$$

4. In order to show the sequence of the sums of the divisors, I add the following table[3] containing the sums of the divisors of all integers from 1 up to 99.

n	0	1	2	3	4	5	6	7	8	9
0	—	1	**3**	**4**	7	**6**	12	**8**	15	13
10	18	**12**	28	**14**	24	24	31	**18**	39	**20**
20	42	32	36	**24**	60	31	42	40	56	**30**
30	72	**32**	63	48	54	48	91	**38**	60	56
40	90	**42**	96	**44**	84	78	72	**48**	124	57
50	93	72	98	**54**	120	72	120	80	90	**60**
60	168	**62**	96	104	127	84	144	**68**	126	96
70	144	**72**	195	**74**	114	124	140	96	168	**80**
80	186	121	126	**84**	224	108	132	120	180	**90**
90	234	112	168	128	144	120	252	**98**	171	156

[3] The number in the intersection of the row marked 60 and the column marked 7, that is, 68, is $\sigma(67)$. If p is prime, $\sigma(p)$ is in heavy print. This arrangement of the table is a little more concise than the arrangement in the original.

If we examine a little the sequence of these numbers, we are almost driven to despair. We cannot hope to discover the least order. The irregularity of the primes is so deeply involved in it that we must think it impossible to disentangle any law governing this sequence, unless we know the law governing the sequence of the primes itself. It could appear even that the sequence before us is still more mysterious than the sequence of the primes.

5. Nevertheless, I observed that this sequence is subject to a completely definite law and could even be regarded as a *recurring* sequence. This mathematical expression means that each term can be computed from the foregoing terms, according to an invariable rule. In fact, if we let $\sigma(n)$ denote any term of this sequence, and $\sigma(n-1)$, $\sigma(n-2)$, $\sigma(n-3)$, $\sigma(n-4)$, $\sigma(n-5)$, . . . the preceding terms, I say that the value of $\sigma(n)$ can always be combined from some of the preceding as prescribed by the following formula:

$$\begin{aligned}\sigma(n) = \sigma(n-1) &+ \sigma(n-2) - \sigma(n-5) - \sigma(n-7)\\ &+ \sigma(n-12) + \sigma(n-15) - \sigma(n-22) - \sigma(n-26)\\ &+ \sigma(n-35) + \sigma(n-40) - \sigma(n-51) - \sigma(n-57)\\ &+ \sigma(n-70) + \sigma(n-77) - \sigma(n-92) - \sigma(n-100)\\ &+ \cdots .\end{aligned}$$

On this formula we must make the following remarks.

I. In the sequence of the signs $+$ and $-$, each arises twice in succession.

II. The law of the numbers 1, 2, 5, 7, 12, 15, . . . which we have to subtract from the proposed number n, will become clear if we take their differences:

Nrs.	1,	2,	5,	7,	12,	15,	22,	26,	35,	40,	51,	57,	70,	77,	92,	100, . . .
Diff.	**1,**	3,	**2,**	5,	**3,**	7,	**4,**	9,	**5,**	11,	**6,**	13,	**7,**	15,	**8,** . . .	

In fact, we have here, alternately, all the integers 1, 2, 3, 4, 5, 6, . . . and the odd numbers 3, 5, 7, 9, 11, . . . , and hence we can continue the sequence of these numbers as far as we please.

III. Although this sequence goes to infinity, we must take, in each case, only those terms for which the numbers under the sign σ are still positive and omit the σ for negative values.

IV. If the sign $\sigma(0)$ turns up in the formula, we must, as its value in itself is indeterminate, substitute for $\sigma(0)$ the number n proposed.

6. After these remarks it is not difficult to apply the formula to any given particular case, and so anybody can satisfy himself of its truth by as many examples as he may wish to develop. And since I must admit that I am not in a position to give it a rigorous demonstration, I will justify it by a sufficiently large number of examples.

$$
\begin{array}{lll}
\sigma(1) = \sigma(0) & = 1 & = 1 \\
\sigma(2) = \sigma(1) + \sigma(0) & = 1 + 2 & = 3 \\
\sigma(3) = \sigma(2) + \sigma(1) & = 3 + 1 & = 4 \\
\sigma(4) = \sigma(3) + \sigma(2) & = 4 + 3 & = 7 \\
\sigma(5) = \sigma(4) + \sigma(3) - \sigma(0) & = 7 + 4 - 5 & = 6 \\
\sigma(6) = \sigma(5) + \sigma(4) - \sigma(1) & = 6 + 7 - 1 & = 12 \\
\sigma(7) = \sigma(6) + \sigma(5) - \sigma(2) - \sigma(0) & = 12 + 6 - 3 - 7 & = 8 \\
\sigma(8) = \sigma(7) + \sigma(6) - \sigma(3) - \sigma(1) & = 8 + 12 - 4 - 1 & = 15 \\
\sigma(9) = \sigma(8) + \sigma(7) - \sigma(4) - \sigma(2) & = 15 + 8 - 7 - 3 & = 13 \\
\sigma(10) = \sigma(9) + \sigma(8) - \sigma(5) - \sigma(3) & = 13 + 15 - 6 - 4 & = 18 \\
\sigma(11) = \sigma(10) + \sigma(9) - \sigma(6) - \sigma(4) & = 18 + 13 - 12 - 7 & = 12 \\
\sigma(12) = \sigma(11) + \sigma(10) - \sigma(7) - \sigma(5) + \sigma(0) & = 12 + 18 - 8 - 6 + 12 & = 28 \\
\sigma(13) = \sigma(12) + \sigma(11) - \sigma(8) - \sigma(6) + \sigma(1) & = 28 + 12 - 15 - 12 + 1 & = 14 \\
\sigma(14) = \sigma(13) + \sigma(12) - \sigma(9) - \sigma(7) + \sigma(2) & = 14 + 28 - 13 - 8 + 3 & = 24 \\
\sigma(15) = \sigma(14) + \sigma(13) - \sigma(10) - \sigma(8) + \sigma(3) + \sigma(0) & = 24 + 14 - 18 - 15 + 4 + 15 & = 24 \\
\sigma(16) = \sigma(15) + \sigma(14) - \sigma(11) - \sigma(9) + \sigma(4) + \sigma(1) & = 24 + 24 - 12 - 13 + 7 + 1 & = 31 \\
\sigma(17) = \sigma(16) + \sigma(15) - \sigma(12) - \sigma(10) + \sigma(5) + \sigma(2) & = 31 + 24 - 28 - 18 + 6 + 3 & = 18 \\
\sigma(18) = \sigma(17) + \sigma(16) - \sigma(13) - \sigma(11) + \sigma(6) + \sigma(3) & = 18 + 31 - 14 - 12 + 12 + 4 & = 39 \\
\sigma(19) = \sigma(18) + \sigma(17) - \sigma(14) - \sigma(12) + \sigma(7) + \sigma(4) & = 39 + 18 - 24 - 28 + 8 + 7 & = 20 \\
\sigma(20) = \sigma(19) + \sigma(18) - \sigma(15) - \sigma(13) + \sigma(8) + \sigma(5) & = 20 + 39 - 24 - 14 + 15 + 6 & = 42
\end{array}
$$

I think these examples are sufficient to discourage anyone from imagining that it is by mere chance that my rule is in agreement with the truth.

7. Yet somebody could still doubt whether the law of the numbers 1, 2, 5, 7, 12, 15, . . . which we have to subtract is precisely that one which I have indicated, since the examples given imply only the first six of these numbers. Thus, the law could still appear as insufficiently established and, therefore, I will give some examples with larger numbers.

I. Given the number 101, find the sum of its divisors. We have

$$
\begin{array}{rl}
\sigma(101) = & \sigma(100) + \sigma(99) - \sigma(96) - \sigma(94) \\
& + \ \sigma(89) + \sigma(86) - \sigma(79) - \sigma(75) \\
& + \ \sigma(66) + \sigma(61) - \sigma(50) - \sigma(44) \\
& + \ \sigma(31) + \sigma(24) - \sigma(9) - \sigma(1) \\
= & 217 + 156 - 252 - 144 \\
& + \ 90 + 132 - 80 - 124 \\
& + \ 144 + 62 - 93 - 84 \\
& + \ 32 + 60 - 13 - 1 \\
= & 893 - 791 \\
= & 102
\end{array}
$$

and hence we could conclude, if we would not have known it before, that 101 is a prime number.

II. Given the number 301, find the sum of its divisors. We have

$$
\begin{array}{rccccccccc}
\text{diff.} & & 1 & & 3 & & 2 & & 5 & \\
\sigma(301) = & & \sigma(300) & + & \sigma(299) & - & \sigma(296) & - & \sigma(294) & + \\
& & 3 & & 7 & & 4 & & 9 & \\
& + & \sigma(289) & + & \sigma(286) & - & \sigma(279) & - & \sigma(275) & + \\
& & 5 & & 11 & & 6 & & 13 & \\
& + & \sigma(266) & + & \sigma(261) & - & \sigma(250) & - & \sigma(244) & + \\
& & 7 & & 15 & & 8 & & 17 & \\
& + & \sigma(231) & + & \sigma(224) & - & \sigma(209) & - & \sigma(201) & + \\
& & 9 & & 19 & & 10 & & 21 & \\
& + & \sigma(184) & + & \sigma(175) & - & \sigma(156) & - & \sigma(146) & + \\
& & 11 & & 23 & & 12 & & 25 & \\
& + & \sigma(125) & + & \sigma(114) & - & \sigma(91) & - & \sigma(79) & + \\
& & 13 & & 27 & & 14 & & & \\
& + & \sigma(54) & + & \sigma(41) & - & \sigma(14) & - & \sigma(0). &
\end{array}
$$

We see by this example how we can, using the differences, continue the formula as far as is necessary in each case. Performing the computations, we find

$$\sigma(301) = 4939 - 4587 = 352.$$

We see hence that 301 is not a prime. In fact, $301 = 7 \cdot 43$ and we obtain

$$\sigma(301) = \sigma(7)\sigma(43) = 8 \cdot 44 = 352$$

as the rule has shown.

8. The examples that I have just developed will undoubtedly dispel any qualms which we might have had about the truth of my formula. Now, this beautiful property of the numbers is so much more surprising as we do not perceive any intelligible connection between the structure of my formula and the nature of the divisors with the sum of which we are here concerned. The sequence of the numbers 1, 2, 5, 7, 12, 15, . . . does not seem to have any relation to the matter in hand. Moreover, as the law of these numbers is "interrupted" and they are in fact a mixture of two sequences with a regular law, of 1, 5, 12, 22, 35, 51, . . . and 2, 7, 15, 26, 40, 57, . . . , we would not expect that such an irregularity can turn up in Analysis. The lack of demonstration must increase the surprise still more, since it seems wholly impossible to succeed in discovering such a property without being guided by some reliable method which could take the place of a perfect proof. I confess that I did not hit on this discovery by mere chance, but another proposition opened the path to this beautiful property—another proposition of the same nature which must be accepted as true although I am unable to

prove it. And although we consider here the nature of integers to which the Infinitesimal Calculus does not seem to apply, nevertheless I reached my conclusion by differentiations and other devices. I wish that somebody would find a shorter and more natural way, in which the consideration of the path that I followed might be of some help, perhaps.

9. In considering the partitions of numbers, I examined, a long time ago, the expression

$$(1-x)(1-x^2)(1-x^3)(1-x^4)(1-x^5)(1-x^6)(1-x^7)(1-x^8)\ldots,$$

in which the product is assumed to be infinite. In order to see what kind of series will result, I multiplied actually a great number of factors and found

$$1-x-x^2+x^5+x^7-x^{12}-x^{15}+x^{22}+x^{26}-x^{35}-x^{40}+\ldots.$$

The exponents of x are the same which enter into the above formula; also the signs $+$ and $-$ arise twice in succession. It suffices to undertake this multiplication and to continue it as far as it is deemed proper to become convinced of the truth of this series. Yet I have no other evidence for this, except a long induction which I have carried out so far that I cannot in any way doubt the law governing the formation of these terms and their exponents. I have long searched in vain for a rigorous demonstration of the equation between the series and the above infinite product $(1-x)(1-x^2)(1-x^3)\ldots$, and I have proposed the same question to some of my friends with whose ability in these matters I am familiar, but all have agreed with me on the truth of this transformation of the product into a series, without being able to unearth any clue of a demonstration. Thus, it will be a known truth, but not yet demonstrated, that if we put

$$s=(1-x)(1-x^2)(1-x^3)(1-x^4)(1-x^5)(1-x^6)\ldots$$

the same quantity s can also be expressed as follows:

$$s=1-x-x^2+x^5+x^7-x^{12}-x^{15}+x^{22}+x^{26}-x^{35}-x^{40}+\ldots.$$

For each of us can convince himself of this truth by performing the multiplication as far as he may wish; and it seems impossible that the law which has been discovered to hold for 20 terms, for example, would not be observed in the terms that follow.

10. As we have thus discovered that those two infinite expressions are equal even though it has not been possible to demonstrate their equality, all the conclusions which may be deduced from it will be of the same nature, that is, true but not demonstrated. Or, if one of these conclusions could be demonstrated, one could reciprocally obtain a clue to the demonstration of that equation; and it was with this purpose in mind that I maneuvered those two expressions in many ways, and so I was led among other discoveries to that which I explained above; its truth, therefore, must be as certain as

that of the equation between the two infinite expressions. I proceeded as follows. Being given that the two expressions

$$\text{I.}\ s = (1-x)(1-x^2)(1-x^3)(1-x^4)(1-x^5)(1-x^6)(1-x^7)\ldots$$

$$\text{II.}\ s = 1 - x - x^2 + x^5 + x^7 - x^{12} - x^{15} + x^{22} + x^{26} - x^{35} - x^{40} + \ldots$$

are equal, I got rid of the factors in the first by taking logarithms

$$\log s = \log(1-x) + \log(1-x^2) + \log(1-x^3) + \log(1-x^4) + \ldots.$$

In order to get rid of the logarithms, I differentiate and obtain the equation

$$\frac{1}{s}\frac{ds}{dx} = -\frac{1}{1-x} - \frac{2x}{1-x^2} - \frac{3x^2}{1-x^3} - \frac{4x^3}{1-x^4} - \frac{5x^4}{1-x^5} - \ldots$$

or

$$-\frac{x}{s}\frac{ds}{dx} = \frac{x}{1-x} + \frac{2x^2}{1-x^2} + \frac{3x^3}{1-x^3} + \frac{4x^4}{1-x^4} + \frac{5x^5}{1-x^5} + \ldots.$$

From the second expression for s, as infinite series, we obtain another value for the same quantity

$$-\frac{x}{s}\frac{ds}{dx} = \frac{x + 2x^2 - 5x^5 - 7x^7 + 12x^{12} + 15x^{15} - 22x^{22} - 26x^{26} + \ldots}{1 - x - x^2 + x^5 + x^7 - x^{12} - x^{15} + x^{22} + x^{26} - \ldots}.$$

11. Let us put

$$-\frac{x}{s}\frac{ds}{dx} = t.$$

We have above two expressions for the quantity t. In the first expression, I expand each term into a geometric series and obtain

$$\begin{aligned} t = x + x^2 + x^3 + x^4 + x^5 + x^6 + x^7 + x^8 &+ \ldots \\ + 2x^2 \qquad + 2x^4 \qquad + 2x^6 \qquad + 2x^8 &+ \ldots \\ + 3x^3 \qquad\qquad + 3x^6 \qquad\qquad &+ \ldots \\ + 4x^4 \qquad\qquad + 4x^8 &+ \ldots \\ + 5x^5 \qquad\qquad &+ \ldots \\ + 6x^6 \qquad &+ \ldots \\ + 7x^7 \quad &+ \ldots \\ + 8x^8 &+ \ldots. \end{aligned}$$

Here we see easily that each power of x arises as many times as its exponent has divisors, and that each divisor arises as a coefficient of the same power of x. Therefore, if we collect the terms with like powers, the coefficient of each power of x will be the sum of the divisors of its exponent. And, therefore, using the above notation $\sigma(n)$ for the sum of the divisors of n, I obtain

$$t = \sigma(1)x + \sigma(2)x^2 + \sigma(3)x^3 + \sigma(4)x^4 + \sigma(5)x^5 + \ldots.$$

The law of the series is manifest. And, although it might appear that some induction was involved in the determination of the coefficients, we can easily satisfy ourselves that this law is a necessary consequence.

12. By virtue of the definition of t, the last formula of No. 10 can be written as follows:

$$t(1 - x - x^2 + x^5 + x^7 - x^{12} - x^{15} + x^{22} + x^{26} - \dots)$$
$$- x - 2x^2 + 5x^5 + 7x^7 - 12x^{12} - 15x^{15} + 22x^{22} + 26x^{26} - \dots = 0.$$

Substituting for t the value obtained at the end of No. 11, we find

$$\begin{aligned} 0 = \sigma(1)x + \sigma(2)x^2 + \sigma(3)x^3 + \sigma(4)x^4 + \sigma(5)x^5 + \sigma(6)x^6 + \dots \\ - x - \sigma(1)x^2 - \sigma(2)x^3 - \sigma(3)x^4 - \sigma(4)x^5 - \sigma(5)x^6 - \dots \\ - \quad 2x^2 - \sigma(1)x^3 - \sigma(2)x^4 - \sigma(3)x^5 - \sigma(4)x^6 - \dots \\ + \quad 5x^5 + \sigma(1)x^6 + \dots . \end{aligned}$$

Collecting the terms, we find the coefficient for any given power of x. This coefficient consists of several terms. First comes the sum of the divisors of the exponent of x, and then sums of divisors of some preceding numbers, obtained from that exponent by subtracting successively 1, 2, 5, 7, 12, 15, 22, 26, ... Finally, if it belongs to this sequence, the exponent itself arises. We need not explain again the signs assigned to the terms just listed. Therefore, generally, the coefficient of x^n is

$$\sigma(n) - \sigma(n-1) - \sigma(n-2) + \sigma(n-5) + \sigma(n-7) - \sigma(n-12)$$
$$- \sigma(n-15) + \dots .$$

This is continued as long as the numbers under the sign σ are not negative. Yet, if the term $\sigma(0)$ arises, we must substitute n for it.

13. Since the sum of the infinite series considered in the foregoing No. 12 is 0, whatever the value of x may be, the coefficient of each single power of x must necessarily be 0. Hence we obtain the law that I explained above in No. 5; I mean the law that governs the sum of the divisors and enables us to compute it recursively for all numbers. In the foregoing development, we may perceive some reason for the signs, some reason for the sequence of the numbers

$$1, 2, 5, 7, 12, 15, 22, 26, 35, 40, 51, 57, 70, 77, \dots$$

and, especially, a reason why we should substitute for $\sigma(0)$ the number n itself, which could have appeared the strangest feature of my rule. This reasoning, although still very far from a perfect demonstration, will certainly lift some doubts about the most extraordinary law that I explained here.

3. Transition to a more general viewpoint. Euler's foregoing text is extraordinarily instructive. We can learn from it a great deal about mathematics, or the psychology of invention, or inductive reasoning. The examples and comments at the end of this chapter provide for opportunity to examine some of Euler's mathematical ideas, but now we wish to concentrate on his inductive argument.

The theorem investigated by Euler is remarkable in several respects and is of great mathematical interest even today. However, we are concerned here not so much with the mathematical content of this theorem, but rather with the reasons which induced Euler to believe in the theorem when it was still unproved. In order to understand better the nature of these reasons, I shall ignore the mathematical content of Euler's memoir and give a schematic outline of it, emphasizing a certain general aspect of his inductive argument.

As we shall disregard the mathematical content of the various theorems that we must discuss, we shall find it advantageous to designate them by letters, as T, T^*, C_1, C_2, . . . , C_1^*, C_2^*, ... The reader may ignore the meaning of these letters completely. Yet, in case he wishes to recognize them in Euler's text, here is the key.

T is the theorem

$$(1 - x)(1 - x^2)(1 - x^3) \ldots = 1 - x - x^2 + x^5 + x^7 - x^{12} - x^{15} + \ldots .$$

The law of the numbers 1, 2, 5, 7, 12, 15, . . . is explained in sect. 2, No. 5, II.

C_n is the assertion that the coefficient of x^n is the same on both sides of the foregoing equation. For example, C_6 asserts that expanding the product on the left hand side, we shall find that the coefficient of x^6 is 0. Observe that C_n is a consequence of the theorem T.

C_n^* is the equation

$$\sigma(n) = \sigma(n - 1) + \sigma(n - 2) - \sigma(n - 5) - \sigma(n - 7) + \ldots$$

explained at length in sect. 2, No. 5. For example, C_6^* asserts that

$$\sigma(6) = \sigma(5) + \sigma(4) - \sigma(1).$$

T^* is the "most extraordinary law," asserting that C_1^*, C_2^*, C_3^*, . . . are all true. Observe that C_n^* is a consequence (a particular case) of the theorem T^*.

4. Schematic outline of Euler's memoir.[4] *Theorem T is of such a nature that we can be assured of its truth without giving it a perfect demonstration. Nevertheless, I shall present such evidence for it as might be regarded as almost equivalent to a rigorous demonstration.*

Theorem T includes an infinite number of particular cases: C_1, C_2, C_3, ... Conversely, the infinite set of these particular cases C_1, C_2, C_3, . . . is equivalent to theorem T. We can find out by a simple calculation whether C_1 is true or not.

---

[4] This outline was first published in my paper, "Heuristic Reasoning and the Theory of Probability," *Amer. Math. Monthly*, vol. 48, 1941, p. 450–465. The italics indicate phrases which are not due to Euler.

Another simple calculation determines whether C_2 is true or not, and similarly for C_3, and so on. I have made these calculations and I find that $C_1, C_2, C_3, \ldots, C_{40}$ are all true. It suffices to undertake *these calculations* and to continue *them* as far as is deemed proper to become convinced of the truth of this *sequence continued indefinitely.* Yet I have no other evidence for this, except a long induction which I have carried out so far that I cannot in any way doubt the law of which $C_1, C_2, \ldots$ *are the particular cases.* I have long searched in vain for a rigorous demonstration of *theorem T*, and I have proposed the same question to some of my friends with whose ability in these matters I am familiar, but all have agreed with me on the truth of *theorem T* without being able to unearth any clue of a demonstration. Thus it will be a known truth, but not yet demonstrated; for each of us can convince himself of this truth by the *actual calculation of the cases* $C_1, C_2, C_3, \ldots$ as far as he may wish; and it seems impossible that the law which has been discovered to hold for 20 terms, for example, would not be observed in the terms that follow.

As we have thus discovered the truth of *theorem T* even though it has not been possible to demonstrate it, all the conclusions which may be deduced from it will be of the same nature, that is, true but not demonstrated. Or, if one of these conclusions could be demonstrated, one could reciprocally obtain a clue to the demonstration of *theorem T*; and it was with this purpose in mind that I maneuvered *theorem T* in many ways and so discovered among others *theorem* T^* whose truth must be as certain as that of *theorem T*.

Theorems T and T^* are equivalent; they are both true or false; they stand or fall together. Like T, theorem T^* includes an infinity of particular cases $C_1^*, C_2^*, C_3^*, \ldots$, and this sequence of particular cases is equivalent to theorem T^*. Here again, a simple calculation shows whether C_1^* is true or not. Similarly, it is possible to determine whether C_2^* is true or not, and so on. It is not difficult to apply theorem T^* to any given particular case, and so anybody can satisfy himself of its truth by as many examples as he may wish to develop. And since I must admit that I am not in a position to give it a rigorous demonstration, I will justify it by a sufficiently large number of examples, by $C_1^*, C_2^*, \ldots, C_{20}^*$. I think these examples are sufficient to discourage anyone from imagining that it is by mere chance that my rule is in agreement with the truth.

If one still doubts that the law is precisely that one which I have indicated, I will give some examples with larger numbers. *By examination, I find that C_{101}^* and C_{301}^* are true, and so I find that theorem T^* is valid even for these cases which are far removed from those which I examined earlier.* These examples which I have just developed undoubtedly will dispel any qualms which we might have had about the truth of *theorems T and* T^*.

EXAMPLES AND COMMENTS ON CHAPTER VI

In discovering his "Most Extraordinary Law of the Numbers" Euler "reached his conclusion by differentiations and other devices" although "the

Infinitesimal Calculus does not seem to apply to the nature of integers." In order to understand Euler's method, we apply it to similar examples. We begin by giving a name to his principal "device" or mathematical tool.

1. *Generating functions.* We restate the result of No. 11 of Euler's memoir in modern notation:

$$\sum_{n=1}^{\infty} \frac{nx^n}{1 - x^n} = \sigma(1)x + \sigma(2)x^2 + \ldots + \sigma(n)x^n + \ldots .$$

The right hand side is a power series in x. The coefficient of x^n in this power series is $\sigma(n)$, the sum of the divisors of n. Both sides of the equation represent the same function of x. The expansion of this function in powers of x "generates" the sequence $\sigma(1)$, $\sigma(2)$, . . . $\sigma(n)$, . . . and so we call this function the *generating function* of $\sigma(n)$. Generally, if

$$f(x) = a_0 + a_1x + a_2x^2 + \ldots + a_nx^n + \ldots$$

we say that $f(x)$ is the generating function of a_n, or the function generating the sequence $a_0, a_1, a_2, \ldots a_n, \ldots$.

The name "generating function" is due to Laplace. Yet, without giving it a name, Euler used the device of generating functions long before Laplace, in several memoirs of which we have seen one in sect. 2. He applied this mathematical tool to several problems in Combinatory Analysis and the Theory of Numbers.

A generating function is a device somewhat similar to a bag. Instead of carrying many little objects detachedly, which could be embarrassing, we put them all in a bag, and then we have only one object to carry, the bag. Quite similarly, instead of handling each term of the sequence $a_0, a_1, a_2, \ldots a_n, \ldots$ individually, we put them all in a power series Σa_nx^n, and then we have only one mathematical object to handle, the power series.

2. Find the generating function of n. Or, what is the same, find the sum of the series Σnx^n.

3. Being given that $f(x)$ generates the sequence $a_0, a_1, a_2, \ldots a_n, \ldots$ find the function generating the sequence

$$0a_0,\ 1a_1,\ 2a_2,\ \ldots\ na_n,\ \ldots .$$

4. Being given that $f(x)$ generates the sequence $a_0, a_1, a_2, \ldots a_n, \ldots$, find the function generating the sequence

$$0,\ a_0,\ a_1,\ \ldots\ a_{n-1},\ \ldots .$$

5. Being given that $f(x)$ is the generating function of a_n, find the generating function of

$$s_n = a_0 + a_1 + a_2 + \ldots + a_n.$$

6. Being given that $f(x)$ and $g(x)$ are the generating functions of a_n and b_n, respectively, find the generating function of

$$c_n = a_0 b_n + a_1 b_{n-1} + a_2 b_{n-2} + \ldots + a_n b_0.$$

7. *A combinatorial problem in plane geometry.* A convex polygon with n sides is dissected into $n - 2$ triangles by $n - 3$ diagonals; see fig. 6.1. Call D_n the number of different dissections.

Find D_n for $n = 3, 4, 5, 6$.

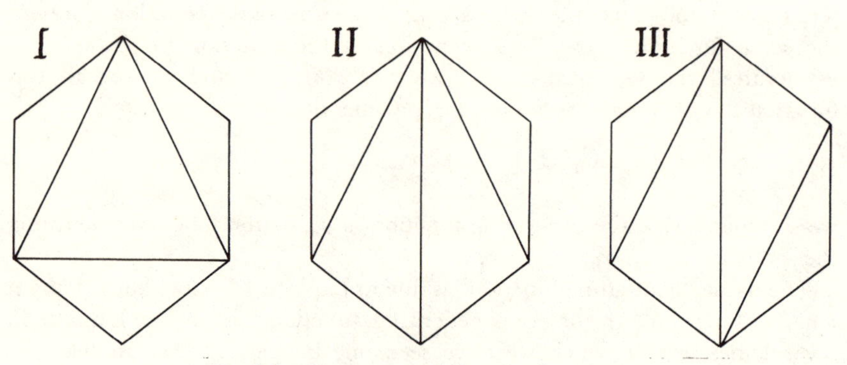

Fig. 6.1. Three types of dissection for a hexagon.

8 (continued). It is not easy to guess a general, explicit expression for D_n on the basis of the numerical values considered in ex. 7. Yet the sequence $D_3, D_4, D_5, \ldots$ is a "recurring" sequence in the following, very general sense: each term can be computed from the foregoing terms according to an invariable rule, a "recursion formula." (See Euler's memoir, No. 5.)

Define $D_2 = 1$, and show that for $n \geqq 3$

$$D_n = D_2 D_{n-1} + D_3 D_{n-2} + D_4 D_{n-3} + \ldots + D_{n-1} D_2.$$

[Check the first cases. Refer to fig. 6.2.]

9 (continued). The derivation of an explicit expression for D_n from the recursion formula of ex. 8 is not obvious. Yet consider the generating function

$$g(x) = D_2 x^2 + D_3 x^3 + D_4 x^4 + \ldots + D_n x^n + \ldots.$$

Show that $g(x)$ satisfies a quadratic equation and derive hence that for $n = 3, 4, 5, 6, \ldots$

$$D_n = \frac{2}{2}\,\frac{6}{3}\,\frac{10}{4}\,\frac{14}{5} \cdots \frac{4n - 10}{n - 1}.$$

10. *Sums of squares.* Recall the definition of $R_k(n)$ (ex. 4.1), extend it to $n = 0$ in setting $R_k(0) = 1$ (a reasonable extension), introduce the generating function

$$\sum_{n=0}^{\infty} R_k(n)x^n = R_k(0) + R_k(1)x + R_k(2)x^2 + \ldots ,$$

and show that

$$\sum_{n=0}^{\infty} R_3(n)x^n = (1 + 2x + 2x^4 + 2x^9 + \ldots)^3.$$

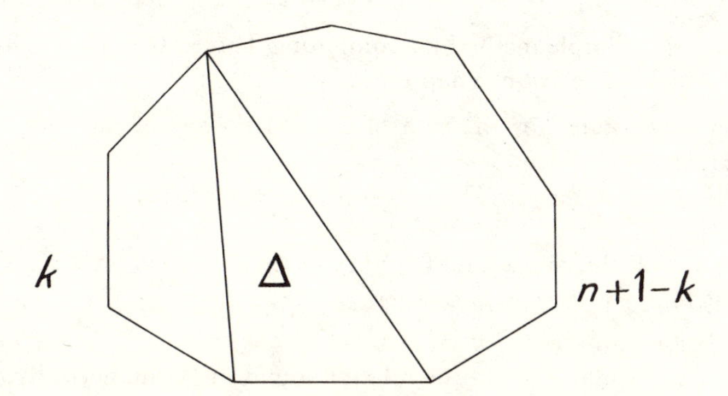

Fig. 6.2. Starting the dissection of a polygon with n sides.

[What is $R_3(n)$? The number of solutions of the equation

$$u^2 + v^2 + w^2 = n$$

in integers u, v, and w, positive, negative, or 0.

What may be the rôle of the series on the right-hand side of the equation that you are required to prove?

$$1 + 2x + 2x^4 + 2x^9 + \ldots = \sum_{u=-\infty}^{\infty} x^{u^2} = \sum_{n=0}^{\infty} R_1(n)x^n.$$

How should you conceive the right-hand side of the equation you are aiming at? Perhaps so:

$$\Sigma x^{u^2} \cdot \Sigma x^{u^2} \cdot \Sigma x^{u^2}.]$$

11. Generalize the result of ex. 10.

12. Recall the definition of $S_k(n)$ (ex. 4.1) and express the generating function

$$\sum_{n=1}^{\infty} S_k(n)x^n.$$

13. Use ex. 11 to prove that, for $n \geq 1$, $R_2(n)$ is divisible by 4, $R_4(n)$ by 8, and $R_8(n)$ by 16. (The result was already used in ch. IV, Tables II and III.)

14. Use ex. 12 to prove that

$$S_2(n) = 0 \text{ if } n \text{ is not of the form } 8m + 2,$$
$$S_4(n) = 0 \text{ if } n \text{ is not of the form } 8m + 4,$$
$$S_8(n) = 0 \text{ if } n \text{ is not of the form } 8m.$$

15. Use ex. 11 to prove that

$$R_{k+l}(n) = R_k(0)R_l(n) + R_k(1)R_l(n-1) + \dots + R_k(n)R_l(0).$$

16. Prove that

$$S_{k+l}(n) = S_k(1)S_l(n-1) + S_k(2)S_l(n-2) + \dots + S_k(n-1)S_l(1).$$

17. Propose a simple method for computing Table III of ch. IV from the Tables I and II of the same chapter.

18. Let $\sigma_k(n)$ stand for the sum of the kth powers of the divisors of n. For example,

$$\sigma_3(15) = 1^3 + 3^3 + 5^3 + 15^3 = 3528;$$

$\sigma_1(n) = \sigma(n)$.

(1) Show that the conjectures found in sect. 4.6 and ex. 4.23 imply

$$\sigma(1)\sigma(2u-1) + \sigma(3)\sigma(2u-3) + \dots + \sigma(2u-1)\sigma(1) = \sigma_3(u)$$

where u denotes an odd integer.

(2) Test particular cases of the relation found in (1) numerically.

(3) How does such a verification influence your confidence in the conjectures from which the relation verified has been derived?

19. *Another recursion formula.* We consider the generating functions

$$G = \sum_{m=1}^{\infty} S_1(m)x^m, \qquad H = \sum_{m=1}^{\infty} S_4(m)x^m.$$

We set

$$S_4(4u) = s_u$$

where u is an odd integer. Then

$$G = x + x^9 + x^{25} + \dots + x^{(2n-1)^2} + \dots,$$
$$H = s_1x^4 + s_3x^{12} + s_5x^{20} + \dots + s_{2n-1}x^{8n-4} + \dots,$$
$$G^4 = H$$

by ex. 14 and 12. We derive from the last equation, by taking the logarithms and differentiating, that

$$4 \log G = \log H,$$
$$\frac{4G'}{G} = \frac{H'}{H},$$
$$G \cdot xH' = 4 \cdot xG' \cdot H,$$

$$(x + x^9 + x^{25} + \dots)(4s_1x^4 + 12s_3x^{12} + 20s_5x^{20} + \dots)$$
$$= 4(x + 9x^9 + 25x^{25} + \dots)(s_1x^4 + s_3x^{12} + s_5x^{20} + \dots).$$

Comparing the coefficients of x^5, x^{13}, x^{21}, . . . on both sides of the foregoing equation, we find, after some elementary work, the following relations:

$$\begin{aligned}
&0s_1 = 0\\
&1s_3 - 4s_1 = 0\\
&2s_5 - 3s_3 = 0\\
&3s_7 - 2s_5 - 12s_1 = 0\\
&4s_9 - 1s_7 - 11s_3 = 0\\
&5s_{11} \qquad\quad - 10s_5 = 0\\
&6s_{13} + 1s_{11} - 9s_7 - 24s_1 = 0\\
&7s_{15} + 2s_{13} - 8s_9 - 23s_3 = 0\\
&8s_{17} + 3s_{15} - 7s_{11} - 22s_5 = 0\\
&9s_{19} + 4s_{17} - 6s_{13} - 21s_7 = 0\\
&10s_{21} + 5s_{19} - 5s_{15} - 20s_9 - 40s_1 = 0\\
&11s_{23} + 6s_{21} - 4s_{17} - 19s_{11} - 39s_3 = 0\\
&\ldots\ldots\ldots\ldots\ldots\ldots\ldots\ldots\ldots
\end{aligned}$$

The very first equation of this system is vacuous and is displayed here only to emphasize the general law. Yet we know that $s_1 = 1$. Knowing this, we obtain from the next equation s_3. Knowing s_3, we obtain from the following equation s_5. And so on, we can compute from the system the terms of the sequence $s_1, s_3, s_5, \ldots$ as far as we wish, one after the other, *recurrently*.

The system has a remarkable structure. There is 1 equation containing 1 of the quantities $s_1, s_3, s_5, \ldots$, 2 equations containing 2 of them, 3 equations containing 3 of them, and so on. The coefficients in each column are increased by 1 and the subscripts by 2 as we pass from one row to the next. The subscript at the head of each column is 1 and the coefficient is -4 multiplied by the first coefficient in the same row.

We can concentrate the whole system in one equation (recursion formula); write it down.

20. *Another Most Extraordinary Law of the Numbers Concerning the Sum of their Divisors.* If the conjecture of sect. 4.6 stands

$$s_{2n-1} = S_4(4(2n-1)) = \sigma(2n-1)$$

and so ex. 19 yields a recursion formula connecting the terms of the sequence $\sigma(1), \sigma(3), \sigma(5), \sigma(7), \ldots$ which is in many ways strikingly similar to Euler's formula.

Write out in detail and verify numerically the first cases of the indicated recursion formula.

21. For us there is also a heuristic similarity between Euler's recursion formula for $\sigma(n)$ (sect. 2) and the foregoing recursion formula for $\sigma(2n-1)$ (ex. 20). For us this latter is a conjecture. We derived this conjecture, as Euler has derived his, "by differentiation and other devices" from another conjecture.

Show that the recursion formula for $\sigma(2n-1)$ indicated by ex. 20 is equivalent to the equation

$$S_4(4(2n-1)) = \sigma(2n-1)$$

to which we arrived in sect. 4.6. That is, if one of the two assertions is true, the other is necessarily also true.

22. Generalize ex. 19.

23. Devise a method for computing $R_8(n)$ independently of $R_4(n)$.

24. *How Euler missed a discovery.* The method illustrated by ex. 19 and ex. 23, and generally stated in ex. 22, is due to Euler.[5] In inventing his method, Euler aimed at the problem of four squares and some related problems. In fact, he applied his method to the problem of four squares and investigated inductively the number of representations, but failed to discover the remarkable law governing $R_4(n)$, which is after all not so difficult to discover inductively (ex. 4.10–4.15). How did it happen?

In examining the equation

$$n = x^2 + y^2 + z^2 + w^2$$

we may choose various standpoints, especially the following:

(1) We admit for x, y, z, and w only non-negative integers.

(2) We admit for x, y, z, and w all integers, positive, negative, and null.

The second standpoint may be less obvious, but leads to $R_4(n)$ and to the remarkable connection between $R_4(n)$ and the divisors of n. The first standpoint is more obvious, but the number of solutions does not seem to have any simple remarkable property. Euler chose the standpoint (1), not the standpoint (2), he applied his method explained in ex. 22 to

$$(1 + x + x^4 + x^9 + \ldots)^4,$$

not to

$$(1 + 2x + 2x^4 + 2x^9 + \ldots)^4,$$

and so he bypassed a great discovery. It is instructive to compare two lines of inquiry which look so much alike at the outset, but one of which is wonderfully fruitful and the other almost completely barren.

---

[5] *Opera Omnia*, ser. 1, vol. 4, p. 125–135.

The properties of $R_4(n)$, $S_4(n)$, $R_8(n)$, and $S_8(n)$ investigated in ch. IV (ex. 4.10–4.15, sect. 4.3–4.6, ex. 4.18–4.23) have been discovered by Jacobi, not inductively, but as incidental corollaries of his researches on elliptic functions. Several proofs of these theorems have been found since, but no known proof is quite elementary and straightforward.[6]

25. *A generalization of Euler's theorem on $\sigma(n)$.* Given k, set

$$\prod_{n=1}^{\infty} (1 - x^n)^k = 1 - \sum_{n=1}^{\infty} a_n x^n$$

and show that, for $n = 1, 2, 3, \ldots$

$$\sigma(n) = \sum_{m=1}^{n-1} a_m \sigma(n - m) + n a_n / k.$$

Which particular case yields Euler's theorem of sect. 2?

---

[6] See also for further references G. H. Hardy and E. M. Wright, *An introduction to the theory of numbers*, Oxford, 1938, chapter XX.

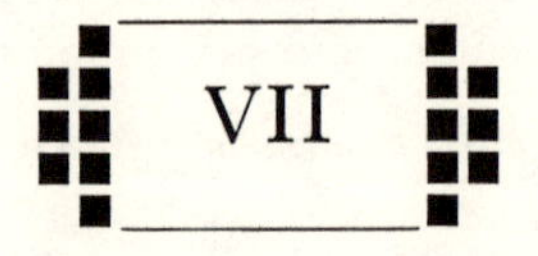

MATHEMATICAL INDUCTION

Jacques Bernoulli's method is important also to the naturalist. We find what seems to be a property A of the concept B by observing the cases C_1, C_2, C_3, ... We learn from Bernoulli's method that we should not attribute such a property A, found by incomplete, non-mathematical induction, to the concept B, unless we perceive *that A is* linked *to the characteristics of B and is* independent *of the variation of the cases. As in many other points, mathematics offers here a model to natural science.*—ERNST MACH[1]

1. The inductive phase. Again, we begin with an example.

There is little difficulty in finding the sum of the first n integers. We take here for granted the formula

$$1 + 2 + 3 + \ldots + n = \frac{n(n+1)}{2}$$

which can be discovered and proved in many ways.[2] It is harder to find a formula for the sum of the first n squares

$$1 + 4 + 9 + 16 + \ldots + n^2.$$

There is no difficulty in computing this sum for small values of n, but it is not so easy to disentangle a rule. It is quite natural, however, to seek some sort of parallelism between the two sums and to observe them together:

n	1	2	3	4	5	6	...
$1 + 2 + \ldots + n$	1	3	6	10	15	21	...
$1^2 + 2^2 + \ldots + n^2$	1	5	14	30	55	91	... .

How are the last two rows related? We may hit upon the idea of examining their ratio:

n	1	2	3	4	5	6	...
$\dfrac{1^2 + 2^2 + \ldots + n^2}{1 + 2 + \ldots + n}$	1	$\frac{5}{3}$	$\frac{7}{3}$	3	$\frac{11}{3}$	$\frac{13}{3}$	...

[1] *Erkenntnis und Irrtum*, 4th ed., 1920, p. 312.

[2] See *How to Solve It*, p. 107.

Here the rule is obvious and it is almost impossible to miss it if the foregoing ratios are written as follows:

$$\frac{3}{3} \quad \frac{5}{3} \quad \frac{7}{3} \quad \frac{9}{3} \quad \frac{11}{3} \quad \frac{13}{3}.$$

We can hardly refrain from formulating the conjecture that

$$\frac{1^2 + 2^2 + \ldots + n^2}{1 + 2 + \ldots + n} = \frac{2n + 1}{3}.$$

Using the value of the denominator on the left-hand side, which we took for granted, we are led to restating our conjecture in the form

$$1^2 + 2^2 + 3^2 + \ldots + n^2 = \frac{n(n + 1)\,(2n + 1)}{6}.$$

Is this true? That is, is it generally true? The formula is certainly true in the particular cases $n = 1, 2, 3, 4, 5, 6$ which suggested it. Is it true also in the next case $n = 7$? The conjecture leads us to predicting that

$$1 + 4 + 9 + 16 + 25 + 36 + 49 = \frac{7 \cdot 8 \cdot 15}{6}$$

and, in fact, both sides turn out to be equal to 140.

We could, of course, go on to the next case $n = 8$ and test it, but the temptation is not too strong. We are inclined to believe anyhow that the formula will be verified in the next case too, and so this verification would add but little to our confidence—so little that going through the computation is hardly worth while. How could we test the conjecture more efficiently?

If the conjecture is true at all, it should be independent of the variation of the cases, it should hold good in the transition from one case to another. Supposedly,

$$1 + 4 + \ldots + n^2 = \frac{n(n + 1)\,(2n + 1)}{6}.$$

Yet, if this formula is generally true, it *should hold also in the next case*: we should have

$$1 + 4 + \ldots + n^2 + (n + 1)^2 = \frac{(n + 1)\,(n + 2)\,(2n + 3)}{6}.$$

Here is an opportunity to check efficiently the conjecture: by subtracting the upper line from the lower we obtain

$$(n + 1)^2 = \frac{(n + 1)\,(n + 2)\,(2n + 3)}{6} - \frac{n(n + 1)\,(2}{6}$$

Is this consequence of the conjecture true?

An easy rearrangement of the right hand side yields

$$\begin{aligned}
&\frac{n+1}{6}[(n+2)(2n+3)-n(2n+1)] \\
&=\frac{n+1}{6}[2n^2+3n+4n+6-2n^2-n] \\
&=\frac{n+1}{6}[6n+6] \\
&=(n+1)^2.
\end{aligned}$$

The consequence examined is incontestably true, the conjecture passed a severe test.

2. The demonstrative phase. The verification of any consequence increases our confidence in the conjecture, but the verification of the consequence just examined can do more: it can *prove* the conjecture. We need only a little change of our viewpoint and a little reshuffling of our remarks.

It is *supposedly* true that

$$1^2+2^2+\ldots+n^2=\frac{n(n+1)(2n+1)}{6}.$$

It is *incontestably* true that

$$(n+1)^2=\frac{(n+1)(n+2)(2n+3)}{6}-\frac{n(n+1)(2n+1)}{6}.$$

It is *consequently* true that

$$1^2+2^2+\ldots+n^2+(n+1)^2=\frac{(n+1)(n+2)(2n+3)}{6}$$

(we added the two foregoing equations). This means: If our conjecture is true for a certain integer n, it remains necessarily true for the next integer $n+1$.

Yet we know that the conjecture is true for $n = 1, 2, 3, 4, 5, 6, 7$. Being true for 7, it must be true also for the next integer 8; being true for 8, it must be true for 9; since true for 9, also true for 10, and so also for 11, and so on. The conjecture is true for all integers; we succeeded in proving it in full generality.

3. Examining transitions. The last reasoning of the foregoing section can be simplified a little. It is enough to know two things about the conjecture:

It is true for $n = 1$.

Being true for n, it is also true for $n + 1$.

Then the conjecture is true for all integers: true for 1, therefore, also for 2; true for 2, therefore, also for 3; and so on.

We have here a fundamentally important procedure of demonstration. We could call it "passing from n to $n + 1$," but it is usually called" mathematical induction." This usual designation is a very inappropriate name for a procedure of demonstration, since induction (in the meaning in which the term is most frequently used) yields only a plausible, and not a demonstrative, inference.

Has mathematical induction anything to do with induction? Yes, it has, and we consider it here for this reason and not only for its name.

In our foregoing example, the demonstrative reasoning of sect. 2 naturally completes the inductive reasoning of sect. 1, and this is typical. The demonstration of sect. 2 appears as a "mathematical complement to induction," and if we take "mathematical induction" as an abbreviation in this sense, the term may appear quite appropriate, after all. (Therefore, let us take it in this sense—no use quarreling with established technical terms.) Mathematical induction often arises as the finishing step, or last phase, of an inductive research, and this last phase often uses suggestions which turned up in the foregoing phases.

Another and still better reason to consider mathematical induction in the present context is hinted by the passage quoted from Ernst Mach at the beginning of this chapter.[3] Examining a conjecture, we investigate the various cases to which the conjecture is supposed to apply. We wish to see whether the relation asserted by the conjecture is *stable*, that is, independent of, and undisturbed by, the variation of the cases. Our attention turns so naturally to the *transition* from one such case to another. "That by means of centripetal forces the planets may be retained in certain orbits, we may easily understand, if we consider the motions of projectiles" says Newton, and then he imagines a stone that is projected with greater and greater initial velocity till its path goes round the earth as the path of the moon; see ex. 2.18 (4). Thus Newton visualizes a continuous transition from the motion of a projectile to the motion of a planet. He considers the transition between two cases to which the law of universal gravitation, that he undertook to prove, should equally apply. Any beginner, who uses mathematical induction in proving some elementary theorem, acts like Newton in this respect: he considers the transition from n to $n + 1$, the transition between two cases to which the theorem that he undertook to prove should equally apply.

4. The technique of mathematical induction. To be a good mathematician, or a good gambler, or good at anything, you must be a good guesser. In order to be a good guesser, you should be, I would think, naturally clever to begin with. Yet to be naturally clever is certainly not

[3] Mach believed that Jacques Bernoulli invented the method of mathematical induction, but most of the credit for its invention seems to be due to Pascal. Cf. H. Freudenthal, *Archives internationales d'histoire des sciences*, no. 22, 1953, p. 17–37. Cf. also *Jacobi Bernoulli Basileensis Opera*, Geneva 1744, vol. I, p. 282–283.

enough. You should examine your guesses, compare them with the facts, modify them if need be, and so acquire an extensive (and intensive) experience with guesses that failed and guesses that came true. With such an experience in your background, you may be able to judge more competently which guesses have a chance to turn out correct and which have not.

Mathematical induction is a demonstrative procedure often useful in verifying mathematical conjectures at which we arrived by some inductive procedure. Therefore, if we wish to acquire some experience in inductive mathematical research, some acquaintance with the technique of mathematical induction is desirable.

The present section and the following examples and comments may give a little help in acquiring this technique.

(1) *The inductive phase.* We begin with an example very similar to that discussed in sect. 1 and 2. We wish to express in some shorter form another sum connected with the first n squares,

$$\frac{1}{3}+\frac{1}{15}+\frac{1}{35}+\dots+\frac{1}{4n^2-1}=$$

$$\frac{1}{4\cdot 1^2-1}+\frac{1}{4\cdot 2^2-1}+\frac{1}{4\cdot 3^2-1}+\dots+\frac{1}{4n^2-1}.$$

We compute this sum in the first few cases and tabulate the results:

$$n \qquad\qquad = 1,\ 2,\ 3,\ 4,\ \dots$$

$$\frac{1}{3}+\frac{1}{15}+\dots+\frac{1}{4n^2-1}=\frac{1}{3},\ \frac{2}{5},\ \frac{3}{7},\ \frac{4}{9},\ \dots\ .$$

There is an obvious guess:

$$\frac{1}{3}+\frac{1}{15}+\frac{1}{35}+\dots+\frac{1}{4n^2-1}=\frac{n}{2n+1}.$$

Profiting from our experience with our former similar problem, we test our conjecture right away as efficiently as we can: we test the transition from n to $n+1$. If our conjecture is generally true, it must be true both for n and for $n+1$:

$$\frac{1}{3}+\frac{1}{15}+\dots+\frac{1}{4n^2-1}=\frac{n}{2n+1},$$

$$\frac{1}{3}+\frac{1}{15}+\dots+\frac{1}{4n^2-1}+\frac{1}{4(n+1)^2-1}=\frac{n+1}{2n+3}.$$

By subtracting we obtain

$$\frac{1}{4(n+1)^2-1}=\frac{n+1}{2n+3}-\frac{n}{2n+1}.$$

Is this consequence of our conjecture true? We transform both sides, trying to bring them nearer to each other:

$$\frac{1}{(2n+2)^2-1}=\frac{2n^2+3n+1-2n^2-3n}{(2n+3)\,(2n+1)}.$$

Very little algebra is enough to see that the two sides of the last equation are actually identical. The consequence examined is incontestably true.

(2) *The demonstrative phase.* Now we reshuffle our remarks, as in our foregoing example, sect. 2.

Supposedly

$$\frac{1}{3}+\frac{1}{15}+\dots+\frac{1}{4n^2-1}=\frac{n}{2n+1}.$$

Incontestably

$$\frac{1}{4(n+1)^2-1}=\frac{n+1}{2n+3}-\frac{n}{2n+1}.$$

Consequently

$$\frac{1}{3}+\frac{1}{15}+\dots+\frac{1}{4n^2-1}+\frac{1}{4(n+1)^2-1}=\frac{n+1}{2n+3}.$$

The conjecture, supposed to be true for n, turns out to be true, in consequence of this supposition, also for $n+1$. As it is true for $n=1$, it is generally true.

(3) *Shorter.* We could have spent a little less time on the inductive phase of our solution. Having conceived the conjecture, we could have suspected that mathematical induction may be appropriate to prove it. Then, without any testing, we could have tried to apply mathematical induction directly, as follows.

Supposedly

$$\frac{1}{3}+\frac{1}{15}+\dots+\frac{1}{4n^2-1}=\frac{n}{2n+1}.$$

Consequently

$$\begin{aligned}\frac{1}{3}+\frac{1}{15}+\dots+\frac{1}{4n^2-1}+\frac{1}{4(n+1)^2-1}&=\frac{n}{2n+1}+\frac{1}{4(n+1)^2-1}\\&=\frac{n}{2n+1}+\frac{1}{(2n+2)^2-1}\\&=\frac{n(2n+3)+1}{(2n+1)\,(2n+3)}\\&=\frac{2n^2+3n+1}{(2n+1)\,(2n+3)}\\&=\frac{(2n+1)\,(n+1)}{(2n+1)\,(2n+3)}\\&=\frac{n+1}{2n+3}\end{aligned}$$

and so we succeeded in deriving for $n + 1$ the relation that we have supposed for n. This is exactly what we were required to do, and so we proved the conjecture.

This variant of the solution is less repetitious, but perhaps also a trifle less natural, than the first, presented under (1) and (2).

(4) *Still shorter.* We can see the solution almost at a glance if we notice that

$$\frac{1}{4n^2 - 1} = \frac{1}{(2n-1)(2n+1)} = \frac{1}{2}\left(\frac{1}{2n-1} - \frac{1}{2n+1}\right).$$

(We are led to this formula quite naturally, if we are familiar with the decomposition of rational functions in partial fractions.) Putting $n = 1, 2, 3, \ldots, n$ and adding, we obtain

$$\frac{1}{4-1} + \frac{1}{16-1} + \frac{1}{36-1} + \cdots + \frac{1}{4n^2-1}$$

$$= \frac{1}{1 \cdot 3} + \frac{1}{3 \cdot 5} + \frac{1}{5 \cdot 7} + \cdots + \frac{1}{(2n-1)(2n+1)}$$

$$= \frac{1}{2}\left[\left(\frac{1}{1} - \frac{1}{3}\right) + \left(\frac{1}{3} - \frac{1}{5}\right) + \left(\frac{1}{5} - \frac{1}{7}\right) + \cdots + \left(\frac{1}{2n-1} - \frac{1}{2n+1}\right)\right]$$

$$= \frac{1}{2}\left[1 - \frac{1}{2n+1}\right]$$

$$= \frac{n}{2n+1}.$$

What has just happened happens not infrequently. A theorem proved by mathematical induction can often be proved more shortly by some other method. Even the careful examination of the proof by mathematical induction may lead to such a shortcut.

(5) *Another example.* We examine two numbers, a and b, subject to the inequalities

$$0 < a < 1, \quad 0 < b < 1.$$

Then, obviously,

$$(1 - a)(1 - b) = 1 - a - b + ab > 1 - a - b.$$

A natural generalization leads us to suspect the following statement: *If* $n \geqq 2$ *and* $0 < a_1 < 1, 0 < a_2 < 1, \ldots 0 < a_n < 1$, *then*

$$(1 - a_1)(1 - a_2) \ldots (1 - a_n) > 1 - a_1 - a_2 - \ldots - a_n.$$

We use mathematical induction to prove this. We have seen that the inequality is true in the first case to which it is asserted to apply, for $n = 2$. Therefore, supposing it to be true for n, where $n \geqq 2$, we have to derive it for $n + 1$.

Supposedly

$$(1 - a_1) \dots (1 - a_n) > 1 - a_1 - \dots - a_n$$

and we know that

$$0 < a_{n+1} < 1.$$

Consequently

$$(1 - a_1) \dots (1 - a_n)(1 - a_{n+1}) > (1 - a_1 - \dots - a_n)(1 - a_{n+1})$$
$$= 1 - a_1 - \dots - a_n - a_{n+1} + (a_1 + \dots + a_n)a_{n+1}$$
$$> 1 - a_1 - \dots - a_n - a_{n+1}.$$

We derived for $n + 1$ what we have supposed for n: the proof is complete.

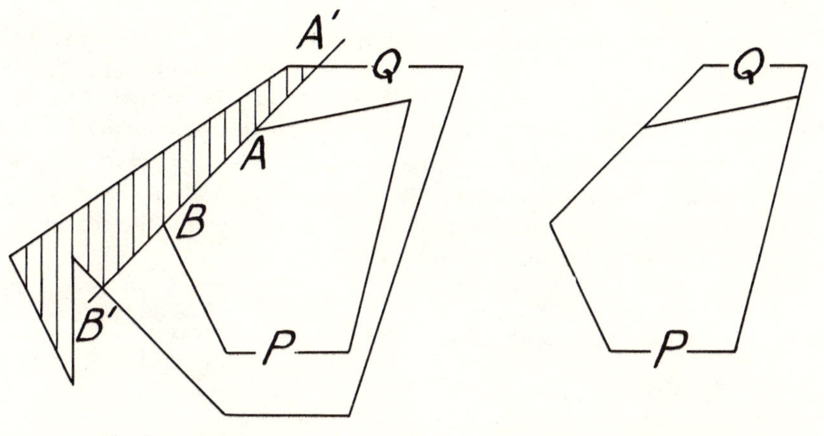

Fig. 7.1. From n to $n + 1$.

Fig. 7.2. The case $n = 1$.

Let us note that mathematical induction may be used to prove propositions which apply, not to all positive integers absolutely, but to all positive integers from a certain integer onwards. For example, the theorem just proved is concerned only with values $n \geqq 2$.

(6) *What is n?* We discuss now a theorem of plane geometry.

If the polygon P is convex and contained in the polygon Q, the perimeter of P is shorter than the perimeter of Q.

That the area of the inner polygon P is less than the area of the outer polygon Q is obvious. Yet the theorem stated is not quite so obvious; without the restriction that P is convex, it would be false.

Fig. 7.1 shows the essential idea of the proof. We cut off the shaded piece from the outer polygon Q; there remains a new polygon Q', a part of Q, which has two properties:

First, Q' still contains the convex polygon P which, being convex, lies fully on one side of the straight line $A'B'$ into which the side AB of P has been produced.

Second, the perimeter of Q' is shorter than that of Q. In fact, the perimeter of Q' differs from that of Q in so far as the former contains the straight line-segment joining the points A' and B', and the latter contains a broken line instead, joining the same points (on the far side of the shaded piece). Yet the straight line is the shortest distance between the points A' and B'.

As we passed from Q to Q', so we can pass from Q' to another polygon Q''. We thus obtain a sequence of polygons $Q, Q', Q'', \ldots$. Each polygon is included in, and has a shorter perimeter than, the foregoing, and the last polygon in this sequence is P. Therefore, the perimeter of P is shorter than that of Q.

We should recognize the nature of the foregoing proof: it is, in fact, a proof by mathematical induction. But what is n? With respect to which quantity is the induction performed?

This question is serious. Mathematical induction is used in various domains and sometimes in very difficult and intricate questions. Trying to find a hidden proof, we may face a crucial decision: What should be n? With respect to what quantity should we try mathematical induction?

In the foregoing proof it is advisable to choose as n the *number of those sides of the inner convex polygon which do not belong entirely to the perimeter of the outer polygon.* Fig. 7.2 illustrates the case $n = 1$. I leave to the reader to find out what is advisable to call n in fig. 7.1.

EXAMPLES AND COMMENTS ON CHAPTER VII

1. Observe that

$$\begin{aligned} 1 &= 1 \\ 1 - 4 &= -(1+2) \\ 1 - 4 + 9 &= 1 + 2 + 3 \\ 1 - 4 + 9 - 16 &= -(1 + 2 + 3 + 4). \end{aligned}$$

Guess the general law suggested by these examples, express it in suitable mathematical notation, and prove it.

2. Prove the explicit formulas for P_n, S_n and S_n^∞ guessed in ex. 3.13, 3.14 and 3.20, respectively. [Ex. 3.11, 3.12.]

3. Guess an expression for

$$1^3 + 2^3 + 3^3 + \ldots + n^3$$

and prove it by mathematical induction. [Ex. 1.4.]

4. Guess an expression for

$$\left(1 - \frac{1}{4}\right)\left(1 - \frac{1}{9}\right)\left(1 - \frac{1}{16}\right)\cdots\left(1 - \frac{1}{n^2}\right)$$

valid for $n \geqq 2$ and prove it by mathematical induction.

5. Guess an expression for

$$\left(1-\frac{4}{1}\right)\left(1-\frac{4}{9}\right)\left(1-\frac{4}{25}\right)\cdots\left(1-\frac{4}{(2n-1)^2}\right)$$

valid for $n \geqq 1$ and prove it by mathematical induction.

6. Generalize the relation

$$\frac{x}{1+x}+\frac{2x^2}{1+x^2}+\frac{4x^4}{1+x^4}+\frac{8x^8}{1+x^8}=\frac{x}{1-x}-\frac{16x^{16}}{1-x^{16}}$$

and prove your generalization by mathematical induction.

7. We consider the operation that consists in passing from the sequence

$$a_1, a_2, a_3, \ldots, a_n, \ldots$$

to the sequence

$$s_1, s_2, s_3, \ldots, s_n, \ldots$$

with the general term

$$s_n = a_1 + a_2 + a_3 + \ldots + a_n.$$

We shall call this operation (the formation of the sequence $s_1, s_2, s_3, \ldots$) "summing the sequence $a_1, a_2, a_3, \ldots$". With this terminology, we can express a fact already observed (in ex. 1.3) as follows.

You can pass from the sequence of all positive integers 1, 2, 3, 4, . . . to the sequence of the squares 1, 4, 9, 16 . . . in two steps: (1) leave out each second term (2) sum the remaining sequence. In fact, see the table:

1	2	3	4	5	6	7	8	9	10	11	12	13	. . .
1		3		5		7		9		11		13	. . .
1		4		9		16		25		36		49	. . .

Prove this assertion by mathematical induction.

8 (continued). You can pass from the sequence of all positive integers 1, 2, 3, 4, . . . to the sequence of the cubes 1, 8, 27, 64, . . . in four steps: (1) leave out each third term (2) sum the remaining sequence (3) leave out each second term (4) sum the remaining sequence. Prove this by mathematical induction, after having examined the table:

1	2	3	4	5	6	7	8	9	10	11	12	13	. . .
1	2		4	5		7	8		10	11		13	. . .
1	3		7	12		19	27		37	48		61	. . .
1			7			19			37			61	. . .
1			8			27			64			125	. . .

9 (continued). You can pass from the sequence of all positive integers 1, 2, 3, 4, . . . to the sequence of the fourth powers 1, 16, 81, 256, . . . in six steps, visible from the table:

1	2	3	4	5	6	7	8	9	10	11	12	13	. . .
1	2	3		5	6	7		9	10	11		13	. . .
1	3	6		11	17	24		33	43	54		67	. . .
1	3			11	17			33	43			67	. . .
1	4			15	32			65	108			175	. . .
1				15				65				175	. . .
1				16				81				256	. . .

What do these facts suggest?

10. Observing that

$$\begin{aligned}
&1 &&= 1\\
&1 - 5 &&= -4\\
&1 - 5 + 10 &&= 6\\
&1 - 5 + 10 - 10 &&= -4\\
&1 - 5 + 10 - 10 + 5 &&= 1\\
&1 - 5 + 10 - 10 + 5 - 1 &&= 0
\end{aligned}$$

we are led to the general statement

$$\binom{n}{0} - \binom{n}{1} + \binom{n}{2} - \binom{n}{3} + \dots + (-1)^k \binom{n}{k} = (-1)^k \binom{n-1}{k}$$

for $0 < k < n$, $n = 1, 2, 3, \dots$.

In proving this by mathematical induction, would you prefer to proceed from n to $n + 1$ or rather from k to $k + 1$?

11. In a tennis tournament there are $2n$ participants. In the first round of the tournament each participant plays just once, so there are n games, each occupying a pair of players. Show that the pairing for the first round can be arranged in exactly

$$1 \cdot 3 \cdot 5 \cdot 7 \cdots (2n - 1)$$

different ways.

12. *To prove more may be less trouble.* Let

$$\frac{1}{1 - x} = f_0(x)$$

and define the sequence $f_0(x), f_1(x), f_2(x), \ldots$ by the condition that

$$f_{n+1}(x) = x\frac{df_n(x)}{dx}$$

for $n = 0, 1, 2, 3, \ldots$. (Such a definition is called *recursive*: to find f_{n+1} we have to go back to f_n.) Observing that

$$f_1(x) = \frac{x}{(1-x)^2}, \quad f_2(x) = \frac{x+x^2}{(1-x)^3}, \quad f_3(x) = \frac{x+4x^2+x^3}{(1-x)^4},$$

prove by mathematical induction that, for $n \geq 1$, the numerator of $f_n(x)$ is a polynomial, the constant term of which is 0 and the other coefficients positive integers.

13 (continued). Find by induction and prove by mathematical induction further properties of $f_n(x)$.

14. *Balance your theorem.* The typical proposition A accessible to proof by mathematical induction has an infinity of cases $A_1, A_2, A_3, \ldots A_n, \ldots$. The case A_1 is often easy; at any rate, A_1 has to be handled by specific means. Once A_1 is established, we have to prove A_{n+1} assuming A_n. A proposition A' stronger than A may be easier to prove than A.[4] In fact, let A' consist of the cases A'_1 $A'_2, \ldots, A'_n, \ldots$. In passing from A to A' we make the burden of the proof heavier: we have to prove the stronger A'_{n+1} instead of A_{n+1}. Yet we make also the support of the proof stronger: we may use the more informative A'_n instead of A_n.

The solution of ex. 12 provides an illustration. Yet we would have made this solution uselessly cumbersome by including the materials treated in ex. 13 which are more conveniently handled by additional remarks, as a corollary.

In general, in trying to devise a proof by mathematical induction, you may fail for two opposite reasons. You may fail because you try to prove too much: your A_{n+1} is too heavy a burden. Yet you may also fail because you try to prove too little: your A_n is too weak a support. You have to balance the statement of your theorem so that the support is just enough for the burden. And so the machinery of the proof edges you towards a more balanced, better adapted view of the facts. This may be typical of the rôle of proofs in building up science.

15. *Outlook.* More intricate problems in more difficult domains demand a more sophisticated technique of mathematical induction and lead to various modifications of this important method of proof. The theory of groups provides some of the most remarkable examples. An interesting variant is the "backward mathematical induction" or "inference from n to

---

[4] This is the "inventor's paradox"; see *How to Solve It*, p. 110.

$n - 1$"; for an interesting elementary example cf. G. H. Hardy, J. E. Littlewood, and G. Pólya, *Inequalities*, p. 17 and p. 20.

16. Being given that $Q_1 = 1$ and

$$Q_{n-1}Q_n = \frac{1}{2^{n(n-1)}} \frac{n!(n+1)!(n+2)!\ldots(2n-1)!}{0!\quad 1!\quad 2!\quad \ldots(n-1)!}$$

for $n = 2, 3, \ldots$, find, and prove, a general expression for Q_n.

17. *Are any n numbers equal?* You would say, No. Yet we can try to prove the contrary by mathematical induction. It may be more attractive however, to prove the assertion: "Any n girls have eyes of the same color."

For $n = 1$ the statement is obviously (or "vacuously") true. It remains to pass from n to $n + 1$. For the sake of concreteness, I shall pass from 3 to 4 and leave the general case to you.

Let me introduce you to any four girls, Ann, Berthe, Carol, and Dorothy, or A, B, C, and D, for short. Allegedly ($n = 3$) the eyes of A, B, and C are of the same color. Also the eyes of B, C, and D are of the same color, allegedly ($n = 3$). Consequently, the eyes of all four girls, A, B, C, and D, must be of the same color; for the sake of full clarity, you may look at the diagram:

$$\rlap{\overbrace{}}A,\ \underbrace{B,\ C,\ D}.$$

This proves the point for $n + 1 = 4$, and the passage from 4 to 5, for example, is, obviously, not more difficult.

Explain the paradox. You may try the experimental approach by looking into the eyes of several girls.

18. If parallel lines are regarded as intersecting (at infinity) the statement "Any n lines in a plane have a common point" is true for $n = 1$ (vacuously) and for $n = 2$ (thanks to our interpretation). Construct a (paradoxical) proof by mathematical induction.

MAXIMA AND MINIMA

Since the fabric of the world is the most perfect and was established by the wisest Creator, nothing happens in this world in which some reason of maximum or minimum would not come to light.—EULER

1. Patterns. Problems concerned with greatest and least values, or maximum and minimum problems, are more attractive, perhaps, than other mathematical problems of comparable difficulty, and this may be due to a quite primitive reason. Everyone of us has his personal problems. We may observe that these problems are very often maximum or minimum problems of a sort. We wish to obtain a certain object at the lowest possible price, or the greatest possible effect with a certain effort, or the maximum work done within a given time and, of course, we wish to run the minimum risk. Mathematical problems on maxima and minima appeal to us, I think, because they idealize our everyday problems.

We are even inclined to imagine that Nature acts as we would like to act, obtaining the greatest effect with the least effort. The physicists succeeded in giving clear and useful forms to ideas of this sort; they describe certain physical phenomena in terms of "minimum principles." The first dynamical principle of this kind (the "Principle of Least Action" which usually goes under the name of Maupertuis) was essentially developed by Euler; his words, quoted at the beginning of this chapter, describe vividly a certain aspect of the problems on minima and maxima which may have appealed to many scientists in his century.

In the next chapter we shall discuss a few problems on minima and maxima arising in elementary physics. The present chapter prepares us for the next.

The Differential Calculus provides a general method for solving problems on minima and maxima. We shall not use this method here. It will be more instructive to develop a few "patterns" of our own instead.

Having solved a problem with real insight and interest, you acquire a precious possession: a pattern, a model, that you can imitate in solving similar problems. You develop this pattern if you try to follow it, if you score a success in following it, if you reflect upon the reasons of your success, upon the analogy of the problems solved, upon the relevant circumstances that make a problem accessible to this kind of solution, etc. Developing such a pattern, you may finally attain a real discovery. At any rate, you have a chance to acquire some well ordered and readily available knowledge.

2. Example. *Given two points and a straight line, all in the same plane, both points on the same side of the line. On the given straight line, find a point from which the segment joining the two given points is seen under the greatest possible angle.*

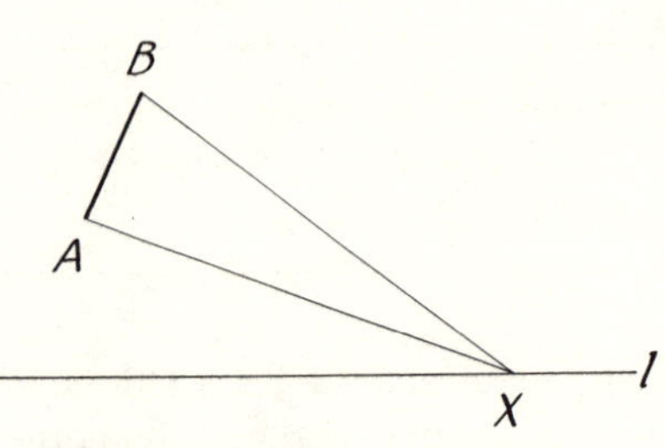

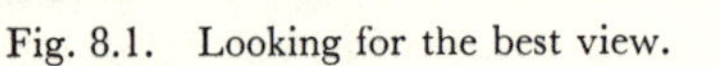
Fig. 8.1. Looking for the best view.

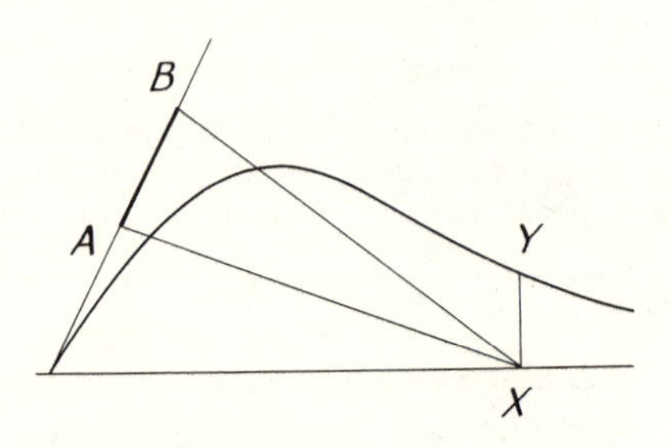

Fig. 8.2. The variation of the angle may look like this.

This is the problem that we wish to solve. We draw a figure (fig. 8.1) and introduce suitable notation. Let

A and B denote the two given points,

l the given straight line,

X a variable point of the line l.

We consider $\angle AXB$, the angle subtended by the given segment AB at the variable point X. We are required to find that position of X on the given line l for which this angle attains its maximum.

Imagine that l is a straight road. If from some point of the road l you wish to fire a shot on a target stretching from A to B, you should choose the point that we are seeking; it gives you the best chance to hit. If you have the more peaceful intention to take, from the road l, a snapshot of a façade the corners of which are at A and at B, you should again choose the point that we are seeking; it gives you the most extensive view.

The solution of our problem is not quite immediate. But, even if we do not know yet *where* the maximum is attained, we do not doubt that it *is* attained somewhere. Why is this so plausible?

We can account for the plausibility if we visualize the variation of the angle the maximum of which we are trying to find. Let us imagine that we walk along the line l and look at the segment AB. Let us start from the point at which the line l and the line through A and B intersect and proceed to the right. At the start, the angle under which AB appears vanishes; then the angle increases; yet finally, when we are very far from AB, it must decrease again since it vanishes at infinite distance.[1] Between the two extreme cases in which the angle vanishes it must become a maximum somewhere. But where?

This question is not easy to answer, although we could point out long stretches of the line l where the maximum is probably not attained. Let us choose any point on the line and call it X. This point, chosen at random, is very likely not in the maximum position that we are trying to find. How could we decide quite clearly whether it is in the maximum position or not?

There is a fairly easy remark.[2] If a point is *not* in the maximum position, there must be another point, on the other side of the maximum position, at which the angle in question has the same value. Is there any other point X' on the line l seen from which the segment AB appears under the same angle as it does from X? Here is, at last, a question that we can readily answer: both X and X' (if there is an X') must be on the same circle passing through the points AB by virtue of a familiar property of the angles inscribed in a circle (Euclid III, 21).

And now, the idea may appear. Let us draw several circles passing through the given points A and B. If such a circle intersects the line in two points, as X and X' in fig. 8.3, the segment AB is seen from both points X and X' under the same angle, but this angle is not the greatest possible: a circle that intersects l between X and X' yields a greater angle. Intersecting circles cannot do the trick: the vertex of the maximum angle is the point at which a circle through A and B *touches* the line l (the point M in fig. 8.3).

3. The pattern of the tangent level line. Let us look back at the solution that we have just found. What can we learn from it? What is essential in it? Which features are capable of an appropriate generalization?

The step which appears the most essential after some reflection is not too conspicuous. I think that the decisive step was to broaden our viewpoint. to step out of the line l, to consider the values of the quantity to be maximized (the angle subtended by AB) at points of the plane outside l. We considered the variation of this angle when its vertex moved in the plane, we considered the dependence of this angle on the position of its vertex. In short, we conceived this angle as the *function of a variable point* (its vertex), and regarded this point (the vertex) as varying in the *plane*.

The angle remains unchanged when its vertex moves along an arc of circle joining the points A and B. Let us call such an arc of circle a *level line*. This expression underscores the general viewpoint that we are about to attain. The lines along which a function of a variable point remains constant are usually called the level lines of that function.

---

[1] If we consider $\angle AXB$ as function of the distance measured along the line l, we can graph it (represent it in rectangular coordinates) in the usual manner. Fig. 8.2 gives a qualitative sketch of the graph; $\angle AXB$ is represented by the ordinate XY.

[2] Very easy if we look at fig. 8.2.

Yet, let us not forget the unknown of our problem. We were required to find the maximum of the angle (of this function of a variable point) when its vertex (the variable point) cannot move freely in the plane, but is restricted to a *prescribed path*, the line l. At which point of the prescribed path is the maximum attained?

We know the answer already, but let us understand it better, let us look at it from a more general viewpoint. Let us consider an analogous, fairly general and very intuitive, example.

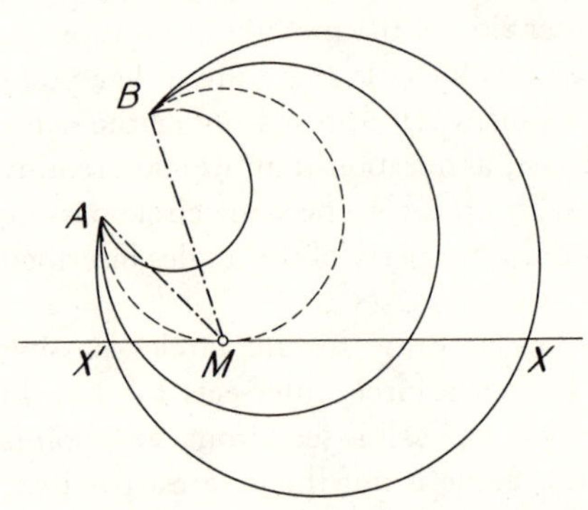

Fig. 8.3. A tangent level line.

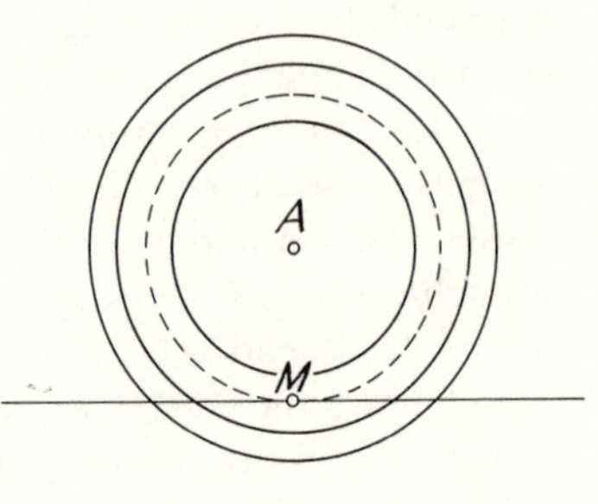

Fig. 8.4. Another tangent level line.

You know what "level lines" or "contour lines" are on a map or in the terrain (let us think of a hilly country) that the map represents. They are the lines of constant elevation; a level line connects those points on the map which represent points upon the earth's surface of equal height above sea level. If you imagine the sea rising 100 feet, a new coastline with bays intruding into the valleys appears at the new sea level. This new coastline is the level line of elevation 100. The map-maker plots only a few level lines, at constant intervals, for example, at elevations 100, 200, 300, . . . ; yet we can think a level line, there *is* a level line, at each elevation, through each point of the terrain. The function of the variable point that is important for the map-maker, or for you when you move about in the terrain, is the elevation above sea level; this function remains constant along each level line.

Now, here is a problem analogous to our problem just discussed (in sect. 2). You walk along a road, a prescribed path. At which point of the road do you attain the maximum elevation?

It is very easy to say where you do not attain it. A point which you pass going up or down is certainly not the point of maximum elevation, nor the point of minimum elevation. At such a point, your road crosses a level line: *the maximum* (or the minimum) *can* NOT *be attained at a point where the prescribed path crosses a level line.*

With this essential remark, let us return to our example (sect. 2, fig. 8.1, 8.2, 8.3). Let us consider the whole prescribed path: the line l from its intersection with the line through A and B to infinite distance (to the right). At each of its points, this prescribed path intersects a level line (an arc of circle with endpoints A and B) except in just one point, where it is tangent to a level line (to such a circle). If the maximum is anywhere, it must be at that point: *at the point of maximum the prescribed path is tangent to a level line.*

This hints very strongly the general idea behind our example. Yet let us examine this hint. Let us apply it to a simple analogous case and see how it works. Here is an easy example.

On a given straight line find the point which is at minimum distance from a given point.

Let us introduce suitable notation:

A is the given point,

a is the given straight line.

It is understood that the given point A is not on the given line a. We have to find the shortest distance from A to a.

Everybody knows the solution. Imagine that you are swimming in a calm sea; in this moment you are at the point A; the line a marks a straight uniform beach. Suddenly you are scared, you wish to reach firm ground as quickly as possible. Where is the nearest point of the beach? You know it without reflection. A dog would know it. A dog or a cow, thrown into the water, would start swimming without delay along the perpendicular from A to a.

Yet our purpose is not just to find the solution, but to examine a general idea in finding it. The quantity that we wish to minimize is the distance of a variable point from the given point A. This distance depends on the position of the variable point. The level lines of the distance are obviously concentric circles with A as their common center. The "prescribed path" is the given straight line a. The minimum is not attained at a point where the prescribed path crosses a level line. In fact, it is attained at the (only) point at which the prescribed path is tangent to a level line (at the point M of fig. 8.4). The shortest distance from the point A to the line a is the radius of the circle with center A that is tangent to a—as we knew from the start. Still, we learned something. The general idea appears now more clearly and it may be left to the reader to clarify it completely.

With the essential common features of the foregoing problems clearly in mind, we are naturally looking out for analogous problems to which we could apply the same pattern of solution. In the foregoing, we considered a point variable in a plane and sought the minimum or maximum of a function of such a point along a prescribed path. Yet we could consider a point variable in space and seek the minimum or maximum of a function of such a point along a prescribed path, or on a prescribed surface. In the plane, the tangent level lines played a special rôle. Analogy prompts us to expect that the tangent level surfaces will play a similar rôle in space.

4. Examples. We discuss two examples which can be treated by the same method, but have very little in common otherwise.

(1) *Find the minimum distance between two given skew straight lines.*

Let us call

a and *b* the two given skew lines,

X a variable point on *a*,

Y a variable point on *b*;

see fig. 8.5. We are required to determine that position of the line-segment *XY* in which it is the shortest.

The distance *XY* depends on the position of its two end-points, *X* and *Y*, which are both variable. There are two variable points, not just one, and this may be the characteristic difficulty of the problem. If one of the two points were given, fixed, invariable, and only the other variable, the problem would be easy. In fact, it would not even be new; it would be identical with a problem just solved (sect. 3).

Let us *fix for a moment* one of the originally variable points, say, *Y*. Then the segment *XY* is in the plane through the fixed point *Y* and the given line *a*, and only one of its end points, *X*, is variable, running along the line *a*. Obviously, *XY* becomes a minimum when it becomes perpendicular to *a* (by sect. 3, fig. 8.4).

Yet we could interchange the rôles of the two points, *X* and *Y*. Let us now, for a change, fix *X* and make *Y* alone variable. Obviously, the segment *XY* becomes shortest when perpendicular to *b*.

The minimum position of *XY*, however, is independent of our whims and of our choice of rôles for *X* and *Y*, and so we are led to suspect that it is perpendicular both to *a* and to *b*. Yet let us look more closely at the situation.

In fact, the foregoing argument shows directly where the minimum position can *not* be (and only indirectly where it should be). I say that a position in which the segment *XY* is not perpendicular to the line *a* at the point *X*, is *not* the minimum position. In fact, I fix the point *Y* and move *X* to another position where *XY* becomes perpendicular to *a* and so doing I make *XY* shorter (by sect. 3). This reasoning obviously applies just as well to *Y* as to *X*, and so we see: *the length of the segment XY cannot be a minimum unless this segment is perpendicular both to a and to b.* If there is a shortest distance, it must be along the common perpendicular to the two given lines.

We need not take anything for granted. In fact, we can see at a glance that the common perpendicular is actually the shortest distance. Let us assume that, in fig. 8.5, the plane of the drawing is parallel to both given lines *a* and *b* (*a* above, *b* below). We may consider any point or line in space as represented in fig. 8.5 by its orthogonal projection. The true length of the segment *XY* is the hypotenuse of a right triangle of which another side is the orthogonal projection of *XY* seen in fig. 8.5; the third side is the shortest distance of two parallel planes, one through a, the other through b, both parallel to the plane of the drawing to which that third side is perpendicular. Therefore, the shorter the projection of XY shown in fig. 8.5, the shorter is XY itself. The projection of XY reduces to a point, its length to nil, and so the length of XY is a minimum if, and only if, XY is perpendicular to the plane of the drawing and so to both lines a and b.

And so we have verified directly what we discovered before by another method.

(2) *Find the maximum of the area of a polygon inscribed in a given circle and having a given number of sides.*

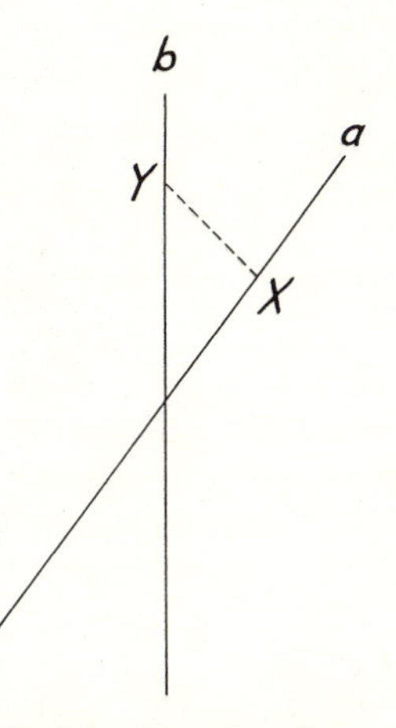

Fig. 8.5. Two skew lines.

The circle is given. On this circle, we have to choose the n vertices $U, \ldots, W, X, Y$, and Z of a polygon so that the area becomes a maximum. Just as in the foregoing problem, under (1), the main difficulty seems to be that there are many variables (the vertices $U, \ldots, W, X, Y$, and Z). We should, perhaps, try the method that worked in the foregoing problem. What is the essential point of this method?

Let us take the problem as *almost solved.* Let us imagine that we have obtained already the desired position of all vertices except one, say, X. The $n - 1$ other points, $U, \ldots, W, Y$, and Z, are already fixed, each in the position where it should be, but we have yet to choose X so that the area becomes a maximum. The whole area consists, however, of two parts: the polygon $U \ldots WYZ$ with $n - 1$ fixed vertices which is independent of X and the triangle WXY which depends on X. We focus our attention upon this triangle which must become a maximum when the whole area becomes a maximum; see fig. 8.6. The base WY of ΔWXY is fixed. If the vertex X moves along a parallel to the base WY, the area remains constant: such parallels to WY are the level lines. We pick out the tangent level line: the tangent to the circle parallel to WY. Its point of contact is obviously the position of X that renders the area of ΔWXY a maximum. With X in this position the triangle is isosceles, $WX = XY$. These two adjacent sides must be equal, if the area of the polygon is a maximum. Yet the same reasoning applies to any pair of adjacent sides: all sides must be equal when the maximum of the area is attained, and so the inscribed polygon with maximum area must be *regular*.

5. The pattern of partial variation. Comparing the two examples discussed in the foregoing section (sect. 4), we easily recognize some common features and a common pattern of solution. In both problems we seek the extremum (minimum or maximum) of a quantity depending on *several* variable elements. In both solutions we fix for a moment all originally

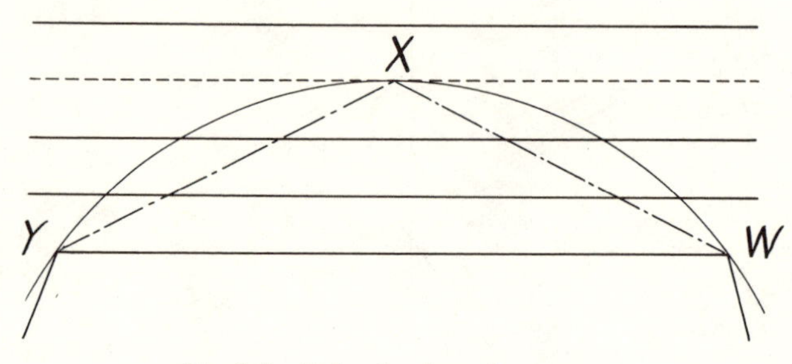

Fig. 8.6. Triangle of maximum area.

variable elements except one and study the effect of the variation of this single element. The simultaneous variation of all variable elements, or *total* variation, is not so easy to survey. We treated our examples with good results in studying the *partial* variation when only a single element varies and the others are fixed. The principle underlying our procedure seems to be: *a function f of several variables cannot attain a maximum with respect to all its variables jointly, unless it attains a maximum with respect to each single variable.*

This statement is fairly general, although unnecessarily restricted in one respect: it clings too closely to the foregoing examples in which we varied just one element at a time. Yet we can imagine that in other examples it may be advantageous to vary just two elements at a time and fix the others, or perhaps just three, and so on. In such cases we could still speak appropriately of "partial variation." The general idea appears now fairly clearly and after one more example the reader may undertake by himself to clarify it completely.

A line of length l is divided into n parts. Find the maximum of the product of these n parts.

Let $x_1, x_2, \ldots x_n$ denote the lengths of the n parts; $x_1, x_2, \ldots x_n$ are positive numbers with a given sum

$$x_1 + x_2 + \ldots + x_n = l.$$

We are required to make $x_1 x_2 \ldots x_n$ a maximum.

We examine first the simplest special case: Being given the sum $x_1 + x_2$ of two positive quantities, find the maximum of their product x_1x_2. We can interpret x_1 and x_2 as adjacent sides of a rectangle and restate the problem in the following more appealing form: Being given L, the length of the perimeter of a rectangle, find the maximum of its area. In fact, the sum of the two sides just mentioned is given:

$$x_1 + x_2 = \frac{L}{2}.$$

There is an obvious guess: the area becomes a maximum when the rectangle becomes a square. This guess cannot be hard to verify. Each side of the square with perimeter L is equal to

$$\frac{L}{4} = \frac{x_1 + x_2}{2}.$$

We have to verify that the area of the square is greater than the area of the rectangle, or, which is the same, that their difference

$$\left(\frac{x_1 + x_2}{2}\right)^2 - x_1x_2$$

is positive. Is that so? Very little algebra is enough to see that

$$\left(\frac{x_1 + x_2}{2}\right)^2 - x_1x_2 = \left(\frac{x_1 - x_2}{2}\right)^2.$$

This formula shows the whole situation at a glance. The right hand side is positive, unless $x_1 = x_2$ and the rectangle is a square.

In short, the area of a rectangle with given perimeter becomes a maximum when the rectangle is a square; the product of two positive quantities with given sum becomes a maximum when these two quantities are equal.

Let us try to use the special case just solved as a stepping stone to the solution of the general problem. Let us take the problem as almost solved. Let us imagine that we have obtained already the desired values of all parts, except those of the first two, x_1 and x_2. Thus, we regard x_1 and x_2 as variables, but $x_3, x_4, \ldots x_n$ as constants. The sum of the two variable parts is constant,

$$x_1 + x_2 = l - x_3 - x_4 - \ldots - x_n.$$

Now, the product of all parts

$$x_1x_2(x_3x_4 \ldots x_n)$$

cannot become a maximum unless the product x_1x_2 of the first two parts becomes a maximum, too. This requires, however, that $x_1 = x_2$. Yet there is no reason why any other pair of parts should behave differently.

The desired maximum of the product cannot be attained unless all the quantities with given sum are equal. We quote Colin Maclaurin (1698–1746) to whom the foregoing reasoning is due: "If the Line *AB* is divided into any Number of Parts *AC*, *CD*, *DE*, *EB*, the Product of all those Parts multiplied into one another will be a *Maximum* when the Parts are equal amongst themselves."

The reader can learn a great deal in clarifying the foregoing proof. Is it quite satisfactory?

6. The theorem of the arithmetic and geometric means and its first consequences. Let us reconsider the result of the foregoing section: If

$$x_1 + x_2 + x_3 + \ldots + x_n = l$$

then

$$x_1 x_2 x_3 \ldots x_n < \left(\frac{l}{n}\right)^n$$

unless $x_1 = x_2 = x_3 \ldots = x_n = l/n$. Eliminating l, we can restate this result in the form:

$$x_1 x_2 \ldots x_n < \left(\frac{x_1 + x_2 + \ldots + x_n}{n}\right)^n$$

or

$$\sqrt[n]{x_1 x_2 \ldots x_n} < \frac{x_1 + x_2 + \ldots + x_n}{n}$$

unless all the positive quantities $x_1, x_2, \ldots x_n$ *are equal*; if these quantities are equal, the inequality becomes an equation. The left-hand side of the above inequality is called the geometric, the right-hand side the arithmetic, mean of $x_1, x_2, \ldots x_n$. The theorem just stated is sometimes called the "theorem of the arithmetic and geometric means" or, shortly, the "theorem of the means."

The theorem of the means is interesting and important in many respects. It is worth mentioning that it can be stated in two different ways:

The product of n positive quantities with a given sum becomes a maximum when these quantities are all equal.

The sum of n positive quantities with a given product becomes a minimum when these quantities are all equal.

The first statement is concerned with a maximum, the second with the corresponding minimum. The derivation of the foregoing section is aimed at the first statement. Changing this derivation systematically we could arrive at the second statement. It is simpler, however, to observe that the inequality between the means yields impartially both statements: to obtain one or the other, we have to regard one or the other side of the inequality as given. We may call these two (essentially equivalent) statements *conjugate statements*.

The theorem of the means yields the solution of many geometric problems on minima and maxima. We discuss here just one example (several others can be found at the end of this chapter).

Being given the area of the surface of a box, find the maximum of its volume.

I use the word "box" instead of "rectangular parallelepiped" because "box" is expressive enough and so much shorter than the official term.

The solution is easily foreseen and, once foreseen, it is easily reduced to the theorem of the means as follows. Let

a, b, c denote the lengths of the three edges of the box drawn from the same vertex,

S the area of the surface,

V the volume.

Obviously

$$S = 2(ab + ac + bc), \qquad V = abc.$$

Observing that the sum of the three quantities ab, ac, and bc is $S/2$ and their product V^2, we naturally think of the theorem of the means which yields

$$V^2 = (abc)^2 < \left(\frac{ab + ac + bc}{3}\right)^3 = \left(\frac{S}{6}\right)^3$$

unless

$$ab = ac = bc$$

or, which is the same,

$$a = b = c.$$

That is

$$V < (S/6)^{3/2}$$

unless the box is a cube, when equality occurs. We can express the result in two different (although essentially equivalent) forms.

Of all boxes with a given surface area the cube has the maximum volume.

Of all boxes with a given volume the cube has the minimum surface area.

As above, we may call these two statements conjugate statements. As above, one of the two conjugate statements is concerned with a maximum, the other with a minimum.

The preceding application of the theorem of the means has its merits. We may regard it as a pattern and collect cases to which the theorem of the means can be similarly applied.

EXAMPLES AND COMMENTS ON CHAPTER VIII

First Part

1. *Minimum and maximum distances in plane geometry.* Find the minimum distance between (1) two points, (2) a point and a straight line, (3) two parallel straight lines.

Find the minimum, and the maximum, distance between (4) a point and a circle, (5) a straight line and a circle, (6) two circles.

The solution is obvious in all cases. Recall the elementary proof at least in some cases.

2. *Minimum and maximum distances in solid geometry.* Find the minimum distance between (1) two points, (2) a point and a plane, (3) two parallel planes, (4) a point and a straight line, (5) a plane and a parallel straight line, (6) two skew straight lines.

Find the minimum, and the maximum, distance between (7) a point and a sphere, (8) a plane and a sphere, (9) a straight line and a sphere, (10) two spheres.

3. *Level lines in a plane.* Consider the distance of a variable point from a given (1) point, (2) straight line, (3) circle. What are the level lines?

4. *Level surfaces in space.* Consider the distance of a variable point from a given (1) point, (2) plane, (3) straight line, (4) sphere. What are the level surfaces?

5. Answer the questions of ex. 1 using level lines.

6. Answer the questions of ex. 2 using level surfaces.

7. Given two sides of a triangle, find the maximum of the area using level lines.

8. Given one side and the length of the perimeter of a triangle, find the maximum of the area using level lines.

9. Given the area of a rectangle, find the minimum of the perimeter using level lines. (In a rectangular coordinate system, let $(0,0)$, $(x,0)$, $(0,y)$, (x,y) be the four vertices of the rectangle and use analytic geometry.)

10. Examine the following statement: "The shortest distance from a given point to a given curve is perpendicular to the given curve."

11. *The principle of the crossing level line.* We consider a function f of a point which varies in a plane, the maximum and the minimum of f along a prescribed path, and a level line of f which separates two regions of the plane; in one of these regions f takes higher, in the other lower, values than on the level line itself.

If the prescribed path crosses the level line, neither the maximum nor the minimum of f is attained at the point of crossing.

12. The contour-map of fig. 8.7 shows a peak P and a pass (or saddle-point with horizontal tangent plane) S. Hiking through such a country, do you necessarily attain the highest point of your path at a point where the path is tangent to a level line?.

13. Let A and B denote two given points and X a variable point in a plane. The angle at X subtended by the segment AB ($\angle AXB$) which can take any value between 0° and 180° (limits inclusively) is a function of the variable point X.

(1) Give a full description of the level lines.

(2) Of two different level lines, which one corresponds to a higher value of the angle?

You may use figs. 8.1 and 8.3, but you should realize that now you may look at the segment AB from *both* sides.

14. Consider figs. 8.1, 8.2, 8.3, take $\angle AXB$ as in ex. 13, and find its minimum along l. Does the result conform to the principle of ex. 11?

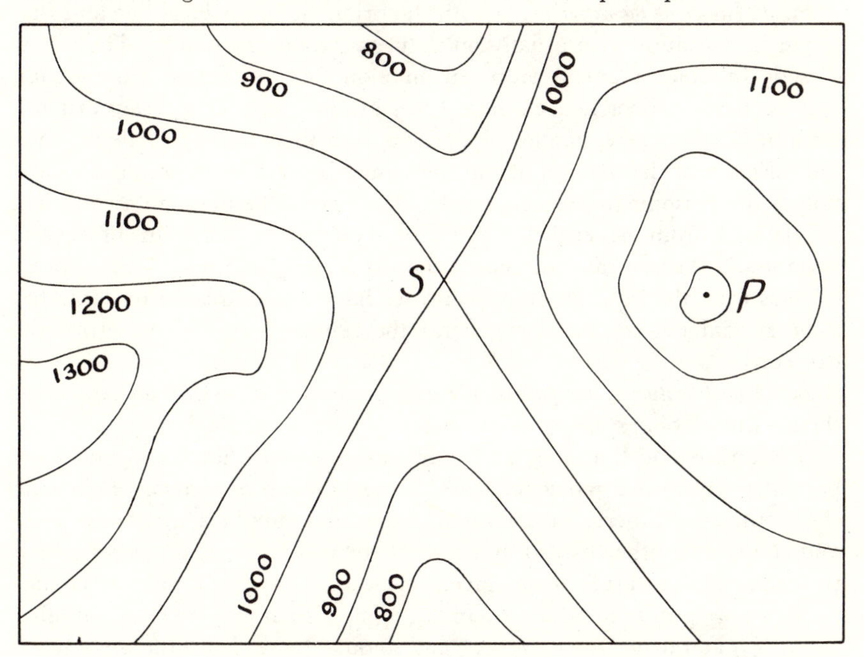

Fig. 8.7. Level lines on a contour map.

15. Given the volume of a box (rectangular parallelepiped), find the minimum of the area of its surface, using partial variation.

16. Of all triangles with a given perimeter, which one has the maximum area? [Ex. 8.]

17. Of all tetrahedra inscribed in a given sphere, which one has the maximum volume? [Do you know a related problem?]

18. Given a, b, and c, the lengths of three edges of a tetrahedron drawn from the same vertex, find the maximum of the volume of the tetrahedron. [Do you know an analogous problem?]

19. Find the shortest distance between a sphere and a cylinder. (Cylinder means: infinite cylinder of revolution.)

20. Find the shortest distance between two cylinders with skew axes.

21. Examine the following statement: "The shortest distance between two given surfaces is perpendicular to both."

22. *The principle of partial variation.* The function $f(X,Y,Z,\ldots)$ of several variables $X,Y,Z,\ldots$ attains its maximum for $X = A$, $Y = B$, $Z = C, \ldots$. Then the function $f(X,B,C,\ldots)$ of the single variable X attains its maximum for $X = A$; and the function $f(X,Y,C,\ldots)$ of the two variables X and Y attains its maximum for $X = A$, $Y = B$; and so on.

A function of several variables cannot attain a maximum with respect to all its variables jointly, unless it attains a maximum with respect to any subset of variables.

23. *Existence of the extremum.* Both the principle of the level line and that of partial variation give usually only "negative information." They show directly in which points a proposed function f can *not* attain a maximum, and we have to infer hence where f *may* attain one. That f *must* attain a maximum somewhere, cannot be derived from these principles alone. Yet the existence of the maximum can sometimes be derived by some modification of the reasoning. Moreover, the existence of the maximum can often be derived from general theorems on continuous functions of several variables.[3] At any rate, whenever the existence of the maximum appears obvious from the intuitive standpoint, we have a good reason to hope that some special device or some general theorem will apply and prove the existence.

24. *A modification of the pattern of partial variation: An infinite process.* Find the maximum of xyz, being given that $x + y + z = l$.

It is understood that x, y, and z are positive, and that l is given. The present problem is a particular case of the problem of sect. 5. Following the method used there, we keep one of the three numbers x, y, and z fixed and change the other two so that they become equal, which increases their product. Let us start from any given system (x, y, z); performing the change indicated we pass to another system (x_1, y_1, z_1); then we pass to still another (x_2, y_2, z_2) and hence to (x_3, y_3, z_3), and so on. Let us leave the three terms unchanged in turn: first the x-term, then y, then z, then again x, then y, then z, then again x, and so on. Thus, we set

$$x_1 = x, \quad y_1 = z_1 = \frac{y + z}{2},$$

$$y_2 = y_1, \quad z_2 = x_2 = \frac{z_1 + x_1}{2},$$

$$z_3 = z_2, \quad x_3 = y_3 = \frac{x_2 + y_2}{2},$$

$$x_4 = x_3, \quad y_4 = z_4 = \frac{y_3 + z_3}{2},$$

$$\cdots\cdots\cdots\cdots\cdots$$

[3] A function of several variables, continuous in a closed set, attains there its lower and upper bounds. This generalizes G. H. Hardy, *Pure Mathematics*, p. 194, theorem 2.

Each step leaves the sum unchanged, but increases the product:

$$x + y + z = x_1 + y_1 + z_1 = x_2 + y_2 + z_2 = \dots$$

$$xyz < x_1 y_1 z_1 < x_2 y_2 z_2 < \dots .$$

We assumed that $y \neq z$ and that $x_1 \neq z_1$. (This is the non-exceptional case; in the exceptional case we attain our end more easily.) We naturally expect that the three numbers x_n, y_n, and z_n become less and less different from each other as n increases. *If* we can prove that finally

$$\lim_{n\to\infty} x_n = \lim_{n\to\infty} y_n = \lim_{n\to\infty} z_n$$

we can immediately infer that

$$xyz < \lim_{n\to\infty} x_n y_n z_n = (l/3)^3.$$

We obtain so this result at considerable expense, yet *without assuming the existence of the maximum* from the outset.

Prove that $\lim_{n\to\infty} x_n = \lim_{n\to\infty} y_n = \lim_{n\to\infty} z_n$.

25. *Another modification of the pattern of partial variation: A finite process.* We are still concerned with the problem of ex. 24, but we use now a more sophisticated modification of the method of sect. 5.

Let $l = 3A$; thus, A is the arithmetic mean of x, y, and z, and we have

$$(x - A) + (y - A) + (z - A) = 0.$$

It may happen that $x = y = z$. If this is not the case, one of the differences on the left hand side of our equation must be negative and another positive. Let us choose the notation so that

$$y < A < z.$$

We pass now from the system (x, y, z) to (x', y', z'), putting

$$x' = x, \quad y' = A, \quad z' = y + (z - A);$$

we left the first quantity unchanged. Then

$$x + y + z = x' + y' + z'$$

and

$$\begin{aligned} y'z' - yz &= A(y + z - A) - yz \\ &= (A - y)(z - A) > 0 \end{aligned}$$

so that

$$xyz < x'y'z'.$$

It may happen that $x' = y' = z'$. If this is not the case, we pass from $(x', y', z',)$ to (x'', y'', z''), putting

$$y'' = y', \qquad z'' = x'' = \frac{z' + x'}{2}$$

which gives

$$x'' = y'' = z'' = A$$

and again increases the product (as we know from sect. 5) so that

$$xyz < x'y'z' < x''y''z'' = A^3$$
$$= \left(\frac{x + y + z}{3}\right)^3.$$

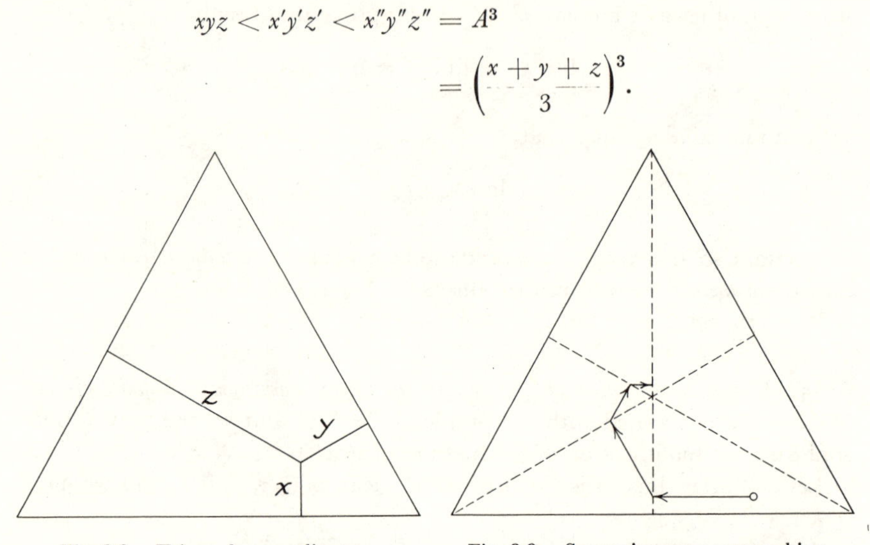

Fig. 8.8. Triangular coordinates.

Fig. 8.9. Successive steps approaching the center.

We proved the desired result, *without assuming the existence of the maximum* and *without considering limits.*

By a suitable extension of this procedure, prove the theorem of the means (sect. 6) generally for n quantities.

26. *Graphic comparison.* Let P be a point in the interior of an equilateral triangle with altitude l, and x, y, and z the distances of P from the three sides of the triangle; see fig. 8.8. Then

$$x + y + z = l.$$

(Why?) The numbers x, y, and z are the *triangular coordinates* of the point P. Any system of three positive numbers x, y, and z with the sum l can be interpreted as the triangular coordinates of a uniquely determined point inside the triangle.

The sequence

$$(x, y, z), \quad (x_1, y_1, z_1), \quad (x_2, y_2, z_2), \quad \ldots$$

considered in ex. 24 is represented by a sequence of points in fig. 8.9. The segments joining the consecutive points are parallel to the various sides of the triangle in succession, to the first, the second, and the third side, respectively, then again to the first side, and so on; each segment ends on an altitude of the triangle. (Why?) The procedure of ex. 25 is represented by three points and two segments. (How?)

27. Reconsider the argument of sect. 4(2) and modify it, taking first ex. 24, then ex. 25 as a model.

28. A necessary condition for a maximum or a minimum value of a function $f(x,y,z)$ at the point (a,b,c) is that the partial derivatives

$$\frac{\partial f}{\partial x}, \quad \frac{\partial f}{\partial y}, \quad \frac{\partial f}{\partial z}$$

vanish for $x = a$, $y = b$, $z = c$.

The usual proof of this theorem exemplifies one of our patterns. Which one?

29. State the well-known necessary condition (in terms of partial derivatives) for a maximum or minimum value of the function $f(x,y)$ under the side-condition (or subsidiary condition) that x and y are linked by the equation $g(x,y) = 0$. Explain the connection with the pattern of the tangent level line.

30. Reexamine the cases mentioned in the solution of ex. 12 in the light of the condition mentioned in ex. 29. Is there any contradiction?

31. State the well-known necessary condition for a maximum or minimum value of a function $f(x,y,z)$ under the side-condition that $g(x,y,z) = 0$. Explain the connection with the pattern of the tangent level surface.

32. State the well-known necessary condition for a maximum or minimum value of a function $f(x,y,z)$ under the two simultaneous side-conditions that $g(x,y,z) = 0$ and $h(x,y,z) = 0$. Explain the connection with the pattern of the tangent level surface.

Second Part

The terminology and notation used in the following are explained in ex. 33, which should be read first.

33. *Polygons and polyhedra. Area and perimeter. Volume and surface.* Dealing with polygons, we shall use the following notation most of the time:

A for the area, and

L for the length of the perimeter.

Dealing with polyhedra, we let

V denote the volume, and

S the area of the surface.

We shall discuss problems of maxima and minima concerned with A and L, or V and S. Such problems were known to the ancient Greeks.[4] We shall discuss mainly problems treated by Simon Lhuilier and Jacob Steiner.[5] Elementary algebraic inequalities, especially the theorem of the means (sect. 6), will turn out useful in solving the majority of the following problems.

Most of the time these problems deal only with the simplest polygons (triangles and quadrilaterals) and the simplest polyhedra (prisms and pyramids). We have to learn a few less usual terms.

Two pyramids, standing on opposite sides of their common base, form jointly a *double pyramid.* If the base has n sides, the double pyramid has $2n$ faces, $n+2$ vertices, and $3n$ edges. The base is *not* a face of the double pyramid.

If all lateral faces are perpendicular to the base, we call the prism a *right prism.*

If the base of a pyramid is circumscribed about a circle and the altitude meets the base at the center of this circle, we call the pyramid a *right pyramid.*

If the two pyramids forming a double pyramid are right pyramids and symmetrical to each other with respect to their common base, we call the double pyramid a *right double pyramid.*

If a prism, pyramid, or double pyramid is not "right," we call it "oblique." Among the five regular solids, there is just one prism, just one pyramid, and just one double pyramid: the cube, the tetrahedron, and the octahedron, respectively. Each of these three is a "right" solid of its kind.

We shall consider also cylinders, cones, and double cones; if there is no remark to the contrary, their bases are supposed to be circles.

34. *Right prism with square base.* Of all right prisms with square base having a given volume, the cube has the minimum surface.

Prove this special case of a theorem already proved (sect. 6, ex. 15) directly, using the theorem of the means.

You may be tempted to proceed as follows. Let V, S, x, and y denote the volume, the surface area, the side of the base, and the altitude of the prism, respectively. Then

$$V = x^2y, \qquad S = 2x^2 + 4xy.$$

Applying the theorem of the means, we obtain

$$(S/2)^2 = [(2x^2 + 4xy)/2]^2 \geqq 2x^2 \cdot 4xy = 8x^3y.$$

Yet this has no useful relation to $V = x^2y$—the theorem of the means does not seem to be applicable.

This was, however, a rash, thoughtless, unprofessional application of the theorem. Try again. [What is the desired conclusion?]

---

[4] Pappus, *Collectiones*, Book V.

[5] Simon Lhuilier, *Polygonométrie et Abrégé d'Isopérimétrie élémentaire*, Genève, 1789. J. Steiner, *Gesammelte Werke*, vol. 2, p. 177–308.

35. *Right cylinder.* Observe that, of all the prisms considered in ex. 34, only the cube is circumscribed about a sphere and prove: Of all right cylinders having a given volume, the cylinder circumscribed about a sphere has the minimum surface. [What is the desired conclusion?]

36. *General right prism.* Of a right prism, given the volume and the shape (but not the size) of the base. When the area of the surface is a minimum, which fraction of it is the area of the base? [Do you know a related problem?]

37. *Right double pyramid with square base.* Prove: Of all right double pyramids with square base having a given volume, the regular octahedron has the minimum surface.

38. *Right double cone.* Observe that the inscribed sphere touches each face of the regular octahedron at its center, which divides the altitude of the face in the ratio 1 : 2 and prove: Of all right double cones having a given volume, the minimum of the surface is attained by the double cone the generatrices of which are divided in the ratio 1 : 2 by the points of contact with the inscribed sphere.

39. *General right double pyramid.* Of a right double pyramid, given the volume and the shape (but not the size) of the base. When the area of the surface is a minimum, which fraction of it is the area of the base?

40. Given the area of a triangle, find the minimum of its perimeter. [Could you predict the result? If you wish to try the theorem of the means, you may need the expression of the area in terms of the sides.]

41. Given the area of a quadrilateral, find the minimum of its perimeter. [Could you predict the result? Call a, b, c, and d the sides of the quadrilateral, ε the sum of two opposite angles, and express the area A in terms of a, b, c, d, and ε. This is a generalization of the problem solved by Heron's formula.]

42. A right prism and an oblique prism have the same volume and the same base. Then the right prism has the smaller surface.

A right pyramid and an oblique pyramid have the same volume and the same base. Then the right pyramid has the smaller surface.

A right double pyramid and an oblique double pyramid have the same volume and the same base. Then the right double pyramid has the smaller surface.

In all three statements, the bases of the two solids compared agree both in shape and in size. (The volumes, of course, agree only in size.)

Choose the statement that seems to you the most accessible of the three and prove it.

43. *Applying geometry to algebra.* Prove: If $u_1, u_2, \ldots u_n, v_1, v_2, \ldots v_n$ are real numbers,

$$\sqrt{u_1^2 + v_1^2} + \sqrt{u_2^2 + v_2^2} + \ldots + \sqrt{u_n^2 + v_n^2}$$
$$\geqq \sqrt{(u_1 + u_2 + \ldots + u_n)^2 + (v_1 + v_2 + \ldots + v_n)^2}$$

and equality is attained if, and only if,

$$u_1 : v_1 = u_2 : v_2 = \ldots = u_n : v_n.$$

[Consider $n+1$ points $P_0, P_1, P_2, \ldots P_n$ in a rectangular coordinate system and the length of the broken line $P_0P_1P_2 \ldots P_n$.]

44. Prove the inequality of ex. 43 independently of geometric considerations. [In the geometric proof of the inequality, the leading special case is $n = 2$.]

45. *Applying algebra to geometry.* Prove: Of all triangles with given base and area, the isosceles triangle has the minimum perimeter. [Ex. 43.]

46. Let V, S, A, and L denote the volume, the area of the whole surface, the area of the base, and the length of the perimeter of the base of a pyramid P, respectively. Let V_0, S_0, A_0, and L_0 stand for the corresponding quantities connected with another pyramid P_0. Supposing that

$$V = V_0, \quad A = A_0, \quad L \geqq L_0$$

and that P_0 is a right pyramid, prove that

$$S \geqq S_0.$$

Equality is attained if, and only if, $L = L_0$ and P is also a right pyramid. [Ex. 43.]

47. Let V, S, A, and L denote the volume, the area of the surface, the area of the base, and the perimeter of the base of a double pyramid D, respectively. Let V_0, S_0, A_0, and L_0 stand for the corresponding quantities connected with another double pyramid D_0. Supposing that

$$V = V_0, \quad A = A_0, \quad L \geqq L_0$$

and that D_0 is a right double pyramid, prove that

$$S \geqq S_0.$$

Equality is attained if, and only if, $L = L_0$ and D is also a right double pyramid. [Ex. 45, 46.]

48. Prove: Of all quadrilateral prisms with a given volume, the cube has the minimum surface. [Compare with ex. 34; which statement is stronger?]

49. Prove: Of all quadrilateral double pyramids with a given volume, the regular octahedron has the minimum surface. [Compare with ex. 37; which statement is stronger?]

50. Prove: Of all triangular pyramids with a given volume, the regular tetrahedron has the minimum surface.

51. *Right pyramid with square base.* Prove: Of all right pyramids with square base having a given volume, the pyramid in which the base is $\frac{1}{4}$ of the total surface has the minimum surface.

52. *Right cone.* Of all right cones having a given volume, the cone in which the base is $\frac{1}{4}$ of the total surface has the minimum surface.

53. *General right pyramid.* Of a right pyramid, given the volume and the shape (but not the size) of the base. When the area of the surface is a minimum, which fraction of it is the area of the base? [Do you know a special case?]

54. Looking back at our various examples dealing with prisms, pyramids, and double pyramids, observe their mutual relations and arrange them in a table so that the analogy of the *results* becomes conspicuous. Point out the gaps which you expect to fill out with further results.

55. *The box with the lid off.* Given S_5, the sum of the areas of five faces of a box. Find the maximum of the volume V. [Do you know a related problem? Could you use the result, or the method?]

56. *The trough.* Given S_4, the sum of the areas of four faces of a right triangular prism; the missing face is a *lateral* face. Find the maximum of the volume V.

57. *A fragment.* In a right prism with triangular base, given S_3, the sum of the areas of three mutually adjacent faces (that is, of two lateral faces and one base). Show that these three faces are of equal area and perpendicular to each other when the volume V attains its maximum. [A fragment—of what?]

58. Given the area of a sector of a circle. Find the value of the angle at the center when the perimeter is a minimum.

59. In a triangle, given the area and an angle. Find the minimum (1) of the sum of the two sides including the given angle, (2) of the side opposite the given angle, (3) of the whole perimeter.

60. Given in position an angle and a point in the plane of the angle, inside the angle. A variable straight line, passing through the given point, cuts off a triangle from the angle. Find the minimum of the area of this triangle.

61. Given E, the sum of the lengths of the 12 edges of a box, find the maximum (1) of its volume V, (2) of its surface S.

62. *A post office problem.* Find the maximum of the volume of a box, being given that the length and girth combined do not exceed l inches.

63. *A problem of Kepler.* Given d, the distance from the center of a generator of a right cylinder to the farthermost point of the cylinder. Find the maximum of the volume of the cylinder.

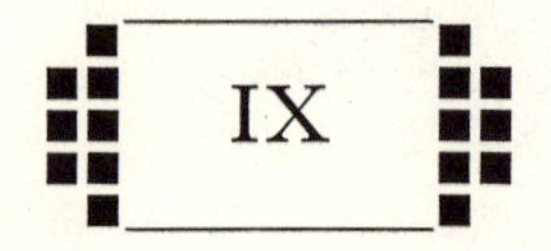

PHYSICAL MATHEMATICS

The science of physics does not only give us [mathematicians] an opportunity to solve problems, but helps us also to discover the means of solving them, and it does this in two ways: it leads us to anticipate the solution and suggests suitable lines of argument.—HENRI POINCARÉ[1]

1. Optical interpretation. Mathematical problems are often inspired by nature, or rather by our interpretation of nature, the physical sciences. Also, the solution of a mathematical problem may be inspired by nature; physics provides us with clues with which, left alone, we had very little chance to provide ourselves. Our outlook would be too narrow without discussing mathematical problems suggested by physical investigation and solved with the help of physical interpretation. Here follows a first, very simple problem of this kind.

(1) *Nature suggests a problem.* The straight line is the shortest path between two given points. Light, travelling through the air from one point to another, chooses this shortest path, so at least our everyday experience seems to show. But what happens when light travels from one point to another not directly, but undergoing a reflection on an interposed mirror? Will light again choose the shortest path? What is the shortest path in these circumstances? By considerations on the propagation of light we are led to the following purely geometrical problem:

Given two points and a straight line, all in the same plane, both points on the same side of the line. On a given straight line, find a point such that the sum of its distances from the two given points be a minimum.

Let (see fig. 9.1)

A and B denote the two given points,

l the given straight line,

X a variable point of the line l.

We consider $AX + XB$, the sum of two distances or, which is the same, the

[1] *La valeur de la science*, p. 152.

length of the path leading from A to X and hence to B. We are required to find that position of X on the given line l for which the length of this path attains its minimum.

We have seen a very similar problem before (sect. 8.2, figs. 8.1, 8.2, 8.3). In fact, both problems have exactly the same data, and even the unknown is of the same nature: Here, as there, we seek the position of a point on a given line for which a certain extremum is attained. The two problems differ only in the nature of this extremum: Here we seek to minimize the sum of two lines; there we sought to maximize the angle included by those two lines.

Still, the two problems are so closely related that it is natural to try the same method. In solving the problem of sect. 8.2, we used level lines; let us use them again.

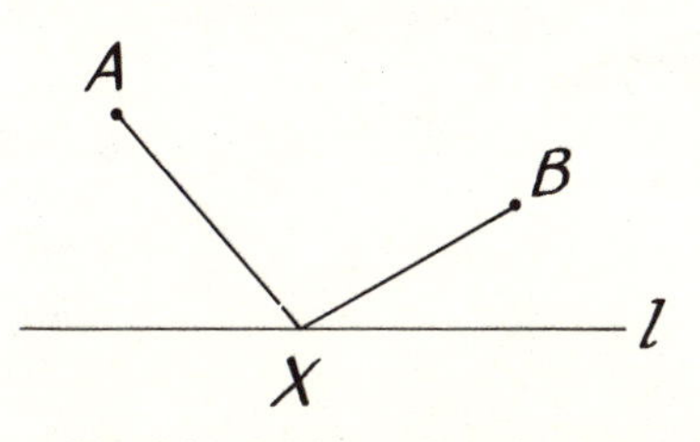

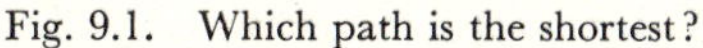
Fig. 9.1. Which path is the shortest?

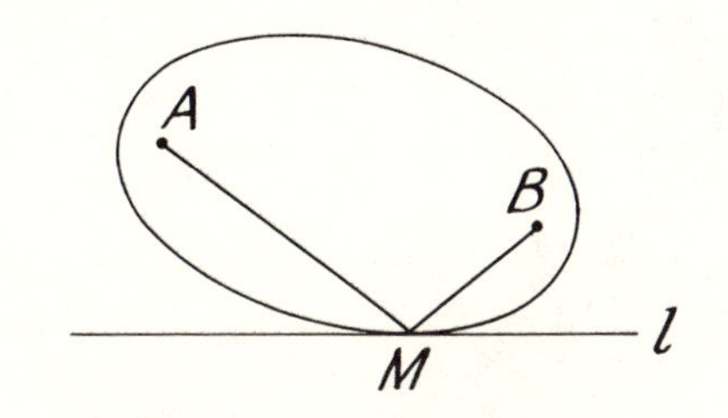

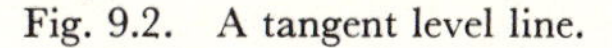
Fig. 9.2. A tangent level line.

Consider a point X that is not bound to the prescribed path but is free to move in the whole plane. If the quantity $AX + XB$ (which we wish to minimize) has a constant value, how can X move? Along an ellipse with foci A and B. Therefore, the level lines are "confocal" ellipses, that is, ellipses with the same foci (the given points A and B). *The desired minimum is attained at the point of contact of the prescribed path l with an ellipse the foci of which are the given points A and B* (see fig. 9.2).

(2) *Nature suggests a solution.* We have found the solution, indeed. Yet, unless we know certain geometric properties of the ellipse, our solution is not of much use. Let us make a fresh start and seek a more informative solution.

Let us visualize the physical situation that suggested our problem. The point A is a source of light, the point B the eye of an observer, and l marks the position of a reflecting plane surface; we may think of the horizontal surface of a quiet pool (which is perpendicular to the plane of fig. 9.1, and intersects it in the line l). The broken line AXB represents, if the point X is correctly chosen, the path of light. We know this path fairly well, by experience. We suspect that the length of the broken line AXB is a minimum when it represents the actual path of the reflected light.

Your eye is in the position B and you look down at the reflecting pool observing in it the image of A. The ray of light that you perceive does not come directly from the object A, but appears to come from a point under the surface of the pool. From which point? From the point A^*, the mirror image of the object A, symmetrical to A with respect to the line l.

Introduce the point A^*, suggested by your physical experience, into the figure! This point A^* changes the face of the problem. We see a host of new relations (fig. 9.3) which we proceed to order and to exploit rapidly. Obviously

$$AX = A^*X.$$

(A^*X is the mirror image of AX. You can also argue from the congruence of the triangles $\triangle ACX$, $\triangle A^*CX$; the line l is the perpendicular bisector of the segment AA^*.) Therefore,

$$AX + XB = A^*X + XB.$$

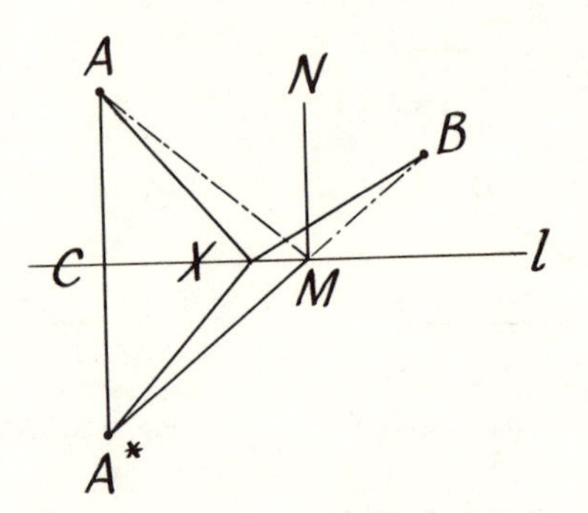

Fig. 9.3. A more informative solution.

Both sides of this equation are minimized by the same position of X. Yet the right-hand side is obviously a *minimum when* A^*, X, *and* B *are on the same straight line*. The straight line is the shortest.

This is the solution (see fig. 9.3). The point M, the minimum position of X, is obtained as intersection of the line l and of the line joining A^* and B. Obviously, AM and MB include the same angle with l. Introducing the line MN, normal to l (parallel to A^*A), we see that

$$\angle AMN = \angle BMN.$$

The equality of these two angles characterizes the shortest path. Yet the very same equality

angle of incidence = angle of reflection

characterizes the actual path of light, as we know by experience. Therefore, in fact, the reflected ray of light takes the shortest possible course between the object and the eye. This discovery is due to Heron of Alexandria.

(3) *Comparing two solutions.* It is often useful to look back at the completed solution. In the present case it should be doubly useful, since we have two solutions which we can compare with each other (under (1) and (2)). Both

methods of solving the problem (figs. 9.2 and 9.3) must yield the same result (imagine the two figures superposed). We can obtain the point M, the solution of our minimum problem, by means of an ellipse tangent to l, or by means of two rays equally inclined to l. Yet these two constructions must agree, whatever the relative position of the data (the points A and B and the line l) may be. The agreement of the two constructions involves a geometric property of the ellipse: *The two straight lines, joining the two foci of an ellipse to any point on the periphery of the ellipse, are equally inclined to the tangent of the ellipse at the point where they meet.*

If we conceive the ellipse as a mirror and take into account the law of reflection (which we have just discussed), we can restate the geometric property in intuitive optical interpretation: *Any ray of light coming from a focus of an elliptic mirror is reflected into the other focus.*

(4) *An application.* Simple as it is, Heron's discovery deserves a place in the history of science. It is the first example of the use of a minimum principle in describing physical phenomena. It is a suggestive example of interrelations between mathematical and physical theories. Much more general minimum principles have been discovered after Heron and mathematical and physical theories have been interrelated on a much grander scale, but the first and simplest examples are in some respects the most impressive.

Looking back at the impressively successful solution under (2), we should ask: Can you use it? Can you use the result? Can you use the method? In fact, there are several openings. We could examine the reflection of light in a curved mirror, or successive reflections in a series of plane mirrors, or combine the result with methods that we learned before, and so on.

Let us discuss here just one example, the problem of the "traffic center." Three towns intend to construct three roads to a common traffic center which should be chosen so that the total cost of road construction is a minimum. If we take all this in utmost simplification, we have the following purely geometric problem: *Given three points, find a fourth point so that the sum of its distances from the three given points is a minimum.*

Let A, B, and C denote the three given points (towns) and X a variable point in the plane determined by A, B, and C. We seek the minimum of $AX + BX + CX$.

This problem seems to be related to Heron's problem. We should bring the two problems together, work out the closest possible relation between them. If, for a moment, we take the distance CX as fixed ($= r$, say), the relation appears very close indeed: Here, as there, we have to find the minimum of $AX + BX$, the sum of the distances of one variable point from two fixed points. The difference is that X is obliged to move along a circle here (with radius r and center C), and along a straight line there. The former problem was about reflection in a plane mirror, the present problem is about reflection in a circular mirror.

Let us rely on the light: it is clever enough to find by itself the shortest path from A to the circular mirror and hence to B. Yet the light moves so that the angle of incidence is equal to the angle of reflection. Therefore, in the desired minimum position, $\angle AXB$ must be *bisected* by the straight line passing through C and X (see fig. 9.4). By the principle of partial variation and the symmetry of the situation, $\angle AXC$ and $\angle BXC$ must be similarly bisected. The three straight lines joining X to A, B, and C dissect the plane into six angles, the common vertex of which is X. Focussing our attention upon the pairs of vertical angles in fig. 9.5, we easily see that all six angles are equal and, therefore, each of them is equal to 60°. *The three roads diverging from the traffic center are equally inclined to each other;* the angle between any two is 120°.

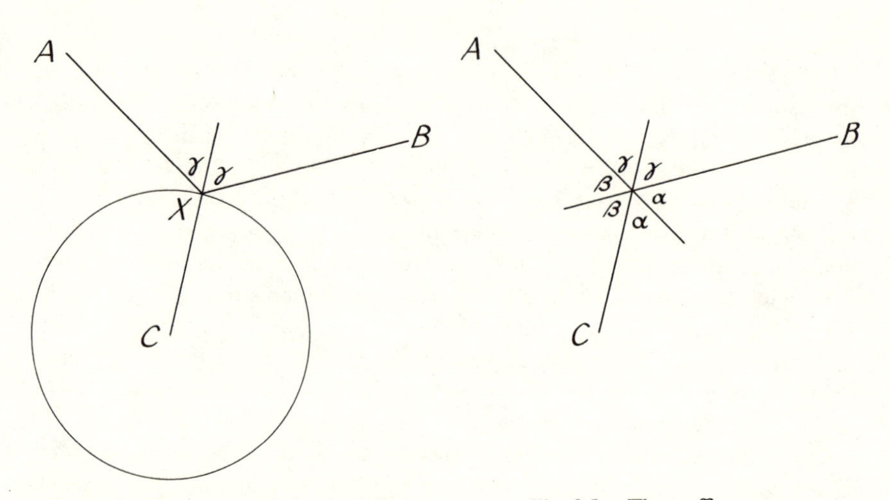

Fig. 9.4. Traffic center and circular mirror.

Fig. 9.5. The traffic center.

(If we remember that the method of partial variation which we have used is subject to certain limitations, we may find a critical reexamination of our solution advisable.)

2. Mechanical interpretation. Mathematical problems and their solutions can be suggested by any sector of our experience, by optical, mechanical, or other phenomena. We shall discuss next how simple mechanical principles can help us to discover the solution.

(1) *A string, of which both extremities are fixed, passes through a heavy ring. Find the position of equilibrium.*

It is understood that the string is perfectly flexible and inextensible, its weight is negligible, the ring slides along the string without friction, and the dimensions of the ring are so small that it can be regarded as a mathematical point.

Let A and B denote the fixed endpoints of the string and X any position of the ring. The string forms the broken line AXB on fig. 9.6.

The proposed problem can be solved by two different methods.

First, the ring must hang as low as possible. (In fact, the ring is heavy; it "wants" to come as close to the ground, or to the center of the earth, as possible.) Both parts of the inextensible string, AX and BX, are stretched, and so the ring, sliding along the string, describes an ellipse with foci A and B. Obviously, the position of equilibrium is at the lowest point M of the ellipse where the tangent is *horizontal.*

Second, the forces acting on the point M of the string must be in equilibrium. The weight of the ring and the tensions in the string act on the point M. The tensions in both parts of the string, MA and MB, are equal and directed along the string to A and B, respectively.

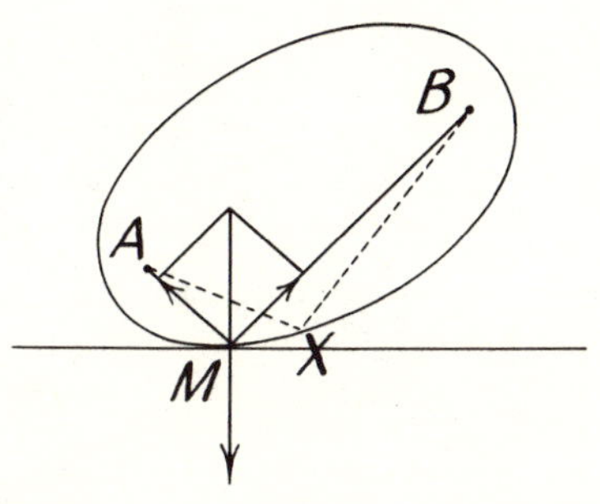

Fig. 9.6. Two conditions of equilibrium.

Their resultant bisects the $\angle AMB$ and, being opposite to the weight of the ring, is *vertical.*

The two solutions, however, must agree. Therefore, the lines MA and MB, equally inclined to the vertical normal of the ellipse, are also equally inclined to its horizontal tangent: *The two straight lines, joining the two foci of an ellipse to any point M on the periphery, are equally inclined to the tangent at the point M.* (By keeping the length AB but changing its angle of inclination to the horizontal, we can bring M in any desired position on one half of the ellipse.)

We derived a former result (sect. 1(3)) by a new method which may be capable of further applications.

(2) We seem to have a surplus of knowledge. Without having learned too much mechanics, we know enough of it, it seems, not only to find a solution of a proposed mechanical problem, but to find two solutions, based on two different principles. These two solutions, compared, led us to an interesting geometrical fact. Could we divert some more of this overflow of mechanical knowledge into other channels?

With a little luck, we can imagine a mechanism to solve the problem of the traffic center considered above (sect. 1 (4)): Three pulleys turn around axles (nails) fixed in a vertical wall at the points A, B, and C; see fig. 9.7.

Three strings, XAP, XBQ, and XCR in fig. 9.7, pass over the pulleys at A, B, and C, respectively. At their common endpoint X the three strings are attached together and each carries a weight, P, Q, and R, respectively, at its other end. These weights P, Q, and R are equal. Our problem is to find the position of equilibrium.

Of course, this problem must be understood with the usual simplifications: The strings are perfectly flexible and inextensible, the friction, the weight of the strings, and the dimensions of the pulleys are negligible (the pulleys are treated as points). As under (1) we can solve the problem by two different methods.

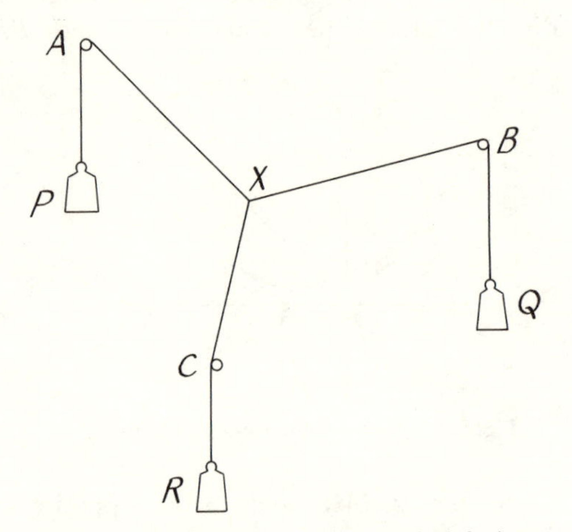

Fig. 9.7. Traffic center by mechanical device.

First, the three weights must hang jointly as low as possible. That is, the sum of their distances from a given horizontal level (the ground) must be a minimum. (That is, the potential energy of the system must be a minimum; remember that the three weights are equal.) Therefore $AP + BQ + CR$ must be a maximum. Therefore, since the length of each string is invariable, $AX + BX + CX$ must be a minimum and so our problem turns out to be identical with the problem of the traffic center of sect. 1 (4), fig. 9.4, 9.5.

Second, the forces acting at the point X must be in equilibrium. The three equal weights pull, each at its own string, with equal force and these forces are transmitted undiminished by the frictionless pulleys. Three equal forces acting at X along the lines XA, XB, and XC, respectively, must be in equilibrium. Obviously, by symmetry, they must be equally inclined to each other; the angle between any two of the three strings meeting at X is 120°. (The triangle formed by the three forces is equilateral, its exterior angles are 120°.)

This confirms the solution of sect. 1 (4). (On the other hand, the mechanical interpretation may emphasize the necessity of some restriction concerning the configuration of the three points A, B, and C.)

3. Reinterpretation. A stick, half immersed in water, appears sharply bent. We conclude hence that the light that follows a straight course in the water as in the air undergoes an abrupt change of direction in emerging from the water into the air. This is the phenomenon of refraction—a phenomenon apparently more complicated and more difficult to understand than reflection. The law of refraction, after unsuccessful efforts by Kepler and others, was finally discovered by Snellius (about 1621) and published by Descartes. Still later came Fermat (1601–1665) who took up the thread of ideas started by Heron.

The light, proceeding from an object A under water to an eye B above the water describes a broken line with an angular point on the surface that separates the air from the water; see fig 9.8. The straight line is, however, the shortest path between A and B, and so the light, in its transition from one medium into another, fails to obey Heron's principle. This is disappointing; we do not like to admit that a simple rule that holds good in two cases (direct propagation and reflection) fails in a third case (refraction). Fermat hit upon an expedient. He was familiar with the idea that the light takes time to travel from one point to another, that it travels with a certain (finite) speed; in fact, Galileo proposed a method for measuring the velocity of light. Perhaps the light that travels with a certain velocity through the air travels with another velocity through the water; such a difference in velocity could explain, perhaps, the phenomenon of refraction. As long as it travels at constant speed, the light, in choosing the shortest course, chooses also the *fastest* course. If the velocity depends on the medium traversed, the shortest course is no more necessarily the fastest. Perhaps the light chooses *always* the fastest course, also in proceeding from the water into the air.

This train of ideas leads to a clear problem of minimum (see fig. 9.8): *Given two points A and B, a straight line l separating A from B, and two velocities u and v, find the minimum time needed in travelling from A to B; you are supposed to travel from A to l with the velocity u, and from l to B with the velocity v.*

Obviously, it is the quickest to follow a straight line from A to a certain point X on l, and some other straight line from X to B. The problem consists essentially in finding the point X. Now, in uniform motion time equals distance divided by speed. Therefore, the time spent in travelling from A to X and hence to B is

$$\frac{AX}{u} + \frac{XB}{v}.$$

This quantity should be made a minimum by the suitable choice of the point X on l. We have to find X, being given A, B, u, v, and l.

To solve this problem without differential calculus is not too easy. Fermat solved it by inventing a method that eventually led to the differential calculus. We rather follow the lead given by the examples of the foregoing section. With a little luck, we can succeed in imagining a mechanism that solves for us the proposed problem of minimum; see fig. 9.9.

A ring X can slide along a fixed horizontal rod l that passes through it. Two strings XAP and XBQ are attached to the ring X. Each of these strings passes over a pulley (at A and at B, respectively) and carries a weight at its other end (at P and at Q, respectively). The main point is to choose the weights. These weights cannot be equal; if they were, the line AXB would be straight in the equilibrium position (this seems plausible at least) and so AXB would not be fit to represent the path of refracted light. Let us postpone the choice of the weights, but let us introduce a suitable notation. We

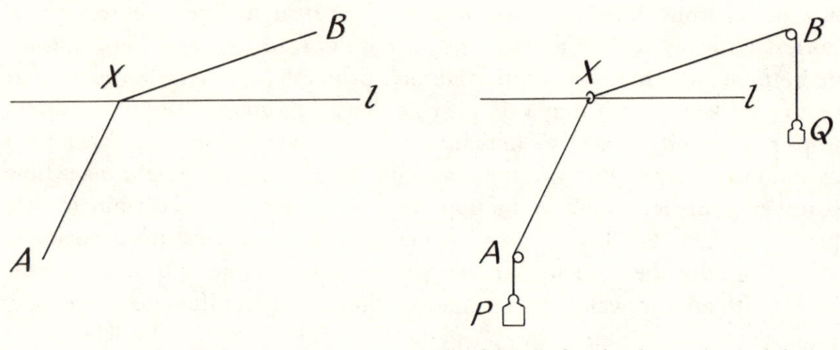

Fig. 9.8. Refraction.

Fig. 9.9. Refraction by mechanical device.

call p the weight at the endpoint P of the first string and q the weight at the endpoint Q of the second string. And now we have to find the position of equilibrium. (We assume the usual simplifications: the rod is perfectly inflexible, the strings perfectly flexible, but also inextensible; we disregard the friction, the weight and stiffness of the strings, the dimensions of the pulleys, and those of the ring.) As in sect. 2, we solve our problem by two different methods.

First, the two weights must hang jointly as low as possible. (That is, the potential energy of the system must be a minimum.) This implies that

$$AP \cdot p + BQ \cdot q$$

must be a maximum. Therefore, since the length of each string is invariable,

$$AX \cdot p + XB \cdot q$$

must be a minimum.

This is very close to Fermat's problem, but not exactly the same. Yet the two problems, the optical and the mechanical, coincide mathematically if we choose

$$p = 1/u, \qquad q = 1/v.$$

Then the problem of equilibrium (fig. 9.9) requires, just as Fermat's problem of the fastest travel, that

$$\frac{AX}{u} + \frac{XB}{v}$$

be a minimum. This we found in looking at the equilibrium of the mechanical system in fig. 9.9 from a first viewpoint.

Second, the forces acting at the point X must be in equilibrium. The pull of the weights is transmitted undiminished by the frictionless pulleys. Two forces, of magnitude $1/u$ and $1/v$, respectively, act on the ring, each pulling it in the direction of the respective string. They cannot move it in the

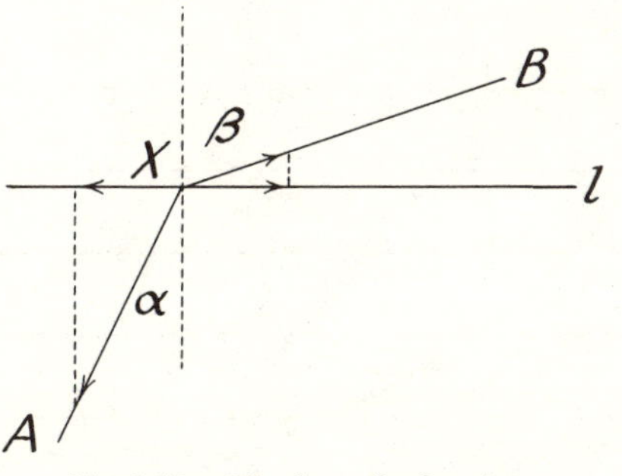

Fig. 9.10. The law of refraction.

vertical direction, because the rod l passing through the ring is perfectly rigid (there is a vertical reaction of unlimited amount due to the rod). Yet the *horizontal components* of the two pulls, which are opposite in direction, must cancel, must be *equal in magnitude*. In order to express this relation, we introduce the angles α and β between the vertical through the point X and the two strings; see fig. 9.10. The equality of the horizontal components is expressed by

$$\frac{1}{u}\sin\alpha = \frac{1}{v}\sin\beta$$

or

$$\frac{\sin\alpha}{\sin\beta} = \frac{u}{v}.$$

This is the *condition of minimum.*

Now, let us return to the optical interpretation. The angle α between the incoming ray and the normal to the refracting surface is called the angle of incidence, and β between the outgoing ray and the normal the angle of refraction. The ratio u/v of the velocities depends on the two media, water and air, but not on geometric circumstances, as the situation of the points A and B. Therefore, the condition of minimum requires that *the sines of the angles of incidence and refraction are in a constant ratio depending on the two media alone* (called nowadays the refractive index). Fermat's "principle of least time" leads to Snellius' law of refraction, confirmed by countless observations

We reconstructed as well as we could the birth of an important discovery. The procedure of solution (that we used instead of Fermat's) is also worth noticing. Our problem had from the start a physical (optical) interpretation. Yet, in order to solve it, we invented *another* physical (mechanical) interpretation. Our solution was a *solution by reinterpretation.* Such solutions may reveal new analogies between different physical phenomena and have a peculiar artistic quality.

4. Jean Bernoulli's discovery of the Brachistochrone. A heavy material point starts from rest at the point A and glides without friction along an inclined plane to a lower point B. The material point starting

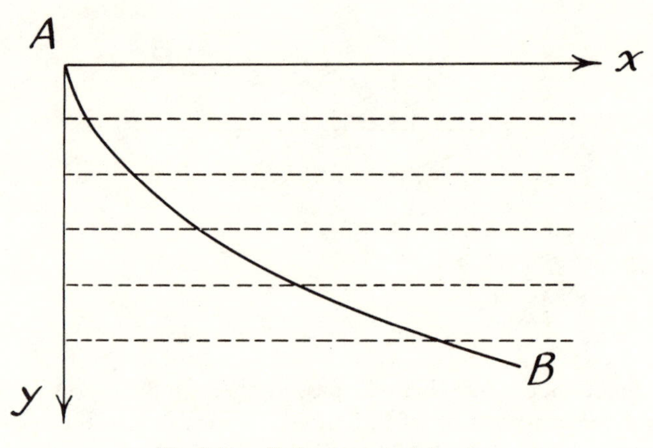

Fig. 9.11. Path of a material point.

from rest could also swing from A to B along a circular arc, as the bob of a pendulum. Which motion takes less time, that along the straight line or that along the circular arc? Galileo thought that the descent along the circular arc is faster. Jean Bernoulli imagined an arbitrary curve in the vertical plane through A and B connecting these two points. There are infinitely many such curves, and he undertook to find the curve that makes the time of descent a minimum; this curve is called the "curve of fastest descent" or the "brachistochrone." We wish to understand Jean Bernoulli's wonderfully imaginative solution of his problem.

We place an arbitrary curve descending from A to B in a coordinate system; see fig. 9.11. We choose A as the origin, the x-axis horizontal, and the y-axis vertically downward. We focus the moment when the material point sliding down the curve passes a certain point (x, y) with a certain velocity v. We have the relation

$$v^2/2 = gy$$

which was perfectly familiar to Bernoulli; we derive it today from the conservation of energy. That is, whatever the path of the descent may be, v, the velocity attained depends only on y, the depth of the descent:

$$v = (2gy)^{1/2}. \tag{1}$$

What does this mean? Let us try to see intuitively the significance of this basic fact.

We draw horizontal lines (see fig. 9.11) dividing the plane in which the material point descends into thin horizontal layers. The descending material point crosses these layers one after the other. Its velocity does not depend on the path that it took, but depends only on the layer that it is just crossing; its velocity varies from layer to layer. Where have we seen such a situation? When the light of the sun comes down to us it crosses several layers of air each of which is of a different density; therefore, the velocity of light varies from layer to layer. The proposed mechanical problem admits an *optical reinterpretation.*

We see now fig. 9.11 in a new context. We regard this figure as representing an optically inhomogeneous medium. This medium is stratified, has strata of different quality; the velocity of light in the horizontal layer at the depth y is $(2gy)^{1/2}$. The light crossing this medium from A to B (from one of the given points to the other) *could* travel along various curves. Yet the light chooses the fastest course; it travels actually along the curve that renders the time of travel a minimum. Therefore, *the actual path of light, traversing the described inhomogeneous, stratified medium from A to B,* **is** *the brachistochrone!* Yet the actual path of light is governed by Snellius' law of refraction: the solution suddenly appears within reach. Jean Bernoulli's imaginative reinterpretation renders accessible a problem that seemed entirely novel and inaccessible.

There still remains some work to do, but it demands incomparably less originality. In order to make Snellius' law applicable in its familiar form (which we have discussed in the preceding sect. 3) we change again our interpretation of fig. 9.11, slightly: the velocity v should not vary continuously with y in infinitesimal steps, but discontinuously, in small steps. We imagine several horizontal layers of transparent matter (several plates of glass) each somewhat different optically from its neighbors. Let v, v', v'', v''', . . . be the velocity of light in the successive layers, and let the light crossing them successively include the angle α, α', α'', α''', . . . with the vertical, respectively; see fig. 9.12. By the law of Snellius (see sect. 3)

$$\frac{\sin\alpha}{v} = \frac{\sin\alpha'}{v'} = \frac{\sin\alpha''}{v''} = \frac{\sin\alpha'''}{v'''} = \cdots.$$

Now we may return from the medium consisting of thin plates to the stratified medium in which v varies continuously with the depth. (Let the plates become infinitely thin.) We see that

$$\frac{\sin \alpha}{v} = \text{const.} \tag{2}$$

along the path of light.

Let β be the angle included by the tangent to the curve with the horizontal. Then

$$\alpha + \beta = 90°, \qquad \tan \beta = dy/dx = y'$$

and so

$$\sin \alpha = \cos \beta = (1 + y'^2)^{-1/2}. \tag{3}$$

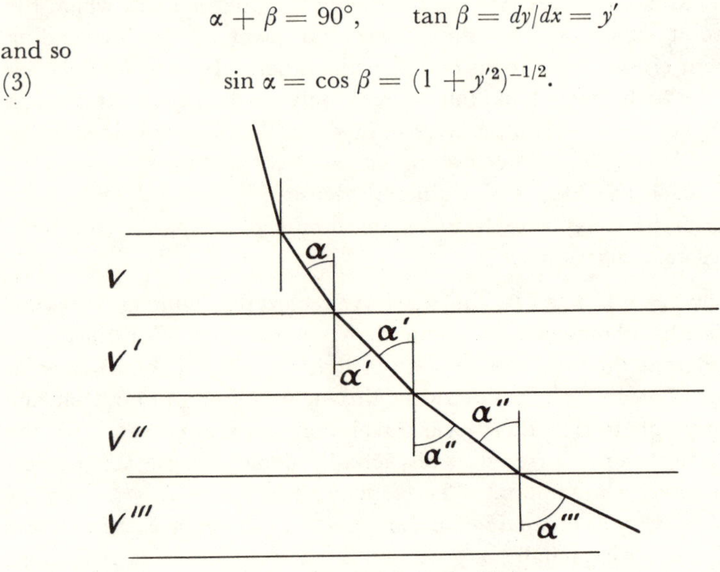

Fig. 9.12. Path of light.

We combine the equations (1), (2), and (3) (derived from mechanics, optics, and the calculus, respectively) introduce a suitable notation for the constant arising in (2), and obtain so

$$y(1 + y'^2) = c;$$

c is a positive constant. We obtained a differential equation of the first order for the brachistochrone. Finding the curves satisfying such an equation was a problem familiar to Bernoulli. We need not go into detail here (see, however, ex. 31): the brachistochrone, determined by the differential equation, turns out to be a *cycloid*. (The cycloid is described by a point in the circumference of a circle that rolls upon a straight line; in our case the straight line is the x-axis and the rolling proceeds upside down: the circle rolls under the x-axis.)

Let us observe, however, that we can see intuitively without resorting to formulas, that Snellius' law implies a differential equation. In fact, this law determines the directions of the successive elements of the path represented in fig. 9.12, and this is precisely what a differential equation does.

Jean Bernoulli's solution of the problem of the brachistochrone, that we have discussed here, has a peculiar artistic quality. Looking at fig. 9.11 or fig. 9.12, we may see intuitively the key idea of the solution. If we can see this idea clearly, without effort, anticipating what it implies, we may notice that there is a real work of art before us.

The key idea of Jean Bernoulli's solution is, of course, reinterpretation. The geometric figure (fig. 9.11 or 9.12) is conceived successively in two different interpretations, is seen in two different "contexts": first in a mechanical context, then in an optical context. Does discovery consist in an unexpected contact and subsequent interpenetration of two different contexts?

5. Archimedes' discovery of the integral calculus. It so happens that one of the greatest mathematical discoveries of all times was guided by physical intuition. I mean Archimedes' discovery of that branch of science that we call today the integral calculus. Archimedes found the area of the parabolic segment, the volume of the sphere and about a dozen similar results by a uniform method in which the idea of equilibrium plays an important role. As he says himself, he "investigated some problems in mathematics by means of mechanics."[2]

If we wish to understand Archimedes' work, we have to know something about the state of knowledge from which he started.

The geometry of the Greeks attained its peak in Archimedes' time; Eudoxus and Euclid were his predecessors, Apollonius his contemporary. We have to mention a few specific points that may have influenced Archimedes' discovery.

As Archimedes himself relates, Democritus found the volume of the cone; he stated that it is one-third of the volume of a cylinder with the same base and the same altitude. We know nothing about Democritus' method, but there seems to be some reason to suspect that he considered what we would call today a variable cross-section of the cone parallel to its base.[3]

Eudoxus was the first to prove Democritus' statement. In proving this and similar results, Eudoxus invented his "method of exhaustion" and set a standard of rigor for Greek mathematics.

We have to realize that the Greeks knew, in a certain sense, "coordinate geometry." They were used to handle loci in a plane by considering the distances of a moving point from two fixed axes of reference. If the sum of the squares of these distances is constant and the axes of reference are perpendicular to each other, the locus is a circle—this proposition belongs to coordinate geometry, but not yet to analytic geometry. Analytic geometry

---

[2] The *Method* of Archimedes, edited by Thomas L. Heath, Cambridge, 1912. Cf. p. 13. This booklet will be quoted as *Method* in the following footnotes. See also *Oeuvres complètes d'Archimède*, translated by P. Ver Eecke, pp. 474–519.

[3] Cf. *Method*, pp. 10–11.

begins in the moment when we express the relation mentioned in algebraic symbols as

$$x^2 + y^2 = a^2.$$

The Mechanics of the Greeks never attained the excellence of their Geometry, and started much later. If we take vague discussions by Aristotle and others for what they are worth, we can say that Mechanics as a science begins with Archimedes. He discovered, as everybody knows, the law of floating bodies. He also discovered the principle of the lever and the main properties of the center of gravity which we shall need in a moment.

Now we are prepared for discussing the most spectacular example of Archimedes' work; we wish to find, with his method, the *volume of the sphere.* Archimedes regards the sphere as generated by a revolving circle, and he regards the circle as a locus, characterized by a relation between the distances of a variable point from two fixed rectangular axes of reference. Written in modern notation, this relation is

$$x^2 + y^2 = 2ax,$$

the equation of a circle with radius a that touches the y-axis at the origin. See fig. 9.13 which differs only slightly from Archimedes' original figure; the circle, revolving about the x-axis, generates a sphere. I think that the use of modern notation does not distort Archimedes' idea. On the contrary, it seems to me that this notation is suggestive. It suggests motives which may lead us to Archimedes' idea today and which are, perhaps, not too different from the motives that led Archimedes himself to his discovery.

In the equation of the circle there is the term y^2. Observe that πy^2 is the area of a *variable cross-section of the sphere.* Yet Democritus found the volume of the cone by examining the variation of its cross-section. This leads us to rewriting the equation of the circle in the form

$$\pi x^2 + \pi y^2 = \pi 2ax.$$

Now we can interpret πx^2 as the variable cross-section of a cone, generated by the rotation of the line $y = x$ about the x-axis, see fig. 9.13. This suggests to seek an analogous interpretation of the remaining term $\pi 2ax$. If we do not see such an interpretation, we may try to rewrite the equation in still other shapes, and so we have a chance to hit upon the form

$$\text{(A)} \qquad 2a(\pi y^2 + \pi x^2) = x\pi(2a)^2.$$

Much is concentrated in this equation (A). Looking at equation (A), noticing the various lengths and areas arising in it and suitably *disposing them in the figure*, we may witness the birth of a great idea; it will be born from the intimate union of formula (A) with fig. 9.13.

We notice the areas of three circular disks, πy^2, πx^2, and $\pi(2a)^2$. The three circles are the intersections of the same plane with three solids of revolution. The plane is perpendicular to the x-axis and at the distance x from the origin O. The three solids of revolution are a sphere, a cone, and a cylinder. They are described by the three lines the equations of which are (A), $y = x$, and $y = 2a$, respectively, when the right hand portion of fig. 9.13 rotates about the x-axis. The cone and the cylinder have the same base and the same altitude. The radius of the common base and the common altitude have the same length $2a$. The vertex of the cone is at the origin O.

Archimedes treats differently the disks the areas of which appear on different sides of the equation (A). He leaves the disk with radius $2a$,

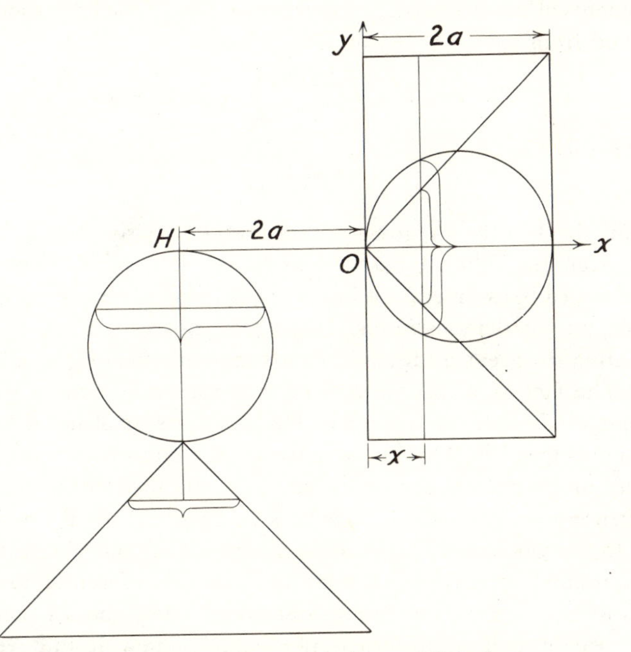

Fig. 9.13. The birth of the Integral Calculus.

cross-section of the cylinder, in its original position, at the distance x from the origin. Yet he removes the disks with radii y and x, cross-sections of the sphere and the cone, respectively, from their original position and transports them to the point H of the x-axis with abscissa $-2a$. We let these disks with radii y and x hang with their center vertically under H, suspended by a string of negligible weight, see fig. 9.13. (This string is an addition, of negligible weight, to Archimedes' original figure.)

Let us regard the x-axis as a *lever*, a rigid bar of negligible weight, and the origin O as its *fulcrum* or point of suspension. Equation (A) deals with *moments*. (A moment is the product of the weight and the arm of the lever.) Equation (A) expresses that the moment of the two disks on the left-hand side equals the moment of the one disk on the right-hand side and so, by the mechanical law discovered by Archimedes, the lever is in equilibrium.

As x varies from 0 to $2a$, we obtain all cross-sections of the cylinder; these cross-sections *fill* the cylinder. To each cross-section of the cylinder there correspond two cross-sections suspended from the point H and these cross-sections *fill* the sphere and the cone, respectively. *As their corresponding cross-sections, the sphere and the cone, hanging from H, are in equilibrium with the cylinder.* Therefore, by Archimedes' mechanical law, the moments must be equal. Let us call V the volume of the sphere, let us recall the expression for the volume of the cone (due to Democritus) and also the volume of the cylinder and the obvious location of its center of gravity. Passing from the moments of the cross-sections to the moments of the corresponding solids, we are led from equation (A) to

$$\text{(B)} \qquad 2a\left(V + \frac{\pi(2a)^2 2a}{3}\right) = a\pi(2a)^2 2a$$

which readily yields[4]

$$V = \frac{4\pi a^3}{3}.$$

Looking back at the foregoing, we see that the decisive step is that from (A) to (B), from the filling cross-sections to the full solids. Yet this step is only heuristically assumed, not logically justified. It is plausible, even very plausible, but not demonstrative. It is a guess, not a proof. And Archimedes, representing the great tradition of Greek mathematical rigor, knows this full well: "The fact at which we arrived is not actually demonstrated by the argument used; but the argument has given a sort of indication that the conclusion is true."[5] This guess, however, is a guess with a prospect. The idea goes much beyond the requirements of the problem at hand, and has an immensely greater scope. The passage from (A) to (B), from the cross-section to the whole solid is, in more modern language, the transition from the infinitesimal part to the total quantity, from the differential to the integral. This transition is a great beginning, and Archimedes, who was a great enough man to see himself in historical perspective, knew it full well: "I am persuaded that this method will be of no little service to mathematics. For I foresee that this method, once understood, will be used to discover other theorems which have not yet occurred to me, by other mathematicians, now living or yet unborn."[6]

EXAMPLES AND COMMENTS ON CHAPTER IX.

1. Given in a plane a point P and two intersecting lines l and m, none of which passes through P. Let Y be a variable point on l and Z a variable

[4] I presented this derivation several times in my classes and once I received a compliment I am proud of. After my usual "Are there any questions?" at the end of the derivation, a boy asked: "Who paid Archimedes for this research?" I must confess that I was not prompt enough to answer: "In those days such research was sponsored only by Urania, the Muse of Science."

[5] *Method*, p. 17.

[6] *Method*, p. 14.

point on m. Determine Y and Z so that the perimeter of $\triangle PYZ$ be a minimum.

Give two solutions, one by physical considerations, the other by geometry.

2. Three circles in a plane, exterior to each other, are given in position. Find the triangle with minimum perimeter that has one vertex on each circle.

Give two different physical interpretations.

3. *Triangle with minimum perimeter inscribed in a given triangle.* Given $\triangle ABC$. Find three points X, Y, and Z on the sides BC, CA, and AB of the triangle, respectively, such that the perimeter of $\triangle XYZ$ is a minimum.

Give two different physical interpretations.

4. Generalize ex. 3.

5. Criticize the solution of ex. 1. Does it apply to all cases?

6. Criticize the solution of ex. 3. Does it apply to all cases?

7. Give a rigorous solution of ex. 3 for acute triangles. [Partial variation, ex. 1, ex. 5.]

8. Criticize the solutions of sect. 1 (4) and sect. 2 (2) for the problem of the traffic center. Do they apply to all cases?

9. *Traffic center of four points in space.* Given a tetrahedron with vertices at the points A, B, C, and D. Assume that there is a point X inside the tetrahedron such that the sum of its distances from the four vertices

$$AX + BX + CX + DX$$

is a minimum. Show that the angles $\angle AXB$ and $\angle CXD$ are equal and are bisected by the same straight line; point out other pairs of angles similarly related. [Do you know a related problem? An analogous problem? Could you use its result, or the method of its solution?]

10. *Traffic center of four points in a plane.* Consider the extreme case of ex. 9 in which the points A, B, C, and D, in the same plane, are the four vertices of a convex quadrilateral $ABCD$. Do the statements of ex. 9 remain valid in this extreme case?

11. *Traffic network for four points.* Let A, B, C, and D be four fixed points, and X and Y two variable points, in a plane. If the minimum of the sum of five distances $AX + BX + XY + YC + YD$ is so attained that all six points A, B, C, D, X, and Y are distinct, the three lines XA, XB, and XY are equally inclined to each other and so are the three lines YC, YD, and YX.

12. *Unfold and straighten.* There is still another useful interpretation of fig. 9.3. Draw l, A^*X, and XB on a sheet of transparent paper, then fold the sheet along the line l: you obtain fig. 9.1 (with A^* instead of A). Imagine the fig. 9.1 originally drawn in this sophisticated manner on a folded transparent sheet. In order to find the position of X that renders $AX + XB$ a minimum, unfold the sheet, draw a straight line from A (or rather A^* in fig. 9.3) to B, and then fold the sheet back.

13. *Billiards.* On a rectangular billiard table there is a ball at the point P. It is required to drive the ball in such a direction that after four successive reflections on the four sides of the rectangle the ball should return to its original position P. [Fig. 9.14.]

14. *Geophysical exploration.* At the point E of the horizontal surface of the earth an explosion takes place. The sound of this explosion is propagated in the interior of the earth and reflected by an oblique plane layer OR which includes the angle α with the earth's surface. The sound coming from E can attain a listening post L at another point of the earth's surface in n

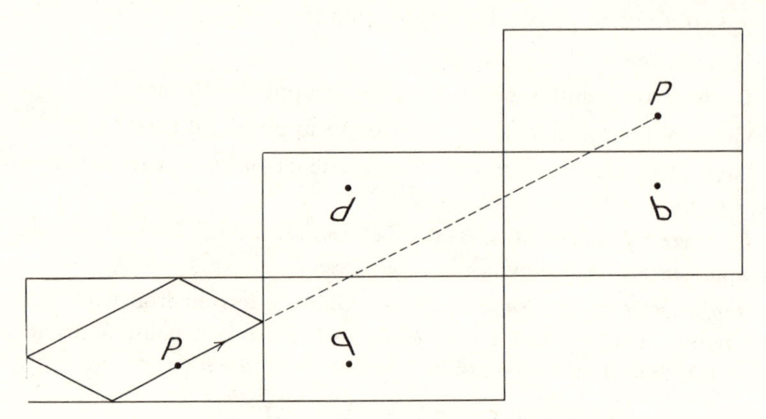

Fig. 9.14. The reflected billiard table.

different ways. (One of the n paths is constructed in fig. 9.15 by the method of ex. 12.) Being given n (observed by suitable apparatus) give limits between which α is included.

15. Given, in space, a straight line l and two points, A and B, not on l. On the line l find a point X such that the sum of its distances from the two given points $AX + XB$, be a minimum.

Do you know a related problem? A more special problem? Could you use its result, or the method of its solution?

16. Solve ex. 15 by using a tangent level surface.

17. Solve ex. 15 by paper-folding.

18. Solve ex. 15 by mechanical interpretation. Is the solution consistent with ex. 16 and 17?

19. Given three skew straight lines in space, a, b, and c. Show that the triangle with one vertex on each given line and minimum perimeter has the following property: the line joining its vertex on the line a to the center of its inscribed circle is perpendicular to a.

20. Consider the particular case of ex. 19 in which the three skew lines are three edges of a cube. Where are the vertices of the desired triangle?

Where is the center of its inscribed circle? What is its perimeter, if the volume of the cube is $8a^3$?

21. Given three skew straight lines in space, a, b, and c. Let X vary along a, Y along b, Z along c, and T freely in space. Find the minimum of $XT + YT + ZT$.

22. Specialize ex. 21 as ex. 20 specializes ex. 19.

23. *Shortest lines on a polyhedral surface.* The end walls of a rectangular room are squares; the room is 20 feet long, 8 feet wide, and 8 feet high. A spider is on one of the end walls, 7 feet above the floor and midway between the side walls. The spider perceives a fly on the opposite wall 1 foot above

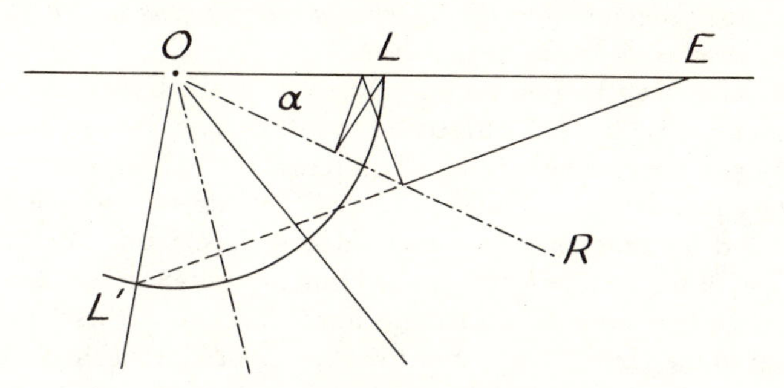

Fig. 9.15. Reflections underground.

the floor and also midway between the side walls. Show that the spider has *less* than 28 feet to travel along the walls, or the ceiling, or the floor, to attain the fly. [Ex. 17.]

24. *Shortest lines (geodesics) on a curved surface.* We regard a curved surface as the limit of a polyhedron. As the polyhedron approaches the curved surface, the number of its faces tends to ∞, the longest diagonal of any face tends to 0, and the faces tend to become tangential to the surface.

On a polyhedral surface, the shortest line between two points is a polygon. It may be a plane polygon all points of which lie in one plane, or it may be a skew polygon the points of which are not contained in one plane. (Both cases can be illustrated by the solution of ex. 23, the first case by (1), the second case by (2) and (3).)

On a curved surface, a shortest line is called a "geodesic" because shortest lines play a role in geodesy, the study of the earth's surface. A geodesic may be a plane curve, fully contained in one plane, or it may be a "space curve" ("tortuous" curve) the points of which cannot be contained in one plane. At any rate, a geodesic must have some intrinsic geometric relation to the surface on which it is a shortest line. What is this relation?

(1) We consider the polygonal line $ABC \ldots L$. Even if $ABC \ldots L$ is a skew polygon, two consecutive segments of it, as HI and IJ, lie in the same

plane. If $ABC \ldots L$ is the shortest line on a polyhedral surface between its endpoints A and L, each of its intermediate vertices $B, C, D, \ldots, H, I, J, \ldots K$ lies on an edge of the polyhedron. The plane that contains the segments HI and IJ contains also the bisector of $\angle HIJ$, and this bisector is perpendicular to the edge of the polyhedron that passes through I; see ex. 16 or ex. 18.

We consider a curve. Even if the curve is tortuous, an infinitesimal (very short) arc of it can be regarded as a plane (almost plane) arc. The plane of the infinitesimal arc is the *osculating plane* at its midpoint. The osculating plane is analogous to the plane in which two successive segments of a skew polygon lie. If the curve is a geodesic, that is, a shortest line on a surface, analogy suggests that *the osculating plane of a geodesic at a point passes through the normal to the surface at that point.*

(2) A geodesic can be interpreted physically as a rubber band stretched along a smooth (frictionless) surface. Let us examine the equilibrium of a small portion of the rubber band. The forces acting on this portion are two tensions of equal amount acting tangentially at the two endpoints of the short arc, and the reaction of the frictionless surface acting normally to it. The reaction of the surface, compounded into a resultant force, and the two tensions at the endpoints are in equilibrium. Therefore, these three forces are parallel to the same plane. Yet two "neighboring tangents" determine the osculating plane which, therefore, contains the normal to the surface.

(3) Each arc of a geodesic is a geodesic. In fact, if in a curve there is a portion that is not the shortest between its endpoints and, therefore, can be replaced by a shorter arc between the same endpoints, the whole curve cannot be a shortest line. Hence it is natural to expect that a geodesic possesses some distinctive property in each of its points. The property suggested by two very different heuristic considerations, (1) and (2), is a property of this kind.

(4) Look out for examples to test the heuristically obtained result. What are the shortest lines on a sphere? Do they have the property suggested? Do other lines on the spherical surface have the same property?

25. A material point moves without friction on a smooth rigid surface. No exterior forces (such as gravitation) act on the point (except, of course, the reaction of the surface). Give reasons why the point should be expected to describe a geodesic.

26. *A construction by paper-folding.* Find a polygon inscribed in a circle if the sides are given in magnitude and in succession.

Let $a_1, a_2, a_3, \ldots a_n$ denote the given lengths. The side of length a_1 is followed by the side of length a_2, this one by the side of length a_3, and so on; the side of length a_n is followed by the side of length a_1. It is understood that any of the lengths $a_1, a_2, \ldots a_n$ is less than the sum of the remaining $n - 1$ lengths.

There is a beautiful solution by paper-folding. Draw $a_1, a_2, \ldots a_n$ on cardboard as successive chords in a sufficiently large circle so that two consecutive chords have a common endpoint. Draw the radii from these endpoints to the center of the circle. Cut out the polygon bounded by the n chords and the two extreme radii, fold the cardboard along the $n - 1$ other radii, and paste together the two radii along which the cardboard was cut. You obtain so an open polyhedral surface; it consists of n rigid isosceles triangles, is bounded by n free edges, of lengths $a_1, a_2, \ldots a_n$, respectively, and has n dihedral angles which can still be varied. (We suppose that $n > 3$.)

What can you do to this polyhedral surface to solve the proposed problem?

27. *The die is cast.* The mass in the interior of a heavy rigid convex polyhedron need not be uniformly distributed. In fact, we can imagine a suitable heterogeneous distribution of mass the center of gravity of which coincides with an arbitrarily assigned interior point of the polyhedron. Thrown on the horizontal floor, the polyhedron will come to rest on one of its faces. This yields a mechanical argument for the following geometrical proposition.

Given any convex polyhedron P and any point C in the interior of P, we can find a face F of P with the following property: the foot of the perpendicular drawn from C to the plane of F is an *interior* point of F.

Find a geometrical proof for this proposition. (Observe that the face F may, but need not, be uniquely determined by the property stated.)

28. *The Deluge.* There are three kinds of remarkable points on a contour map: peaks, passes (or saddle points with a horizontal tangent plane), and "deeps." (On fig. 8.7 P is a peak, S a pass.) A "deep" is the deepest point in the bottom of a valley from which the water finds no outlet. A deep is an "inverted" peak: on the contour-map, regard any level line of elevation h as if it had the elevation $-h$. Then the map is "inverted"; it becomes the map of a landscape under the sea, the peaks become deeps, the deeps become peaks, but the passes remain passes. There is a remarkable connection between these three kinds of points.

Suppose that there are P peaks, D deeps, and S passes on an island. Then

$$P + D = S + 1.$$

In order to derive this theorem intuitively, we imagine that a persistent rain causes the lake around the island to rise till finally the whole island is submerged. We may admit that all P peaks are equally high, and that all D deeps are on, or under, the level of the lake. In fact, we can imagine the peaks raised and the deeps depressed without changing their number. As it starts raining, the water gathers in the deeps; we have, counting the lake,

$D + 1$ sheets of water and 1 island

at the beginning. Just before the island is engulfed, only the peaks show above the water and so we have

1 sheet of water and P islands

at the end. How did the transition take place?

Let us imagine that, at any time, the several sheets of water are at the same elevation. If there is no pass precisely at this elevation, the water can rise a little more without changing the number of the sheets of water, or the number of the islands. When, however, the rising water just attains a pass, the least subsequent rise of its level will either unite two formerly

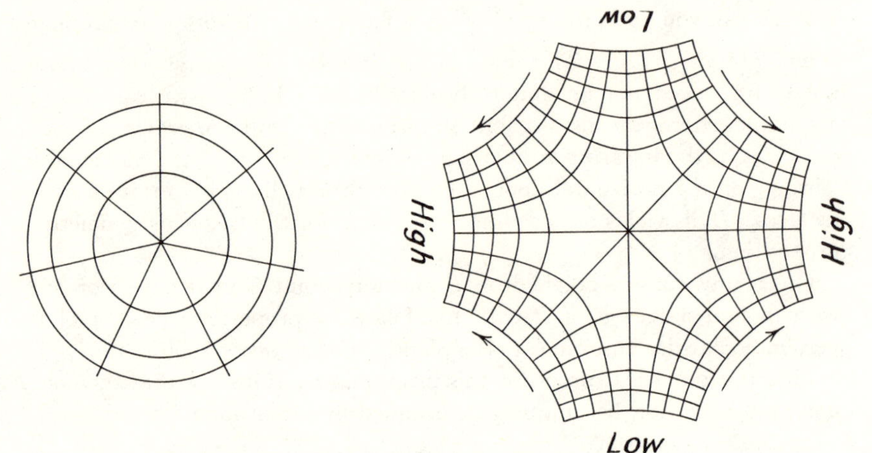

Fig. 9.16. Neighborhood of a peak.

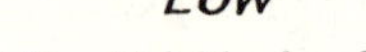

Fig. 9.17. Neighborhood of a pass.

separated sheets of water or isolate a piece of land. Therefore, each pass either decreases the number of the sheets of water by one unit, or increases the number of islands by one unit. Looking at the total change, we obtain

$$(D + 1 - 1) + (P - 1) = S$$

which is the desired theorem.

(a) Suppose now that there are P peaks, D deeps, and S passes on the whole globe (some of them are under water) and show that

$$P + D = S + 2.$$

(b) The last relation reminds us of Euler's theorem (see sect. 3.1–3.7 and ex. 3.1–3.9). Could you use Euler's theorem to construct a geometrical proof for the result just obtained by an intuitive argument? [Figs. 9.16 and 9.17 show important pieces of a more complete map in which not only some level lines are indicated, but also some "lines of steepest descent" which are perpendicular to the level lines. These two kinds of lines subdivide the globe's surface into triangles and quadrilaterals. Cf. ex. 3.2.]

(c) Are there any remarks on the method?

29. *Not so deep as a well.* In order to find d, the depth of a well, you drop a stone into the well and measure the time t between the moment of dropping the stone and the moment when you hear the stone striking the water.

(a) Given g, the gravitational acceleration, and c, the velocity of sound, express d in terms of g, c, and t. (Neglect the resistance of the air.)

(b) If the well is not too deep, even the final velocity of the stone will be a small fraction of the velocity of sound and so we may expect that much the greater part of the measured time t is taken up by the fall of the stone. Hence we should expect that

$$d = gt^2/2 - \text{correction}$$

where the correction is relatively small when t is small.

In order to examine this guess, expand the expression obtained as answer to (a) in powers of t and retain the first two non-vanishing terms.

(c) What would you regard as typical in this example?

30. *A useful extreme case.* An ellipse revolving about its major axis describes a so-called prolate spheroid, or egg-shaped ellipsoid of revolution. The foci of the rotating ellipse do not rotate: they are on the axis of revolution and are also called the foci of the prolate spheroid. We could make an elliptic mirror by covering the inner, concave side of the surface with polished metal; all light coming from one focus is reflected into the other focus by such an elliptic mirror; cf. sect. 1 (3). Elliptic mirrors are very seldom used in practice, but there is a limiting case which is very important in astronomy. What happens if one of the foci of the ellipsoid is fixed and the other tends to infinity?

31. Solve the differential equation of the brachistochrone found in sect. 4.

32. *The Calculus of Variations* is concerned with problems on the maxima and minima of quantities which depend on the shape and size of a variable curve. Such is the problem of the brachistochrone solved in sect. 4 by optical interpretation. The problem of geodesics, or shortest lines on a curved surface, discussed in ex. 24, also belongs to the Calculus of Variations, and the "isoperimetric problem" that will be treated in the next chapter belongs there too. Physical considerations, which can solve various problems on maxima and minima as we have seen, can also solve some problems of the Calculus of Variations. We sketch an example.

Find the curve with given length and given endpoints that has the center of gravity of minimum elevation. It is assumed that the density of heavy matter is constant along the curve which we regard as a uniform cord or chain. When the center of gravity of the chain attains its lowest possible position, the chain is in *equilibrium.* Now we can investigate the equilibrium of the chain in examining the forces acting on it, its weight and its tension. This investigation leads to a differential equation that determines the desired curve, the *catenary.* We do not go into detail. We just wish to remark that the solution sketched has the same basic idea as the mechanical solutions considered in sect. 2.

33. *From the equilibrium of the cross-sections to the equilibrium of the solids.* Archimedes did not state explicitly the general principle of his method, but he applied it to several examples, computing volumes, areas and centers of gravity, and the variety of these applications makes the principle perfectly clear. Let us apply the variant of Archimedes' method that has been presented in sect. 5 to some of his examples.

Prove Proposition 7 of the *Method*: Any segment of the sphere has to the cone with the same base and height the ratio that the sum of the radius of the sphere and the height of the complementary segment has to the height of the complementary segment.

34. Prove Proposition 6 of the *Method*: The center of gravity of a hemisphere is on its axis and divides this axis so that the portion adjacent to the vertex of the hemisphere has to the remaining portion the ratio of 5 to 3.

35. Prove Proposition 9 of the *Method*: The center of gravity of any segment of a sphere is on its axis and divides this axis so that the portion adjacent to the vertex has to the remaining part the ratio that the sum of the axis of the segment and four times the axis of the complementary segment has to the sum of the axis of the segment and double the axis of the complementary segment.

36. Prove Proposition 4 of the *Method*: Any segment of a paraboloid of revolution cut off by a plane at right angles to the axis has the ratio of 3 to 2 to the cone that has the same base and the same height as the segment.

37. Prove Proposition 5 of the *Method*: The center of gravity of a segment of a paraboloid of revolution cut off by a plane at right angles to the axis is on the axis and divides it so that the portion adjacent to the vertex is double the remaining portion.

38. *Archimedes' Method in retrospect.* What was in Archimedes' mind as he discovered his method, we shall never know and we can only vaguely guess. Yet we can set up a clear and fairly short list of such mathematical rules (well known today but unformulated in Archimedes' time) as we need to solve, with contemporary methods, the problems that Archimedes solved with his method. We need:

(1) Two general rules of the integral calculus:

$$\int cf(x)dx = c\int f(x)dx, \qquad \int [f(x) + g(x)]dx = \int f(x)dx + \int g(x)dx;$$

c is a constant, $f(x)$ and $g(x)$ are functions.

(2) The value of four integrals:

$$\int x^n dx = x^{n+1}/(n+1) \text{ for } n = 0, 1, 2, 3.$$

(3) The geometric interpretation of two integrals:

$$\int Q(x)\,dx, \qquad \int xQ(x)\,dx.$$

$Q(x)$ denotes a length in plane geometry and an area in solid geometry; it denotes in both cases the variable cross-section of a figure determined by a plane perpendicular to the x-axis. The first integral expresses an area or a volume, the second integral the moment of a uniformly filled area or volume, according as we consider a problem of plane or solid geometry.

Archimedes did not formulate these rules, although we cannot help thinking that he possessed them, in some form or other. He refrained even from formulating in general terms the underlying process, the passage from the variable cross-section to the area or volume, from the integrand to the integral as we would say today. He described this process in particular cases, he applied it to an admirable variety of cases, he doubtless knew it intimately, but he regarded it as merely heuristic, and he thought this a good enough reason for refraining from stating it generally.

Quote simple geometric facts which can yield intuitively the values of the four integrals listed under (2).[7]

---

[7] For other remarks on Archimedes' discovery cf. B. L. van der Waerden, *Elemente der Mathematik*, vol. 8, 1953, p. 121–129, and vol. 9, 1954, p. 1–9.

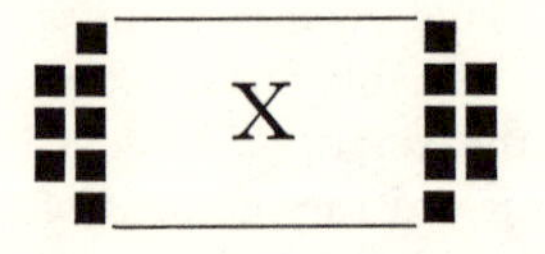

THE ISOPERIMETRIC PROBLEM

The circle is the first, the most simple, and the most perfect figure.—PROCLUS[1]
Lo cerchio è perfettissima figura.—DANTE[2]

1. Descartes' inductive reasons. In Descartes' unfinished work *Regulae ad Directionem Ingenii* (or *Rules for the Direction of the Mind*, which, by the way, must be regarded as one of the classical works on the logic of discovery) we find the following curious passage:[3] "In order to show by enumeration that the perimeter of a circle is less than that of any other figure of the same area, we do not need a complete survey of all the possible figures, but it suffices to prove this for a few particular figures whence we can conclude the same thing, by induction, for all the other figures."

In order to understand the meaning of the passage let us actually perform what Descartes suggests. We compare the circle to a few other figures, triangles, rectangles, and circular sectors. We take two triangles, the equilateral and the isosceles right triangle (with angles 60°, 60°, 60° and 90°, 45°, 45°, respectively). The shape of a rectangle is characterized by the ratio of its width to its height; we choose the ratios 1 : 1 (square), 2 : 1, 3 : 1, and 3 : 2. The shape of a sector of the circle is determined by the angle at the center; we choose the angles 180°, 90°, and 60° (semicircle, quadrant, and sextant). We assume that all these figures have the same area, let us say, 1 square inch. Then we compute the length of the perimeter of each figure in inches. The numbers obtained are collected in the following table; the order of the figures is so chosen that the perimeters increase as we read them down.

[1] Commentary on the first book of Euclid's *Elements*; on Definitions XV and XVI.

[2] *Convivio II*, XIII, 26.

[3] *Oeuvres de Descartes*, edited by Adam and Tannery, vol. 10, 1908, p. 390. The passage is unessentially altered; the property of the circle under consideration is stated here in a different form.

Table I. Perimeters of Figures of Equal Area

Circle	3.55
Square	4.00
Quadrant	4.03
Rectangle 3 : 2	4.08
Semicircle	4.10
Sextant	4.21
Rectangle 2 : 1	4.24
Equilateral triangle	4.56
Rectangle 3 : 1	4.64
Isosceles right triangle	4.84

Of the ten figures listed, which are all of the same area, the circle, listed at the top, has the shortest perimeter. Can we conclude hence by induction, as Descartes seems to suggest, that the circle has the shortest perimeter not only among the ten figures listed but among all possible figures? By no means. But it cannot be denied that our relatively short list suggests very strongly the general theorem. So strongly, indeed, that if we added one or two more figures to the list, the suggestion could not be made much stronger.

I am inclined to believe that Descartes, in writing the passage quoted, thought of this last, more subtle point. He intended to say, I think, that prolonging the list would not have much influence on our belief.

2. Latent reasons. "Of all plane figures of equal area, the circle has the minimum perimeter." Let us call this statement, supported by Table I, the *isoperimetric theorem.*[4] Table I, constructed according to Descartes' suggestion, yields a fairly convincing inductive argument in favor of the isoperimetric theorem. Yet why does the argument appear convincing?

Let us imagine a somewhat similar situation. We choose ten trees of ten different familiar kinds. We measure the specific weight of the wood of each tree, and pick out the tree the wood of which has the least specific weight. Would it be reasonable to believe merely on the basis of these observations that the kind of tree that has the lightest wood among the ten kinds examined has also the lightest wood among all existing kinds of trees? To believe this would not be reasonable, but silly.

What is the difference from the case of the circle? We are *prejudiced* in favor of the circle. The circle is the most perfect figure; we readily believe that, along with its other perfections, the circle has the shortest perimeter for a given area. The inductive argument suggested by Descartes appears so convincing because it corroborates a conjecture plausible from the start.

"The circle is the most perfect figure" is a traditional phrase. We find it in the writings of Dante (1265–1321), of Proclus (410–485), and of still earlier writers. The meaning of the sentence is not clear, but there may be something more behind it than mere tradition.

[4] An explanation of the name and equivalent forms will be given later (sect. 8).

3. Physical reasons. "Of all solids of equal volume, the sphere has the minimum surface." We call this statement the "isoperimetric theorem in space."

We are inclined to believe the isoperimetric theorem in space, as in the plane, without any mathematical demonstration. We are prejudiced in favor of the sphere, perhaps even more than in favor of the circle. In fact, nature itself seems to be prejudiced in favor of the sphere. Raindrops, soap bubbles, the sun, the moon, our globe, the planets are spherical, or nearly spherical. With a little knowledge of the physics of surface tension, we could learn the isoperimetric theorem from a soap bubble.

Yet even if we are ignorant of serious physics, we can be led to the isoperimetric theorem by quite primitive considerations. We can learn it from a cat. I think you have seen what a cat does when he prepares himself for sleeping through a cold night: he pulls in his legs, curls up, and, in short, makes his body as spherical as possible. He does so, obviously, to keep warm, to minimize the heat escaping through the surface of his body. The cat, who has no intention of decreasing his volume, tries to decrease his surface. He solves the problem of a body with given volume and minimum surface in making himself as spherical as possible. He seems to have some knowledge of the isoperimetric theorem.

The physics underlying this consideration is extremely crude.[5] Still, the consideration is convincing and even valuable as a sort of provisional support for the isoperimetric theorem. The elusive reasons speaking in favor of the sphere or the circle, hinted above (sect. 2), start condensing. Are they reasons of physical analogy?

4. Lord Rayleigh's inductive reasons. A little more than two hundred years after the death of Descartes, the physicist Lord Rayleigh investigated the tones of membranes. The parchment stretched over a drum is a "membrane" (or, rather, a reasonable approximation to the mathematical idea of a membrane) provided that it is very carefully made and stretched so that it is uniform throughout. Drums are usually circular in shape but, after all, we could make drums of an elliptical, or polygonal, or any other shape. A drum of any form can produce different tones of which usually the deepest tone, called the principal tone, is much the strongest. Lord Rayleigh compared the principal tones of membranes of different shapes, but of equal area and subject to the same physical conditions. He constructed the following Table II, very similar to our Table I in sect. 1. This Table II lists the same shapes as Table I, but in somewhat different order, and gives for each shape the pitch (the frequency) of the principal tone.[6]

---

[5] A better advised cat should not make the surface of his body a minimum, but its thermal conductance or, which amounts to the same, its electrostatic capacity. Yet, by a theorem of Poincaré, this different problem of minimum has the same solution, the sphere. See G. Pólya, *American Mathematical Monthly*, v. 54, 1947, p. 201–206.

[6] Lord Rayleigh, *The Theory of Sound*, 2nd ed., vol. 1, p. 345.

Table II. Principal Frequencies of Membranes of Equal Area

Circle	4.261
Square	4.443
Quadrant	4.551
Sextant	4.616
Rectangle 3 : 2	4.624
Equilateral triangle	4.774
Semicircle	4.803
Rectangle 2 : 1	4.967
Isosceles right triangle	4.967
Rectangle 3 : 1	5.736

Of the ten membranes listed, which are all of the same area, the circular membrane, listed at the top, has the deepest principal tone. Can we conclude hence by induction that the circle has the lowest principal tone of *all* shapes?

Of course, we cannot; induction is never conclusive. Yet the suggestion is very strong, still stronger than in the foregoing case. We know (and Lord Rayleigh and his contemporaries also knew) that of all figures with a given area the circle has the minimum perimeter, and that this theorem can be demonstrated mathematically. With this geometrical minimum property of the circle in our mind, we are inclined to believe that the circle has also the physical minimum property suggested by Table II. Our judgement is influenced by analogy, and analogy has a deep influence.

The comparison of Tables I and II is highly instructive. It yields various other suggestions which we do not attempt to discuss now.

5. Deriving consequences. We have surveyed various grounds in favor of the isoperimetric theorem which are, of course, insufficient to prove it but sufficient to make it a reasonable conjecture. A physicist, examining a conjecture in his science, derives consequences from it. These consequences may or may not agree with the facts and the physicist devises experiments to find out which is the case. A mathematician, examining a conjecture in his science, may follow a similar course. He derives consequences from his conjecture. These consequences may or may not be true and the mathematician tries to find out which is the case.

Let us follow this course in examining the isoperimetric theorem which we state now in the following form: *Of all plane figures of equal perimeter, the circle has the maximum area.* This statement differs from that given above (sect. 2) and not merely verbally. Yet it can be shown that the two statements are equivalent. We postpone the proof (see sect. 8) and hasten to examine consequences.

(1) Dido, the fugitive daughter of a Tyrian king, arrived after many adventures at the coast of Africa where she became later the founder of Carthage and its first legendary queen. Dido started by purchasing from

the natives a piece of land on the seashore "not larger than what an oxhide can surround." She cut the oxhide into fine narrow strips of which she made a very long string. And then Dido faced a geometric problem: what shape of land should she surround with her string of given length in order to obtain the maximum area?

In the interior of the continent the answer would be a circle, of course, but on the seashore the problem is different. Let us solve it, in assuming that the seashore is a straight line. In fig. 10.1 the arc XYZ has a given length. We are required to make a maximum the area between this arc and the straight line XZ (which lies on a given infinite straight line, but can be lengthened or shortened at pleasure).

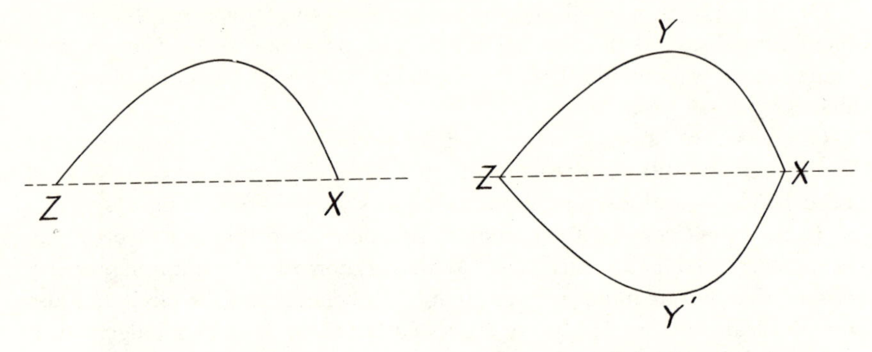

Fig. 10.1. Dido's problem.

Fig. 10.2. Solution by mirror image

To solve this problem we regard the given infinite straight line (the seashore) as a mirror; see fig. 10.2. The line XYZ and its mirror image $XY'Z$ form jointly a *closed* curve $XYZY'$ of given length surrounding an area which is exactly double the one that we have to maximize. This area is a maximum when the closed curve is a circle of which the given infinite straight line (the seashore) is an axis of symmetry. Therefore, the solution of Dido's problem is a semicircle with center on the seashore.

(2) Jakob Steiner derived a host of interesting consequences from the isoperimetric theorem. Let us discuss one of his arguments which is especially striking. Inscribe in a given circle a polygon (fig. 10.3). Regard the segments of the circle (shaded in fig. 10.3) cut off by the sides of the inscribed polygon as rigid (cut out of cardboard). Imagine these rigid segments of the circle connected by flexible joints at the vertices of the inscribed polygon. Deform this articulated system by changing the angles at the joints. After deformation (see fig. 10.4) you obtain a new curve which is not a circle, but consists of successive circular arcs and has the same perimeter as the given circle. Therefore, by the isoperimetric theorem, the area of the new curve must be less than the area of the given circle. Yet the circular segments are rigid (of cardboard), their areas unchanged, and so the deformed polygon must take the blame for lessening the area:

The area of a polygon inscribed in a circle is greater than the area of any other polygon with the same sides. (The sides are the same in length and in order of succession.)

This consequence is elegant, but as yet unproved, insofar as we have not proved yet the isoperimetric theorem itself.

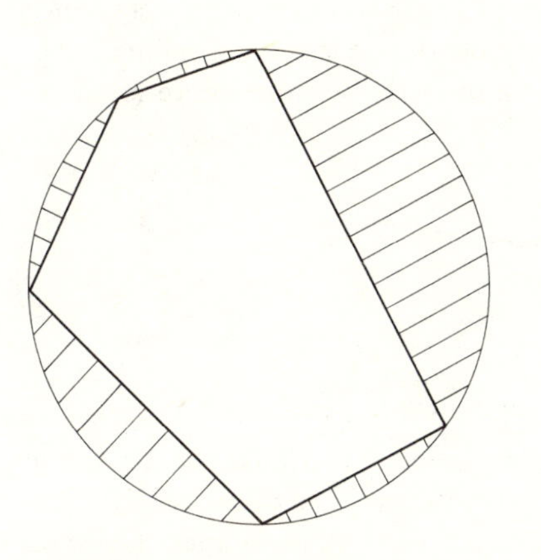

Fig. 10.3. An inscribed polygon.

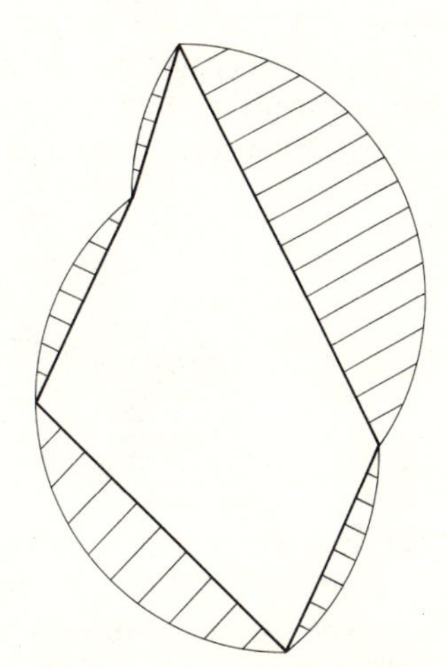

Fig. 10.4. Flexible joints and cardboard segments.

(3) Let us combine Dido's problem with Steiner's method. Inscribe in a given semicircle a polygonal line; see fig. 10.5. Regard the segments cut off by the stretches of the polygonal line from the semicircle (shaded in fig. 10.5) as rigid (of cardboard). Place flexible joints at the vertices of

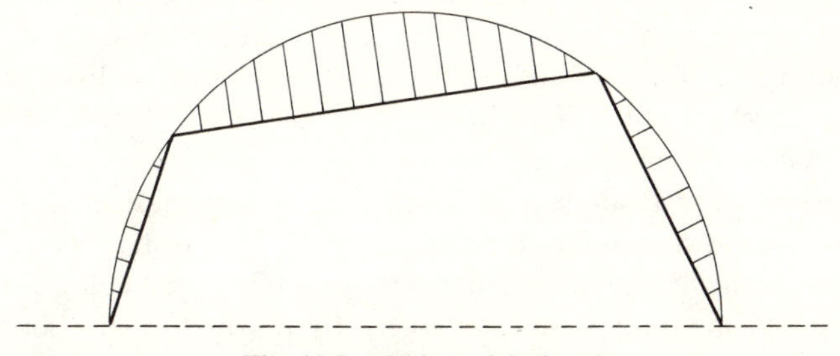

Fig. 10.5. Dido and Steiner.

the polygonal line, vary the angles, and let the two endpoints shift along the line of the diameter, which you regard as given. You obtain so a new curve (fig. 10.6) consisting of circular arcs of the same total length as the semicircle, but including with the given infinite line less area than the semicircle, by virtue of the theorem that we have discussed under (1). Yet the circular segments are rigid (of cardboard) and so the deformed polygon is responsible for the lessening of the area. Hence the theorem: *The sides of a polygon are given, except one, in length and succession. The area becomes a maximum when the polygon is inscribed in a semicircle the diameter of which is the side originally not given.*

6. Verifying consequences. A physicist, having derived various consequences from his conjecture, looks for one that can be conveniently tested by experiments. If the experiments clearly contradict a consequence derived from it, the conjecture itself is exploded. If the experiments verify its consequences, the conjecture gains in authority, becomes more credible.

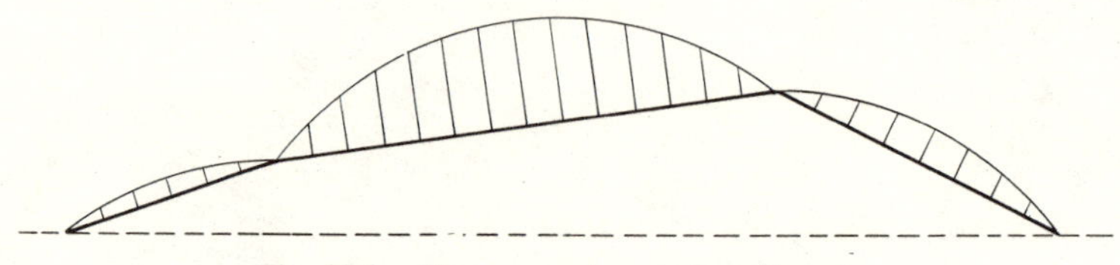

Fig. 10.6. The segments are of cardboard.

The mathematician may follow a similar course. He looks for accessible consequences of his conjecture which he could prove or disprove. A consequence disproved disproves the conjecture itself. A consequence proved renders the conjecture more credible and may hint a line along which the conjecture itself could be proved.

How about our own case? We have derived several consequences of the isoperimetric theorem; which one is the most accessible?

(1) Some of the consequences derived from the isoperimetric theorem in the foregoing section are, in fact, concerned with elementary problems on maxima. Is there any consequence that we could verify? Let us survey the various cases indicated by figs. 10.3–10.6. Which case is the simplest? The complexity of a polygon increases with the number of its sides. Therefore, the simplest polygon of all is the triangle; of course, we like the triangle best because we know the most about it. Now, the problem of figs. 10.3 and 10.4 makes no sense for triangles or, we may say, it is vacuous in the case of a triangle: a triangle with given sides is determined, rigid. There is no transition for a triangle like that from fig. 10.3 to fig. 10.4. Yet the transition from fig. 10.5 to 10.6 is perfectly possible for triangles. This may be the simplest consequence that we have derived so far from the isoperimetric theorem: let us examine it.

The simplest particular case of the result derived in sect. 5 (3) answers the following problem: *Given two sides of a triangle, find the maximum of the area;* see fig. 10.7. Sect. 5 (3) gives this answer: The area is a maximum when the triangle is inscribed in a semicircle, the diameter of which is the side originally not given. This means, however, that the area is a maximum when the two given sides include a right angle which is obvious (ex. 8.7).

We have succeeded in verifying a first consequence of the isoperimetric theorem. Such success naturally raises our spirits. What is behind the fact just verified? Could we generalize it? Could we verify some other consequence?

(2) In generalizing the problem discussed under (1), we arrive at the following: *All successive sides of a polygon are given in magnitude, except one. Find the maximum area.*

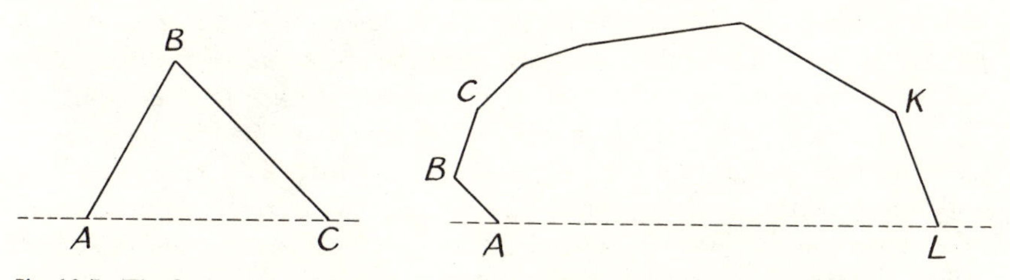

Fig. 10.7. The finger with one joint. Fig. 10.8. The super-finger.

We introduce suitable notation and draw fig. 10.8. The lengths *AB*, *BC*, . . . *KL* are given; the length *LA* is not given. We can imagine the broken line *ABC* . . . *F* . . . *KL* as a sort of "super-finger"; the "bones" *AB*, *BC*, . . . *KL* are of invariable length, the angles at the joints *B*, *C*, . . . *F* . . . *K* variable. We are required to make the area *ABC* . . . *KLA* a maximum.

As in some problems that we have considered some time ago (sect. 8.4, 8.5), the characteristic difficulty seems to be that there are many variables (the angles at *B*, *C*, . . . *F*, . . . and *K*). Yet we have just discussed, under (1), the extreme special case of the problem where there is just one variable angle (just one joint; fig. 10.7). It is natural to hope that we can use this special case as a stepping stone to the solution of the general problem.

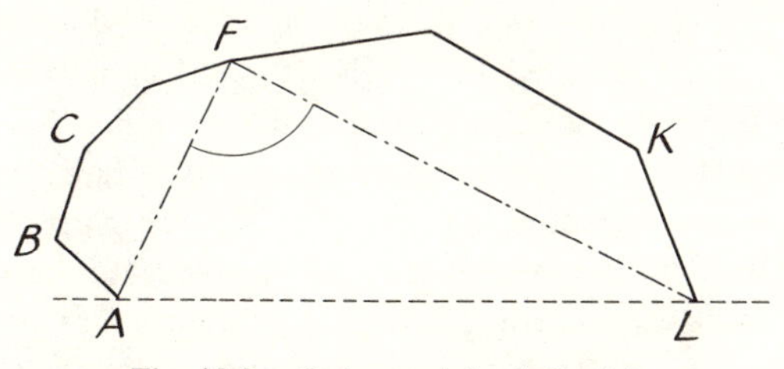

Fig. 10.9. Only one joint is flexible.

In fact, let us take the problem as almost solved. Let us imagine that we have obtained already the desired values of all the angles except one. In fig. 10.9 we regard the angle at *F* as variable, but all the other angles, at *B*, *C*, . . . *K* as fixed; the joints *B*, *C*, . . . *K* are rigid, only *F* is flexible. We join *A* and *L* to *F*. The lengths *AF* and *LF* are invariable. The whole polygon *ABC* . . . *F* . . . *KLA* is decomposed now into three parts, two of which are rigid (of cardboard) and only the third variable. The polygons $ABC \ldots FA$ and $LK \ldots FL$ are rigid. The triangle AFL has two given sides, FA and FL, and a variable angle at F. The area of this triangle, and with it the area of the whole polygon $ABC \ldots F \ldots KLA$, becomes a maximum when $\angle AFL$ is a right angle, as we have said a moment ago, under (1), in discussing fig. 10.7.

This reasoning obviously applies just as well to the other joints, that is, to the angles at B, C, . . . and K (fig. 10.8), and so we see: *the area of the polygon $ABC \ldots KLA$ cannot be a maximum unless the side originally not given, AL, subtends a right angle at each of the vertices not belonging to it, at B, C, . . . F, . . . K.* If there is a maximum area, it must be attained in the situation just described. We may take for granted that there is a maximum area and, remembering a little elementary geometry, describe the situation in other terms, as follows: *the maximum of the area is attained if, and only if, the polygon is inscribed in a semicircle, the diameter of which is the side originally not given.*

We have obtained here exactly the same result as in sect. 5 (3), but we did not use the isoperimetric theorem here and we did there.

(3) We have verified first, under (1), a very special consequence of the isoperimetric theorem, then, under (2), a much broader consequence. We have gathered now, perhaps, enough momentum to attack another broad consequence, derived above, in sect. 5(2).

We compare two polygons $ABC \ldots KL$ and $A'B'C' \ldots K'L'$; see fig. 10.10. The corresponding sides are equal $AB = A'B'$, $BC = B'C'$, $\ldots KL = K'L'$, $LA = L'A'$, but some of the angles are different; $ABC \ldots KL$ is inscribed in a circle, but $A'B'C' \ldots K'L'$ is not.

We join a vertex J of $ABC \ldots KL$ to the center of the circumscribed circle and draw the diameter JM. If, by chance, the point M coincides with a vertex of $ABC \ldots KL$, our task is greatly simplified (we could use then the result under (2) immediately). If not, M lies on the circle between two adjacent vertices of the inscribed polygon, say, A and B. Join MA, MB, consider $\triangle AMB$ (shaded in fig. 10.10) and construct over the base $A'B'$ the $\triangle A'M'B'$ (also shaded) congruent to $\triangle AMB$. Finally, join $J'M'$.

The polygon $AMBC \ldots KL$ is divided into two parts by the line JM (see fig. 10.10; the corresponding polygon is correspondingly divided by $J'M'$). Apply to both parts the theorem proved under (2). The area of the polygon $MBC \ldots J$, inscribed in a semicircle, is not less than the area of $M'B'C' \ldots J'$; in fact, the corresponding sides are all equal, except that MJ, which forms the diameter of the semicircle, may differ from $M'J'$. For the same reason the area of $MALK \ldots J$ is not less than that of $M'A'L'K' \ldots J'$. By adding, we obtain that

$$\text{area } AMBC \ldots KL > \text{area } A'M'B'C' \ldots K'L'.$$

Yet

$$\triangle AMB \cong \triangle A'M'B'.$$

By subtracting, we obtain that

$$\text{area } ABC \ldots KL > \text{area } A'B'C' \ldots K'L'.$$

The area of a polygon inscribed in a circle is greater than the area of any other polygon with the same sides.

We have obtained here exactly the same result as in sect. 5 (2), but we did not use the isoperimetric theorem here and we did there.

(The first inequality, between the areas of the extended polygons, contains the sign $>$ although a conscientious reader may have expected the sign $\geqq$. Let us append the discussion of this somewhat more subtle point. I

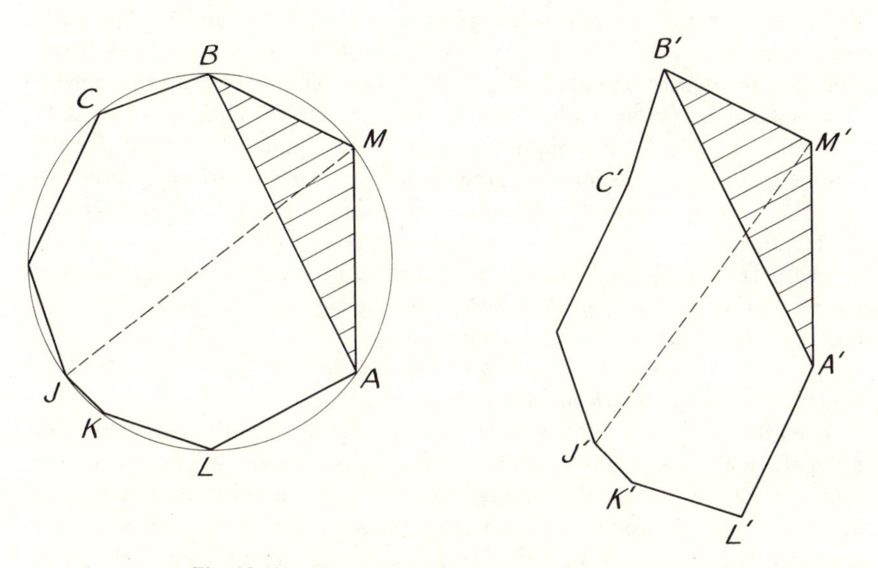

Fig. 10.10. One polygon is inscribed, the other is not.

say that the polygon $A'M'B'C' \ldots K'L'$ is not inscriptible in a circle; otherwise, $A'B'C' \ldots K'L'$ would also be inscriptible, which it is not. I say that the polygons $M'B'C' \ldots J'$ and $M'A'L'K' \ldots J'$ are not both inscriptible in a semicircle with diameter $M'J'$; otherwise, the whole polygon $A'M'B'C' \ldots K'L'$ would be inscriptible in a circle, which it is not. Hence the words "not less," used twice in the derivation of the inequality in question, can be replaced at least once by "greater.")[7]

7. Very close. The consequences that we have succeeded in verifying render the isoperimetric theorem highly plausible. Yet there is more. We may have the feeling that these consequences "contain a lot," that we are "very close" to the final solution, to the complete proof.

[7] The theorems and demonstrations of this section are due to Lhuilier; see footnote 5 of Chapter VIII.

(1) *Find the polygon with a given number of sides and a given perimeter that has the maximum area.*

If there is such a polygon, it must be *inscribed in a circle.* This much we can conclude immediately from our last remark, sect. 6(3).

On the other hand, take the problem as almost solved. Assume that you know already the correct position of all vertices except one, say, X. The $n - 1$ other vertices, say, $U, \dots W, Y$ and Z, are already fixed. The whole polygon $U \dots WXYZ$ consists of two parts: the polygon $U \dots WYZ$ with $n - 1$ already fixed vertices, which is independent of X, and $\triangle WXY$, which depends on X. Of this triangle, $\triangle WXY$, you know the base WY and the sum of the two other sides $WX + XY$; in fact, the remaining $n - 2$ sides of the polygon are supposed to be known, and you actually know the sum of all n sides. The area of $\triangle WXY$ must be a maximum. Yet, as it is almost immediate, the $\triangle WXY$ with known base and perimeter attains its maximum area when it is isosceles (ex. 8.8). That is, $WX = XY$, two adjacent sides of the required polygon are equal. Therefore (by the symmetry of the conditions and the pattern of partial variation) any two adjacent sides are equal. All sides are equal: the desired polygon is *equilateral.*

The desired polygon, which is inscribed in a circle and also equilateral, is necessarily regular: *Of all polygons with a given number of sides and a given perimeter, the regular polygon has the largest area.*

(2) *Two regular polygons, one with n sides and the other with n + 1 sides, have the same perimeter. Which one has the larger area?*

The regular polygon with $n + 1$ sides has a larger area than any irregular polygon with $n + 1$ sides and the same perimeter, as we have just seen, under (1). Yet the regular polygon with n sides, each equal to a, say, can be regarded as an irregular polygon with $n + 1$ sides: $n - 1$ sides are of length a, two sides of length $a/2$, and there is one angle equal to $180°$. (Regard the midpoint of one side of the polygon, conceived in the usual way, as a vertex, and then you arrive at the present less-usual conception.) Therefore, *the regular polygon with n + 1 sides has a larger area than the regular polygon with n sides and the same perimeter.*

(3) *A circle and a regular polygon have the same perimeter. Which one has the larger area?*

Let us realize what the foregoing result, under (2), means. Let us take $n = 3, 4, \dots$ and restate the result in each particular case. In passing from an equilateral triangle to a square with the same perimeter, we find the area increased. In passing from a square to a regular pentagon with the same perimeter, we again find the area increased. And so on, passing from one regular figure to the next, from pentagon to hexagon, from hexagon to heptagon, from n to $n + 1$, we see that the area increases at each step as the perimeter remains unchanged. Ultimately, in the limit, we reach the circle. Its perimeter is still the same, but its area is obviously superior to the area of any regular polygon in the infinite sequence of which it is the limit. *The area of the circle is larger than that of any regular polygon with the same perimeter.*

(4) *A circle and an arbitrary polygon have the same perimeter. Which one has the larger area?*

The circle. This follows immediately from the foregoing (1) and (3).

(5) *A circle and an arbitrary curve have the same perimeter. Which one has the larger area?*

The circle. This follows from the foregoing (4), since any curve is the limit of polygons. We have proved the isoperimetric theorem!

8. Three forms of the Isoperimetric Theorem. In the foregoing (sect. 6 and 7) we have proved the following statement of the isoperimetric theorem:

I. *Of all plane figures of equal perimeter, the circle has the maximum area.*

In sect. 2, however, we discussed another statement.

II. *Of all plane figures of equal area the circle has the minimum perimeter.*

These two statements are different, and different not merely in wording. They need some further clarification.

(1) Two curves are called "isoperimetric" if their perimeters are equal. "Of all isoperimetric plane curves the circle has the largest area"—this is the traditional wording of statement I, which explains the name "isoperimetric theorem."

(2) We may call the two statements of the theorem (I and II) "conjugate statements" (see sect. 8.6). We shall show that these two conjugate statements are equivalent to each other by showing that they are both equivalent to the same third.

(3) Let A denote the area and L the length of the perimeter of a given curve. Let us assume that the given curve and a circle with radius r are isoperimetric: $L = 2\pi r$. Then the first form (statement I) of the isoperimetric theorem asserts that

$$A \leqq \pi r^2.$$

Substituting for r its expression in terms of L, $r = L/2\pi$, we easily transform the inequality into

$$\frac{4\pi A}{L^2} \leqq 1.$$

We call this inequality the *isoperimetric inequality* and the quotient on the left hand side the *isoperimetric quotient.* This quotient depends only on the shape of the curve and is independent of its size. In fact, if, without changing the form, we enlarge the linear dimensions of the curve in the ratio 1 : 2, the perimeter becomes $2L$ and the area $4A$, but the quotient A/L^2 remains unchanged, and the same is true of $4\pi A/L^2$ and of enlargements in any ratio. Some authors call A/L^2 the isoperimetric quotient; we have introduced the factor 4π to make our isoperimetric quotient equal to 1 in the case of the circle. With this terminology, we can say:

III. *Of all plane figures, the circle has the highest isoperimetric quotient.*[8]

This is the third form of the isoperimetric theorem.

(4) We have arrived at the third form of the theorem in coming from figures with equal perimeter. Let us now start from statement III and pass to figures with equal area. Let us assume that a curve with area A and perimeter L has the same area as a circle with radius r. That is, $A = \pi r^2$. Substituting for A this expression, we easily transform the isoperimetric inequality into $L \geqq 2\pi r$. That is, the perimeter of the figure is greater than that of a circle with equal area. We arrive so at the second conjugate form of the theorem, at statement II.

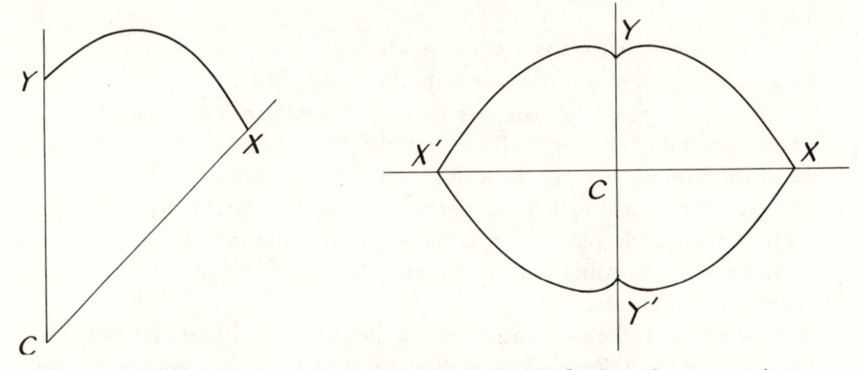

Fig. 10.11. Dido's problem, complicated by a cape.

Fig. 10.12. Reflection solves it sometimes.

(5) We could, of course, proceed along the same line of argument in the opposite direction and, passing through III, derive I from II. And so we can satisfy ourselves that all three forms are equivalent.

9. Applications and questions. If Dido bargained with the natives in the neighborhood of a cape, her problem was, perhaps, more similar to the following than to that discussed in sect. 5 (1).

Given an angle (the infinite part of a plane between two rays drawn from the same initial point). *Find the maximum area cut off from it by a line of given length.*

In fig. 10.11 the vertex of the given angle is called C (cape). The arbitrary line connecting the points X and Y is supposed to have the given length l. We are required to make a maximum the three-cornered area between this curve and the seashore. We may shift the endpoints X and Y of the curve and modify its shape, but cannot change its length l.

The problem is not too easy, but is one of those problems which a particular choice of the data renders more accessible. If the angle at C is a right angle,

[8] Abbreviating "isoperimetric quotient" as I.Q., we could say that the circle has the highest I.Q.

we may take the mirror image of the figure first with respect to one side of the angle, then with respect to the other. We obtain so a new figure, fig. 10.12, and a new problem. The line XY, quadruplicated by reflections, yields a new *closed* line of given length, $4l$. The area to be maximized, quadruplicated by reflections, yields a new area to be maximized, *completely surrounded* by the new given curve. By the isoperimetric theorem, the solution of the new problem is a circle. This circle has two given axes of symmetry, XX' and YY', and so its center is at the intersection of these axes, at the point C. Therefore, the solution of the original problem (Dido's problem) is a *quadrant*: a quarter of a circle with center at the vertex of the given angle.

We naturally recall here the solution of sect. 5 (1) based on fig. 10.2 and observe that it is closely analogous to the present solution. It is easy to see that there are an infinity of further cases in which this kind of solution works. If the given angle at C is $360°/2n = 180°/n$, we can transform, by repeated reflections, the curve XY with given length l into a new closed curve with length $2nl$ and the proposed problem into a new problem, the solution of which is a circle, by virtue of the isoperimetric theorem. The cases treated in sect. 5 (1) and in the present section are just the first two cases in this infinite sequence, corresponding to $n = 1$ and $n = 2$.

That is, if the angle at C is of a special kind ($180°/n$ with integral n) the solution of our problem (fig. 10.11) is a circular arc with center at C. It is natural to expect that this form of the solution is independent of the magnitude of the angle (at least as long as it does not exceed 180°). That is, we conjecture that the solution of the problem of fig. 10.11 is the arc of a circle with center at C, whether the angle at C is, or is not, of the special kind $180°/n$. This conjecture is an inductive conjecture, supported by the evidence of an infinity of particular cases, $n = 1, 2, 3, \ldots$. Is this conjecture true?

The foregoing application of the isoperimetric theorem and the attached question may make us anticipate many similar applications and questions. Our derivation of the theorem raises further questions; its analogues in solid geometry and mathematical physics suggest still other questions. The isoperimetric theorem, deeply rooted in our experience and intuition, so easy to conjecture, but not so easy to prove, is an inexhaustible source of inspiration.

EXAMPLES AND COMMENTS ON CHAPTER X

First Part

1. *Looking back.* In the foregoing (sect. 6–8) we have proved the isoperimetric theorem—have we? Let us check the argument step by step.

There seems to be no objection against the simple result of sect. 6 (1). Yet, in solving the problem of sect. 6 (2), we assumed the existence of the maximum without proof; and we did the same in sect. 7 (1). Do these unproved assumptions invalidate the result?

2. *Could you derive some part of the result differently?* Verify the simplest non-trivial particular case of the result found in sect. 5 (2) directly. That is, prove independently of sect. 6 (3) that the area of a quadrilateral inscribed in a circle is greater than the area of any other quadrilateral with the same sides. [Ex. 8.41.]

3. *Restate with more detail* the argument of sect. 7 (2): construct a polygon with $n+1$ sides that has the same perimeter as, but a greater area than, the regular polygon with n sides.

4. Prove independently of sect. 7 (3) that a circle has a larger area than any regular polygon with the same perimeter.

5. Prove, more generally, that a circle has a larger area than any circumscribable polygon with the same perimeter.

6. Restate with more detail the argument of sect. 7 (5). Does it prove the statement I of sect. 8? Is there any objection?

7. *Can you use the method for some other problem?* Use the method of sect. 8 to prove that the following two statements are equivalent:

"Of all boxes with a given surface area the cube has the maximum volume."

"Of all boxes with a given volume the cube has the minimum surface area."

8. *Sharper form of the Isoperimetric Theorem.* Compare the statements I, II, and III of sect. 8 with the following.

I′. *The area of a circle is larger than that of any other plane curve with the same perimeter.*

II′. *The perimeter of a circle is shorter than that of any other plane curve with the same area.*

III′. *If A is the area of a plane curve and L the length of its perimeter, then*

$$\frac{4\pi A}{L^2} \leqq 1$$

and equality is attained if, and only if, the curve is a circle.

Show that I′, II′, and III′ are equivalent to each other. Have we proved I′?

9. Given a curve C with perimeter L and area A; C is not a circle. Construct a curve C' with the same perimeter L, but with an area A' larger than A.

This problem is important (why?) but not too easy. If you cannot solve it in full generality, solve it in significant special cases; put pertinent questions that could bring you nearer to its general solution; try to restate it; try to approach it in one way or the other.

10. Given a quadrilateral C with a re-entrant angle, the perimeter L and the area A. Construct a triangle C' with the same perimeter L, but with an area A' larger than A.

11. Generalize ex. 10.

12. The information "C is not a circle" is "purely negative." Could you characterize C more "positively" in some manner that would give you a foothold for tackling ex. 9?

[Any three points on any curve are on the same circle, or on a straight line. What about four points?]

13. Given a curve C with perimeter L and area A; there are four points P, Q, R, and S on C which are not on the same circle, nor on the same straight line. Construct a curve C' with the same perimeter L, but with an area A' larger than A. [Ex. 2.]

14. Compare the following two questions.

We consider curves with a given perimeter. If C is such a curve, but not the circle, we can construct another curve C' with a greater area. (In fact, this has been done in exs. 10–13. The condition that C is not a circle is essential; our construction fails to increase the area of the circle.) Can we conclude hence that the circle has the greatest area?

We consider positive integers. If n is such an integer, but not 1, we can construct another integer n' greater than n. (In fact, set $n' = n^2$. The condition $n > 1$ is essential; our construction fails for $n = 1$ as $1^2 = 1$.) Can we conclude hence that 1 is the greatest integer?

Point out the difference if there is any.

15. Prove the statement I' of ex. 8.

Second Part

16. *The stick and the string.* Given a stick and a string, each end of the stick attached to the corresponding end of the string (which, of course, must be longer than the stick). Surround with this contraption the largest possible area.

Lay down the stick. Its endpoints A and B determine its position completely. Yet the string can take infinitely many shapes, forming an arbitrary curve with given length that begins in A and ends in B; see fig. 10.13. One of the possible shapes of the string is a circular arc which includes a segment of a circle with the stick. Complete the circle by adding another segment (shaded in fig. 10.14, I) and add the same segment to the figure included by the stick and an arbitrary position of the string (fig. 10.14, II). The circle I of fig. 10.14 has a larger area than any other curve with the same perimeter, and II of fig. 10.14 *is* such a curve. Subtracting the same (shaded) segment from I and II, we find the result: the area surrounded by the stick and the string is a *maximum when the string forms a circular arc.*

This result remains valid if we add to the variable area of fig. 10.13 any invariable area along its invariable straight boundary line. This remark is often useful.

State the conjugate result. That is, formulate the fact that has the same relation to theorem II of sect. 8 as the fact found in the foregoing has to theorem I of sect. 8.

17. Given an angle (the infinite part of a plane between two rays drawn from the same initial point) and two points, one on each side of the angle. Find the maximum area cut off from the angle by a line of given length

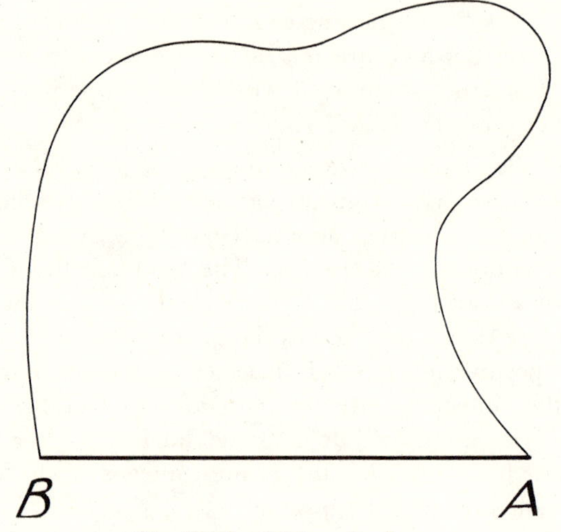

Fig. 10.13. Stick and string.

connecting the two given points. (In fig. 10.11 the points X and Y are given.)

18. Given an angle, with an opening less than 180°, and a point on one of its sides. Find the maximum area cut off from the angle by a line of given length that begins at the given point. (In fig. 10.11 the point X is given, but Y is variable.)

19. Given an angle with an opening less than 180°. Find the maximum area cut off from the angle by a line of given length. (In fig. 10.11, the points X and Y are variable. A conjecture was stated in sect. 9.)

20. Given an angle with an opening less than 180°. Find the maximum area cut off from the angle by a straight line of given length.

21. *Two sticks and two strings.* We have two sticks, AB and CD. A first string is attached to the last point B of the first stick at one end and to the first point C of the second stick at the other end. Another string connects similarly D and A. Surround with this contraption the largest possible area.

22. Generalize.

23. Specialize, and obtain by specializing an elementary theorem that played an important role in the text.

24. Given a circle in space. Find the surface with given area bounded by the given circle that includes the maximum volume with the disk rimmed by the given circle. [Do you know an analogous problem?]

25. *Dido's problem in solid geometry.* Given a trihedral angle (one of the eight infinite parts into which space is divided by three planes intersecting in one point). Find the maximum volume cut off from the trihedral angle by a surface of given area.

This problem is too difficult. You are only asked to pick out a more accessible special case.

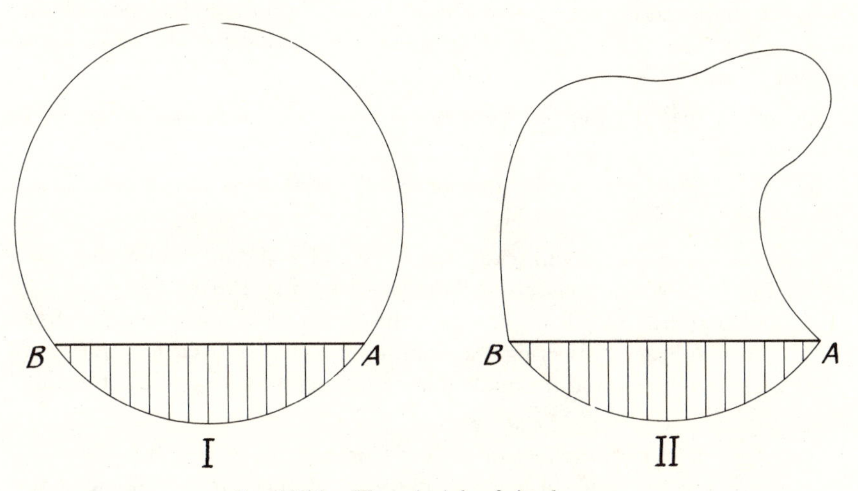

Fig. 10.14. The principle of circular arc.

26. Find a problem analogous to ex. 25 of which you can foresee the result. [Generalize, specialize, pass to the limit, ...]

27. *Bisectors of a plane region.* We consider a plane region surrounded by a curve. An arc that joins two points of the surrounding curve is called a *bisector* of the region if it divides the region into two parts of equal area.

Show that any two bisectors of the same region have at least one common point.

28. Compare two bisectors of a square. One is a straight line parallel to one of the sides that passes through the center of the square. The other is one-quarter of a circle the center of which is a vertex. Which of the two is shorter?

29. Find the shortest straight bisector of an equilateral triangle.

30. Find the shortest bisector of an equilateral triangle.

31. Show that the shortest bisectors of a circle are its diameters.

32. Find the shortest bisector of an ellipse.

33. Try to formulate general theorems covering ex. 28–32.

34. *Bisectors of a closed surface.*[9] A not self-intersecting closed curve on a closed surface is called a *bisector* of the surface if it divides the surface into two parts (open surfaces) of equal area.

Show that any two bisectors of the same surface have at least one common point.

35. A shortest bisector of the surface of a polyhedron consists either of circular arcs with the same radius r or of straight segments ($r = \infty$).

36. A shortest bisector of the surface of a regular solid is a regular polygon. Find its shape and location and the number of solutions for each of the five regular solids. (You may experiment with a model of the solid and a rubber band.)

37. Show that the shortest bisectors of a spherical surface are the great circles.

38. Try to find a generalization of ex. 37 that covers also a substantial part of ex. 36. [Ex. 9.23, 9.24.]

39. Given a sphere S with radius a. We call a *diaphragm* of S that part of a spherical surface intersecting S that is within S. Prove:

(1) All diaphragms passing through the center of S have the same area.

(2) No diaphragm bisecting the volume of S has an area less than πa^2.

The last statement, and the analogous cases discussed, suggest a conjecture. State it. [Ex. 31, 37.]

40. *A figure of many perfections.* We consider a plane region surrounded by a curve. We wish to survey some of the many theorems analogous to the isoperimetric theorem: Of all regions with a given area, the circle has the minimum perimeter.

We met already with a theorem of this kind. In sect. 4, we considered some inductive evidence for the statement: Of all membranes with a given area, the circular membrane emits the deepest principal tone.

Let us now regard the region as a homogeneous plate of uniform thickness. We consider the moment of inertia of this plate about an axis perpendicular to it that passes through its center of gravity. This moment of inertia, which we call the "polar moment of inertia," depends, other things being equal, on the size and shape of the plate. Of all plates with a given area, the circular plate has the minimum polar moment of inertia.

This plate, if it is a conductor of electricity, can also receive an electric charge, proportional to its electrostatic capacity. Also the capacity depends

[9] We consider here only closed surfaces of the "topological type" of the sphere and exclude, for instance, the (doughnut shaped) torus.

on the size and shape of the plate. Of all plates with a given area, the circular plate has the minimum capacity.

Now let the region be a cross-section of a homogeneous elastic beam. If we twist such a beam about its axis, we may observe that it resists the twisting. This resistance, or "torsional rigidity," of the beam depends, other things being equal, on the size and shape of the cross-section. Of all cross-sections with a given area, the circular cross-section has the maximum torsional rigidity.[10]

Why is the circle the solution of so many and so different problems on minima and maxima? What is the "reason"? Is the "perfect symmetry" of the circle the "true reason"? Such vague questions may be stimulating and fruitful, provided that you do not merely indulge in vague talk and speculation, but try seriously to get down to something more precise or more concrete.

41. *An analogous case.* Do you see the analogy between the isoperimetric theorem and the theorem of the means? (See sect. 8.6.)

The length of a closed curve depends in the same manner on each point, or on each element, of the curve. Also the area of the region surrounded by the curve depends in the same manner on each point, or element, of the curve. We seek the maximum of the area when the length is given. As both quantities concerned are of such a nature that no point of the curve plays a favored role in their definition, we need not be surprised that the solution is the only closed curve that contains each of its points in the same way and any two elements of which are superposable: the circle.

The sum $x_1 + x_2 + \dots + x_n$ is a symmetric function of the variables $x_1, x_2, \dots x_n$; that is, it depends in the same manner on each variable. Also the product $x_1 x_2 \dots x_n$ depends in the same manner on each variable. We seek the maximum of the product when the sum is given. As both quantities concerned are symmetric in the n variables, we need not be surprised that the solution requires $x_1 = x_2 = \dots = x_n$.

Besides the area and the length there are other quantities depending on the size and shape of a closed curve which "depend in the same manner on each element of the curve"; we listed several such quantities in ex. 40. We seek the maximum of a quantity of this kind when another quantity of the same kind is given. Is the solution, if there is one, necessarily the circle?

Let us turn to the simpler analogous case for a plausible answer. Let us consider two symmetric functions, $f(x_1, x_2, \dots x_n)$ and $g(x_1, x_2, \dots x_n)$, of n variables and let us seek the extrema of $f(x_1, x_2, \dots x_n)$ when we are given that $g(x_1, x_2, \dots x_n) = 1$. There are cases in which there is no maximum, other cases in which there is no minimum, and still other cases in which neither a maximum nor a minimum exists. The condition

[10] For proofs of the theorems indicated and for similar theorems, see G. Pólya and G. Szegö, *Isoperimetric Inequalities in Mathematical Physics*, Princeton University Press, 1951.

$x_1 = x_2 = \ldots = x_n$ plays an important rôle,[11] yet it need not be satisfied when a maximum or a minimum is attained. There is, however, a simple fact. If

$$x_1 = a_1, \quad x_2 = a_2, \quad x_3 = a_3, \ldots x_n = a_n$$

is a solution, also

$$x_1 = a_2, \quad x_2 = a_1, \quad x_3 = a_3, \ldots x_n = a_n$$

is a solution, by the symmetry of the functions f and g. Therefore, if $a_1 \neq a_2$, there are at least two different solutions. *If there is a unique solution* (that is, if the extremum is attained, and attained for just one system of values of $x_1, x_2, \ldots x_n$) *the solution requires* $x_1 = x_2 = \ldots = x_n$.

"Comparaison n'est pas raison," say the French. Of course, such comparisons as the preceding cannot yield a binding reason, only a heuristic indication. Yet we are quite pleased sometimes to receive such an indication.

Take as an illustration

$$f(x_1, x_2, \ldots x_n) = (x_1 + x_2 + \ldots + x_n)^2,$$

$$g(x_1, x_2, \ldots x_n) = (x_1^2 + x_2^2 + \ldots + x_n^2)/n$$

and find the extrema of f under the condition $g = 1$ considering (1) all real values of $x_1, x_2, \ldots x_n$ and (2) only non-negative real values of these variables.

42. *The regular solids.* Find the polyhedron with a given number n of faces and with a given surface-area that has the maximum volume.

This very difficult problem is suggested by the analogous problem of sect. 7 (1) which also suggests a conjecture: if there is a regular solid with n faces, it yields the maximum volume. However plausible this conjecture may seem, it turned out wrong in two cases out of five. In fact, the conjecture is

correct for $n = 4, 6, 12,$

incorrect for $n = 8, 20.$

What is the difference? Try to observe some simple geometrical property that distinguishes between the two kinds of regular solids.

43. *Inductive reasons.* Let V denote the volume of a solid and S the area of its surface. By analogy, sect. 8 (3) suggests to define

$$\frac{36\pi V^2}{S^3}$$

as the isoperimetric quotient in solid geometry. By analogy, we may conjecture that the sphere has the highest isoperimetric quotient. Table III supports this conjecture inductively.

[11] G. H. Hardy, J. E. Littlewood, and G. Pólya, *Inequalities.* See pp. 109–10, and the theorems there quoted.

Check some of the figures given in Table III and add new material. In particular, try to find a solid with an isoperimetric quotient higher than that of the regular icosahedron.

Table III. The Isoperimetric Quotient $36\pi V^2/S^3$

Sphere	1.0000
Icosahedron	0.8288
Best double cone	0.7698
Dodecahedron	0.7547
Best prism	0.6667
Octahedron	0.6045
Cube	0.5236
Best cone	0.5000
Tetrahedron	0.3023

For "best" double cone, prism, and cone see ex. 8.38, 8.35, and 8.52, respectively.

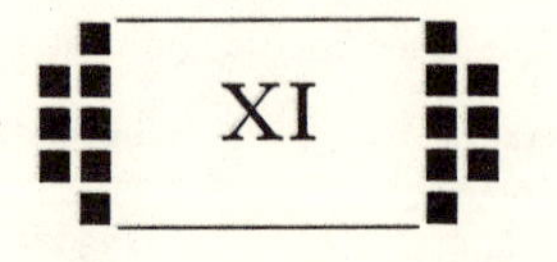

XI

FURTHER KINDS OF PLAUSIBLE REASONS

The most simple relations are the most common, and this is the foundation upon which induction rests.—LAPLACE[1]

1. Conjectures and conjectures. All our foregoing discussions dealt with the rôle of conjectures in mathematical research. Our examples gave us an opportunity to familiarize ourselves with two kinds of plausible arguments speaking for or against a proposed conjecture: we discussed inductive arguments, from the verification of consequences, and arguments from analogy. Are there other kinds of useful plausible arguments for or against a conjecture? The examples of the present chapter aim at clarifying this question.

We should also realize that there are conjectures of various kinds: great and small, original and routine conjectures. There are conjectures which played a spectacular rôle in the history of science, but also the solution of the most modest mathematical problem may need some correspondingly modest conjecture or guess. We begin with examples from the classroom and then proceed to others which are of historical importance.

2. Judging by a related case. Working at a problem, we often try to guess. Of course, we would like to guess the whole solution. If, however, we do not succeed in this, we are quite satisfied if we can guess this or that feature of the solution. At least, we should like to know whether our problem is "reasonable." We ask ourselves: Is our problem reasonable? *Is it possible to satisfy the condition? Is the condition sufficient to determine the unknown? Or is it insufficient? Or redundant? Or contradictory?*[2]

Such questions come naturally and are particularly useful at an early stage of our work when they need not a final answer but just a provisional

[1] Essai philosophique sur les probabilités; see *Oeuvres complètes de Laplace*, vol. 7, p. CXXXIX.

[2] *How to Solve It*, p. 111.

answer, a guess, and there are cases in which we can guess the answer quite reasonably and with very little trouble.

As an illustration we consider an elementary problem in solid geometry. The axis of a cylinder passes through the center of a sphere. The surface of the cylinder intersects the surface of the sphere and divides the solid sphere into two portions: the "perforated sphere" and the "plug". The first portion is outside the cylinder, the second inside. See fig. 11.1, which should be rotated about the vertical line AB. *Given r, the radius of the sphere, and h, the height of the cylindrical hole, find the volume of the perforated sphere.*

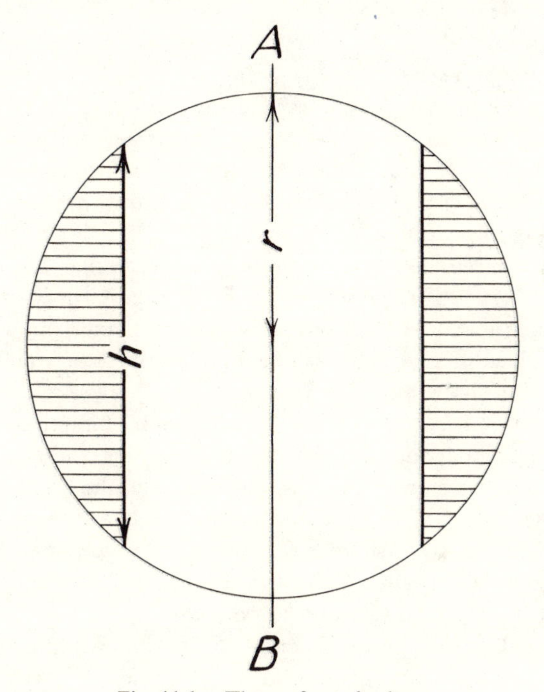

Fig. 11.1. The perforated sphere.

In familiarizing ourselves with the proposed problem, we arrive quite naturally at the usual questions: *Are the data sufficient to determine the unknown? Or are they insufficient? Or redundant?* The data r and h seem to be just enough. In fact, r determines the size of the sphere and h the size of the cylindrical hole. Knowing r and h, we can determine the perforated sphere in shape and size and we also need r and h to determine it so.

Yet, computing the required volume, we find that it is equal to $\pi h^3/6$; see ex. 5. This result looks extremely paradoxical. We have convinced ourselves that we need both r and h to determine the shape and size of the perforated sphere and now it turns out that we do not need r to determine its volume; this sounds quite incredible.

Yet, there is no contradiction. If h remains constant and r increases, the perforated sphere changes considerably in shape: it becomes wider (which tends to increase the volume) but its outer surface becomes flatter (which tends to decrease the volume). Only we did not foresee (and it appears rather unlikely *a priori*) that these two tendencies balance exactly and the volume remains unchanged.

In order to understand both the present particular case and the underlying general idea, we need a distinction. We should distinguish clearly between two related, but different, problems. Being given r and h, we may be required to determine

(a) the volume and

(b) the shape and size

of the perforated sphere. Our original problem was (a). We have seen intuitively that the data r and h are both necessary and sufficient to solve (b). It follows hence that these data are also sufficient to solve (a), but not that they are necessary to solve (a); in fact, they are not.

In answering the question "Are the data necessary?" we *judged by a related case*, we *substituted* (b) for (a), we *neglected the distinction between the original problem* (a) *and the modified problem* (b). From the heuristic viewpoint, such neglect is defensible. We needed only a provisional, but quick, answer. Moreover, such a difference is *usually* negligible: the data which are necessary to determine the shape and size are *usually* also necessary to determine the volume. We became involved into a paradox by forgetting that our conclusion was only heuristic, or by believing in some confused way that the unusual can never happen. And, in our example, the unusual did happen.

Judging a proposed problem by a modified problem is a defensible, reasonable heuristic procedure. We should not forget, however, that the conclusion at which we arrive by such a procedure is only provisional, not final; only plausible, but by no means certain to be true.

3. Judging by the general case. The following problem can be suitably discussed in a class of Algebra for beginners.

The testament of a father of three sons contains the following dispositions. "The part of my eldest son shall be the average of the parts of the two others and three thousand dollars. The part of my second son shall be exactly the average of the parts of the two others. The part of my youngest son shall be the average of the parts of the two others less three thousand dollars." *What are the three parts?*

Is the condition sufficient to determine the unknowns? There is quite a good reason to say yes. In fact, there are three unknowns, say, x, y, and z, the parts of the eldest, the second, and the youngest son, respectively. Each of the three sentences quoted from the testament can be translated into an equation. Now, *in general, a system of three equations with three unknowns determines the unknowns.* Thus we are quite reasonably led to think that the condition of the proposed problem is sufficient to determine the unknowns.

In writing down, however, the three equations, we obtain the following system:

$$x = \frac{y + z}{2} + 3000$$

$$y = \frac{x + z}{2}$$

$$z = \frac{x + y}{2} - 3000.$$

Adding these three equations we obtain

$$x + y + z = x + y + z$$

or

$$0 = 0.$$

Therefore, any equation of the system is a consequence of the other two equations. Our system contains only two *independent* equations and, therefore, in fact, it is *not* sufficient to determine the unknowns.

The problem is essentially modified if the testament contains also the following sentence: "I divide my whole fortune of 15,000 dollars among my three sons." This system adds to the above system the equation

$$x + y + z = 15{,}000.$$

We have now a more comprehensive system of four equations. Yet *in general, a system of four equations with three unknowns is contradictory.* In fact, however, the present system is not contradictory, but just sufficient to determine the unknowns and yields

$$x = 7000, \quad y = 5000, \quad z = 3000.$$

The apparent contradictions of this not too deep example are not too difficult to disentangle, but a careful explanation may be useful.

It is not true that "a system of n equations with n unknowns determines the unknowns." In fact, we have just seen a counter-example with $n = 3$. What matters here, however, is not a mathematical theorem, but a heuristic statement, in fact, the following statement: "A system of n equations with n unknowns determines, *in general*, the unknowns." The term "in general" can be interpreted in various ways. What matters here is a somewhat vague and rough "practical" interpretation: a statement holds "in general" if it holds "in the great majority of such cases as are likely to occur naturally."

Treating a geometrical or physical problem algebraically, we try to express an intuitively submitted condition by our equations. We try to express a different clause of the condition by each equation and we try to cover the whole condition. If we succeed in collecting as many equations as we have unknowns, we hope that we shall be able to determine the unknowns.

Such hope is reasonable. Our equations "occurred naturally"; we may expect that we are "in the general case." Yet the example of the present section did not occur naturally; it was fabricated to show up the lack of absolute certainty in the heuristic statement. Therefore, this example does not invalidate at all the underlying heuristic principle.

We rely on something similar in everyday life. We are, quite reasonably, not too much afraid of things which are very unusual. Letters get lost and trains crash, yet I still send a letter and board a train without hesitation. After all, lost letters and train wrecks are extremely unusual; only to a very small percentage of letters or trains happens such an accident. Why should it happen just now? Similarly, n equations with n unknowns quite naturally obtained may be insufficient to determine the unknowns. Yet, in general, this does not happen; why should it happen just now?

We cannot live and we cannot solve problems without a modicum of optimism.

4. Preferring the simpler conjecture. "Simplex sigillum veri," or "Simplicity is the seal of truth," said the scholastics. Today, as humanity is older and richer with the considerable scientific experience of the intervening centuries, we should express ourselves more cautiously; we know that the truth can be immensely complex. Perhaps the scholastics did not mean that simplicity is a necessary attribute of truth; perhaps they intended to state a heuristic principle: "What is simple has a good chance to be true." It may be even better to say still less and to confine ourselves to the plain advice: "Try the simplest thing first."

This common sense advice includes (somewhat vaguely, it is true) the heuristic moves discussed in the foregoing. That the volume changes when the shape changes is not only the usual case, but also the simplest case. That a system of n equations with n unknowns determines the unknowns is not only the general case, but also the simplest case. It is reasonable to try the simplest case first. Even if we were obliged to return eventually to a closer examination of more complex possibilities, the previous examination of the simplest case may serve as a useful preparation.

Trying the simplest thing first is part of an attitude which is advantageous in face of problems little or great. Let us attempt to imagine (with sweeping simplification and, doubtless, with some distortion) Galileo's situation as he investigated the law of falling bodies. If we wished to count the era of modern science from a definite date, the date of this investigation of Galileo's could be considered as the most appropriate.

We should realize Galileo's position. He had a few forerunners, a few friends sharing his views, but was strenuously opposed by the dominating philosophical school, the Aristotelians. These Aristotelians asked, "Why do the bodies fall?" and were satisfied with some shallow, almost purely verbal explanation. Galileo asked, "How do the bodies fall?" and tried to find an answer from experiment, and a precise answer, expressible in numbers and mathematical concepts. This substitution of "How" for "Why," the search for an answer by experiment, and the search for a mathematical law condensing experimental facts are commonplace in modern science, but they were revolutionary innovations in Galileo's time.

A stone falling from a higher place hits the ground harder. A hammer falling from a higher point drives the stake deeper into the ground. The further the falling body gets from its starting point, the faster it moves—so much is clear from unsophisticated observation. *What is the simplest thing?* It seems simple enough to assume that the velocity of a falling body starting from rest is *proportional to the distance traveled.* "This principle appears very natural," says Galileo, "and corresponds to our experience with machines which operate by percussion." Still, Galileo rejected eventually the proportionality of the velocity to the distance as "not merely false, but impossible."[3]

Galileo's objections against the assumption that appeared so natural to him at first can be more clearly and strikingly formulated in the notation of the calculus. This is, of course, an anachronism; the calculus was invented after Galileo's time and, in part at least, under the impact of Galileo's discoveries. Still, let us use calculus. Let t denote the time elapsed since the beginning of the fall and x the distance traveled. Then the velocity is dx/dt (one of Galileo's achievements was to formulate a clear concept of velocity). Let g be an appropriate positive constant. Then that "simplest assumption," the proportionality of speed to the distance traveled, is expressed by the differential equation

$$\frac{dx}{dt} = gx. \tag{1}$$

We have to add the initial condition

$$x = 0 \quad \text{as} \quad t = 0. \tag{2}$$

From equations (1) and (2) it follows that

$$\frac{dx}{dt} = 0 \quad \text{as} \quad t = 0; \tag{3}$$

this expresses that the falling body starts from rest.

We obtain, however, by integrating the differential equation (1) that

$$\int \frac{dx}{x} = \int g dt$$

$$\log x = gt + \log c$$

[3] See *Le Opere di Galileo Galilei*, edizione nazionale, vol. 8, p. 203, 373, 383.

where c is some positive constant. This yields

$$x = ce^{gt}, \qquad \frac{dx}{dt} = gce^{gt}.$$

We obtain hence, however, that

$$x = c > 0, \qquad \frac{dx}{dt} = gc > 0 \qquad \text{as } t = 0$$

in contradiction to (2) and (3): a motion satisfying the differential equation (1) *cannot start from rest.* And so the assumption that appeared so "natural," just "the simplest thing," is, in fact, self-contradictory: "not merely false, but impossible" as Galileo expressed himself.

Yet what is the "next simplest thing"? It may be to assume that the velocity of a falling body starting from rest is proportional to the time elapsed. This is the well known law at which Galileo eventually arrived. It is expressed in modern notation by the equation

$$\frac{dx}{dt} = gt$$

and a motion satisfying this equation can certainly start from rest.

5. Background. We cannot but admire Galileo's intellectual courage, his freedom from philosophical prejudice and mysticism. Yet we must also admire Kepler's achievements; and Kepler, a contemporary of Galileo, was deeply involved in mysticism and the prejudices of his time.

It is difficult for us to realize Kepler's attitude. The modern reader is amazed by a title as "A prodrome to cosmographic dissertations, containing the COSMIC MYSTERY, on the admirable proportion of celestial orbits and the genuine and proper causes of the number, magnitude, and periodic motions of the heavens, demonstrated by the five regular geometric solids." The contents are still more amazing: astronomy mixed with theology, geometry scrambled with astrology. Yet however extravagant some of the contents may appear, this first work of Kepler marks the beginning of his great astronomical discoveries and gives besides a lively and attractive picture of his personality. His thirst for knowledge is admirable, although it is almost equalled by his hunger for mystery.

As the title of the work quite correctly says, Kepler set out to discover a cause or a reason for the number of the planets, for their distances from the sun, for the period of their revolutions. He asks, in fact: Why are there just six planets? Why are their orbits just so disposed? These questions sound strange to us, but did not sound so to some of his contemporaries.[4]

---

[4] Kepler rejects Rhaeticus' explanation that there are six planets, since 6 is the first "perfect number."

One day he thought that he found the secret and he jotted down in his notebook: "The earth's orbit or sphere is the measure of all. Circumscribe about it a dodecahedron: the sphere surrounding it is Mars. Circumscribe about Mars a tetrahedron: the sphere surrounding it is Jupiter. Circumscribe about Jupiter a cube: the sphere surrounding it is Saturn. Now, inscribe in the Earth an icosahedron: the sphere contained in it is Venus. Inscribe in Venus an octahedron: the sphere contained in it is Mercury. Here you have the reason for the number of the planets."

That is, Kepler imagines 11 concentric surfaces, 6 spheres alternating with the 5 regular solids. The first and outermost surface is a sphere and each surface is surrounded by the preceding. Each sphere is associated with a planet: the radius of the sphere is the distance (mean distance) of the planet from the sun. Each regular solid is inscribed in the preceding, surrounding sphere and circumscribed about the following, surrounded sphere.

And Kepler adds: "I shall never succeed in finding words to express the delight of this discovery."

Kepler (in this respect a modern scientist) carefully compared his conjecture with the facts. He computed a table which is presented here in a slightly modernized form as Table I.

Table I. Kepler's theory compared with the observations

(1) *Planets*	(2) *Copernicus' observation*	(3) *Kepler's theory*	(4) *Regular solids*
Saturn			
	.635	.577	Cube
Jupiter			
	.333	.333	Tetrahedron
Mars			
	.757	.795	Dodecahedron
Earth			
	.794	.795	Icosahedron
Venus			
	.723	.577	Octahedron
Mercury			

Column (1) lists the planets in order of decreasing distances from the sun; it contains six entries, one more than the following columns. Column (2) contains the ratio of the distances of two consecutive planets from the sun, according to Copernicus; each ratio is inserted between the lines in which the names of the respective planets are marked; the distance of the outer planet is the denominator. Column (4) lists the five regular solids in the order chosen by Kepler. Column (3) lists the ratio of the radii of the inscribed and circumscribed spheres for the corresponding regular solid.

The numbers on the same line should agree. In fact, the agreement is good in two cases, and very bad in the remaining three.

Now Kepler (which reminds us in less glorious way of modern scientists) starts shifting his standpoint and modifying his original conjecture. (The main modification is that he compares the distance of Mercury from the sun not to the radius of the sphere inscribed in the octahedron, but to the radius of the circle inscribed in the square in which a certain plane of symmetry intersects the octahedron.) Yet he does not arrive at any startling agreement between conjecture and observation. Still, he sticks to his idea. The sphere is "the most perfect figure," and next to it the five regular solids, known to Plato, are the "noblest figures." Kepler thinks for a moment that the countless crowd of fixed stars may have something to do with the undistinguished multitude of irregular solids. And it seems "natural" to him that the sun and the planets, the most excellent things created, should be somehow related to Euclid's most excellent figures. This could be the secret of the creation, the "Cosmic Mystery."

To modern eyes Kepler's conjecture may look preposterous. We know many relations between observable facts and mathematical concepts, but these relations are of a quite different character. No useful relation is known to us which would have any appreciable analogy to Kepler's conjecture. We find it most strange that Kepler could believe that there is anything deep hidden behind the number of the planets and could ask such a question: Why are there just six planets?

We may be tempted to regard Kepler's conjecture as a queer aberration. Yet we should consider the possibility that some theories which we are respectfully debating today may be considered as queer aberrations in a not far away future, if they are not completely forgotten. I think that Kepler's conjecture is highly instructive. It shows with particular clarity a point that deserves to be borne in mind: the credence that we place in a conjecture is bound to depend on our whole *background*, on the whole *scientific atmosphere* of our time.

6. Inexhaustible. The foregoing example brings into the foreground an important feature of plausible reasoning. Let us try to describe it with some degree of generality.

We have a certain conjecture, say A. That is, A is a clearly formulated, but not proved, proposition. We suspect that A is true, but we do not actually know whether A is true or not. Still we have some confidence in our conjecture A. Such confidence may, but need not, have an articulate basis. After prolonged and apparently unsuccessful work at some problem, there emerges quite suddenly a conjecture A. This conjecture A may appear as the only possible escape from an entangled situation; it may appear as almost certain, although we could not tell why.

After a while, however, some more articulate reasons may occur to us that speak distinctly in favor of A, although they do not prove A: reasons from analogy, from induction, from related cases, from general experience, or from the inherent simplicity of A itself. Such reasons, without providing a strict demonstration, can make A very plausible.

Yet it should be a warning to us that we trusted that conjecture without any of those more distinctly formulated arguments.

And we perceived those arguments successively. There was a first clear point that we succeeded in detaching from an obscure background. Yet there was something more behind this point in the background since afterwards we succeeded in extracting another clear argument. And so there may be something more behind each clarified point. Perhaps that background is inexhaustible. *Perhaps our confidence in a conjecture is never based on clarified grounds alone*; such confidence may need somehow our *whole background* as a basis.

Still, plausible grounds are important, and clarified plausible grounds are particularly important. In dealing with the observable reality, we can never arrive at any demonstrative truth, we have always to rely on some plausible ground. In dealing with purely mathematical questions, we may arrive at a strict demonstration. Yet it may be very difficult to arrive at it, and the consideration of provisional, plausible grounds may give us temporary support and may lead us eventually to the discovery of the definitive demonstrative argument.

Heuristic reasons are important although they prove nothing. To clarify our heuristic reasons is also important although behind each reason so clarified there may be something more—some still obscure and still more important ground, perhaps.[5]

This suggests another remark: If in each concrete case we can clarify only a few of our plausible grounds, and in no concrete case exhaust them, how could we hope to describe exhaustively the kinds of plausible grounds in the abstract?

7. Usual heuristic assumptions. Two of our examples (sect. 2 and 3) bring up another point. Let us recall briefly one of the situations and touch upon a similar situation.

In working at some problem, you obtain from apparently different sources as many equations as you have unknowns. You ought to know that n equations are not always sufficient to determine n unknowns: the equations could be mutually dependent, or contradictory. Still, such a case is exceptional, and so it may be reasonable to hope that your equations will determine your unknowns. Therefore you go ahead, manipulate your equations, and see what follows from them. If there is contradiction or indetermination, it will show itself somehow. On the other hand, if you arrive at a neat result, you may feel more inclined to spend time and effort on a strict demonstration.

In solving another problem you are led to integrate an infinite series

[5] *How to Solve It*, p. 224.

term by term. You ought to know that such an operation is not always permissible, and could yield an incorrect result. Still, such a case is exceptional, and so it may be reasonable to hope that your series will behave. Therefore, it may be expedient to go ahead, see what follows from your formula not completely proved, and postpone worries about a complete proof.

We touched here upon two *usual heuristic assumptions*, one about systems of equations, the other about infinite series. In each branch of mathematics there are such assumptions, and one of the principal assets of the expert in that branch is to know the current assumptions and to know also how he can use them and how far he can trust them.

Of course, you should not trust any guess too far, neither usual heuristic assumptions nor your own conjectures. To believe without proof that your guess is true would be foolish. Yet to undertake some work in the hope that your guess *might* be true, may be reasonable. Guarded optimism is the reasonable attitude.

EXAMPLES AND COMMENTS ON CHAPTER XI

1. Of a triangle, we are given the base a, the altitude h perpendicular to a and the angle α opposite to a. We should (a) construct the triangle, (b) compute its area. Are all the data necessary?

2. Of a trapezoid, we are given the altitude h perpendicular to the two parallel sides, the middle line m which is parallel to the two parallel sides and at the same distance from both, and the angles α and β between one of the two parallel sides and the two remaining (oblique) sides. We should (a) construct the trapezoid, (b) compute its area. Are all the data necessary?

3. A zone is a portion of the surface of the sphere contained between two parallel planes. The altitude of the zone is the distance of the two planes. Given r the radius of the sphere, h the altitude of the zone, and d the distance of that bounding plane from the center of the sphere which is nearer to the center, find the surface of the zone. Any remarks?

4. A first sphere has the radius a. A second sphere, with radius b, intersects the first sphere and passes through its center. Compute the area of that portion of the surface of the second sphere which lies inside the first sphere. Any remarks? Check the extreme cases.

5. Reconsider the example of sect. 2 and prove the solution.

6. A spherical segment is a portion of the sphere contained between two parallel planes. Its surface consists of three parts: of a zone of the sphere and of two circles, called the base and the top of the segment. We use the following notation:

a is the radius of the base,

b the radius of the top,

h the altitude (the distance between the base and the top),

M the area of the middle cross-section (parallel to, and at the same distance from, the base and the top)

V the volume of the segment.

Being given a, b, and h, find $Mh - V$.

Any remarks? Check some extreme cases.

7. The axis of a cone passes through the center of a sphere. The surface of the cone intersects the surface of the sphere in two circles and divides the solid sphere into two portions: the "conically perforated sphere" and the "plug" (see fig. 11.2 which should be rotated about the line AB); the plug is inside the cone. Let r denote the radius of the sphere, c the length

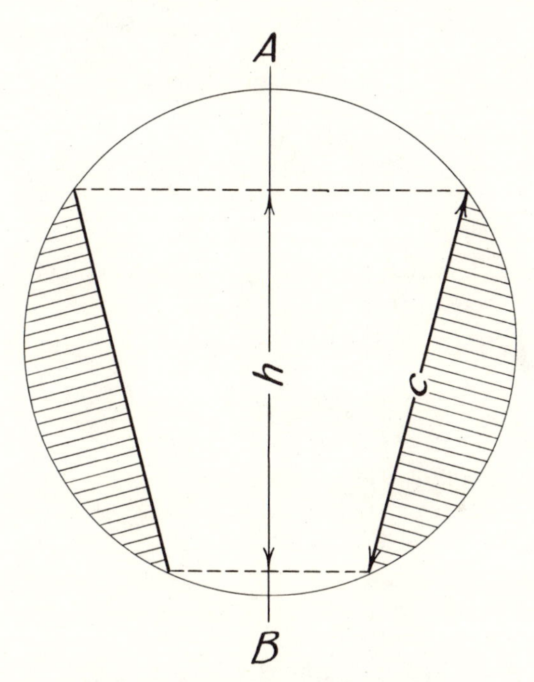

Fig. 11.2. The conically perforated sphere.

of the chord that in rotating generates the conical hole and h (the height of the perforated sphere) the projection of c onto the axis of the cone. Given r, c, and h, find the volume of the conically perforated sphere. Any remarks?

8. The axis of a paraboloid of revolution passes through the center of a sphere and the two surfaces intersect in two circles. Compute the ring-shaped solid between the two surfaces (inside the sphere and outside the paraboloid) being given r the radius of the sphere, h the projection of the ring-shaped solid on the axis of the paraboloid, and d the distance of the center of the sphere from the vertex of the paraboloid. (Rotate fig. 11.3 about OX.) Any remarks?

9. Of a trapezoid, given the lower base a, the upper base b, and the height h, perpendicular to both bases; $a > b$. The trapezoid, revolving about its lower base, describes a solid of revolution (a cylinder topped by two cones) of which find (a) the volume and (b) the surface area. Are the data sufficient to determine the unknown?

10. Ten numbers taken in a definite order, $u_1, u_2, u_3, \ldots u_{10}$, are so connected that, from the third onward, each of them is the sum of the two foregoing numbers:

$$u_n = u_{n-1} + u_{n-2}, \text{ for } n = 3, 4, \ldots 10.$$

Being given u_7, find the sum of all ten numbers $u_1 + u_2 + \ldots + u_{10}$.

Are the data sufficient to determine the unknown?

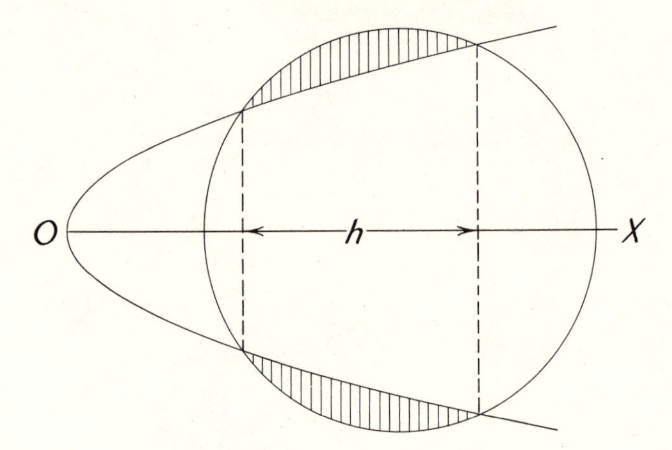

Fig. 11.3. The parabolically perforated sphere.

11. Compute
$$\int_0^\infty \frac{dx}{(1 + x^2)(1 + x^\alpha)}.$$

Any remarks? Check the cases $\alpha = 0$, $\alpha \to \infty$, $\alpha \to -\infty$.

12. Generalize ex. 11. [Try the simplest thing first.]

13. Write one equation with one unknown that does not determine the unknown.

14. One equation may determine several unknowns if the nature of the unknowns is restricted by a suitable additional condition. For example, if x, y, and z are real numbers they are completely determined by the equation

$$x^2 + y^2 + z^2 = 0.$$

Find all systems of positive integers x, y satisfying the equation $x^2 + y^2 = 128$.

15. Find all systems of positive integers x, y, z, w satisfying the equation $x^2 + y^2 + z^2 + w^2 = 64$.

16. *The general case.* Consider the system of three linear equations with three unknowns

$$a_1x + b_1y + c_1z = d_1,$$
$$a_2x + b_2y + c_2z = d_2,$$
$$a_3x + b_3y + c_3z = d_3.$$

We assume that the 12 given numbers $a_1, b_1, c_1, d_1, a_2, \ldots d_3$ are real. The system is called *determinate* if there is just one solution (just one set x, y, z of three numbers satisfying it), *indeterminate* if there are an infinity of solutions, and *inconsistent* if there is no solution. Seen from various standpoints, the case in which the system is determinate appears as the general, usual, normal, regular case and the other cases appear as exceptional, unusual, abnormal, irregular.

(a) Geometrically, we can interpret the set of three numbers x, y, z as a point in a rectangular coordinate system and each equation as the set of points satisfying it, as a plane. (For this interpretation we have to assume, in fact, that on the left-hand side of each equation there is at least one non-vanishing coefficient, but let us assume this.) The system of three equations is determinate if the three planes have just one common point. When they have two common points, they have a straight line in common and so the system is indeterminate. When the three planes are parallel to the same straight line, but have no point common to all three, the system is inconsistent. If the three planes are in a "general position," if they are "chosen at random," they have just one point in common and the system is determinate.

(b) Algebraically, the system of three equations is determinate if, and only if, the determinant of the 9 coefficients on the left-hand sides does not vanish. Therefore, the system is determinate, unless a particular condition or *restriction* is imposed upon the coefficients, in form of an equation.

(c) We may interpret the set of nine (real) coefficients $(a_1, a_2, a_3, b_1, \ldots c_3)$ as a point in nine-dimensional space. The points corresponding to systems that are not determinate (indeterminate or inconsistent) satisfy an equation (the determinant $= 0$) and so they form a manifold of *lower dimension* (an eight-dimensional "hypersurface").

(d) It is *infinitely improbable* that a system of three linear equations with three unknowns given at random is not determinate. Cf. ex. 14.23.

17. For each of the five regular solids, consider the inscribed sphere and the circumscribed sphere and compute the ratio of the radii of these two spheres.

18. Column (3) of Table I would remain unchanged if we interchanged the cube and the octahedron or the dodecahedron and the icosahedron. This would leave Kepler's theory embarrassingly indeterminate. Yet

Kepler displays a singular ingenuity in detecting reasons why one of these five noble solids should be of higher nobility than, and take precedence over, another, as a baron takes precedence over a baronet.

Find some simple geometrical property that distinguishes the three solids that Kepler placed around the Earth's orbit from the two that he placed in this orbit.

19. *No idea is really bad.* "Many a guess has turned out to be wrong but nevertheless useful in leading to a better one." "No idea is really bad, unless we are uncritical. What is really bad is to have no idea at all."[6] I use such sentences almost daily to comfort one or the other student who comes forward with some honest but naïve idea. These sentences apply both to trivial everyday situations and to scientific research. They apply most spectacularly to Kepler's case.

To Kepler himself, with his mind in that singular transition from the medieval to the modern standpoint, his idea of combining the six planets with the five regular solids appeared as brilliant. Yet I cannot imagine that Galileo, Kepler's contemporary, could have conceived such an idea. To a modern mind this idea must appear as pretty bad from the start, because it has so little relation to the rest of our knowledge about nature. Even if it had been in better agreement with the observations, Kepler's conjecture would be weakly supported, because it lacks the support of analogy with what is known otherwise.

Yet Kepler's guess which turned out to be wrong was most certainly useful in leading to a better one. It led Kepler to examine more closely the mean distances of the planets, their orbits, their times of revolution for which he hoped to find some similar "explanation," and so it led finally to Kepler's celebrated laws of planetary motion, to Newton, and to our whole modern scientific outlook.

20. *Some usual heuristic assumptions.* This subject would deserve a fuller treatment, yet we have to restrict ourselves to a very short list and sketchy comments. We must be careful to interpret the words "in general" in a "practical," necessarily somewhat vague, sense.

"If in a system of equations there are as many equations as unknowns, the unknowns are determined, *in general.*"

If, in a problem, there are as many "conditions" as available parameters, it is reasonable to start out with the tentative assumption that the problem has a solution. For instance, a quadratic form of n variables has $n(n+1)/2$ coefficients, and an orthogonal substitution in n variables depends on $n(n-1)/2$ parameters. Therefore, it is pretty plausible from the outset that, by a suitable orthogonal substitution, any quadratic form of n variables can be reduced to the expression

$$\lambda_1 y_1^2 + \lambda_2 y_2^2 + \cdots + \lambda_n y_n^2;$$

[6] *How to Solve It*, pp. 207–8.

$y_1, y_2, \ldots y_n$ are the new variables introduced by the substitution, and $\lambda_1, \lambda_2, \ldots \lambda_n$ suitable parameters. In fact, this expression depends on n parameters and

$$n(n+1)/2 = n(n-1)/2 + n.$$

This remark, coming after a proof of the proposition in the particular cases $n = 2$ and $n = 3$, and an explanation of the geometric meaning of these cases, may create a pretty strong presumption in favor of the general case.

"Two limit operations are, *in general*, commutative."

If one of the limit operations is the summation of an infinite series and the other is integration, we have the case mentioned in sect. 7.[7]

"What is true up to the limit, is true at the limit, *in general*."[8]

Being given that $a_n > 0$ and $\lim_{n\to\infty} a_n = a$, we *cannot* conclude that $a > 0$; merely $a \geqq 0$ is true. We consider a curve as the limit of an inscribed polygon and a surface as the limit of an inscribed polyhedron. Computing the length of the curve as the limit of the length of an inscribed polygon yields the correct result, yet computing the area of the surface as the limit of the area of an inscribed polyhedron may yield an incorrect result.[9] Although it can easily mislead us, the heuristic principle stated is most fertile in inspiring suggestions. See, for instance, ex. 9.24.

"Regard an unknown function, *at first*, as monotonic."

We followed something similar to this advice in sect. 2 as we assumed that with the change of the shape of a body its volume will change, too, and we were misled. Nevertheless, the principle stated is often useful. We may have to prove an inequality of the form

$$\int_a^b f(x)\,dx < \int_a^b g(x)\,dx$$

where $a < b$. We may begin by trying to prove more, namely that

$$f(x) < g(x).$$

This boils down to the initial assumption that the function with derivative $g(x) - f(x)$ is monotonic. (The problem is to compare the values of this function for $x = a$ and $x = b$.) The principle stated is contained in the more general heuristic principle "try the simplest thing first."

"*In general*, a function can be expanded in a power series, the very first term of which yields an acceptable approximation and the more terms we take, the better the approximation becomes."

Without the well-understood restriction "in general" this statement would be monumentally false. Nevertheless, physicists, engineers, and other

[7] See G. H. Hardy, *A Course of Pure Mathematics*, 7th ed., p. 493–496.

[8] Cf. William Whewell, *The Philosophy of the Inductive Sciences*, new ed., vol. I, p. 146.

[9] See H. A. Schwarz, *Gesammelte Mathematische Abhandlungen*, vol. 2, pp. 309–311.

scientists who apply the calculus to their science seem to be particularly fond of it. It includes another principle even more sweeping than the one that we have stated previously: "Regard an unknown function, *at first*, as linear." In fact, if we have the expansion

$$f(x) = a_0 + a_1x + a_2x^2 + a_nx^3 + \ldots,$$

we may take approximately

$$f(x) \sim a_0 + a_1x.$$

(Observe that Galileo, who did not know calculus, had already a strong preference for the linear function; see sect. 4.) The present principle underlies the importance often attributed to the initial term of the relative error; see sect. 5.2. The principle was often useful in suggesting some idea close to the truth, yet it may easily suggest something very far from the truth.

In fact, a physicist (or an engineer, or a biologist) may be led to believe that a physical quantity y depends so on another physical quantity x that there is a differential equation

$$\frac{dy}{dx} = f(y).$$

Now, the integration involved by this equation may be too difficult, or the form of the function $f(y)$ may be unknown. In both cases the physicist expands the function $f(y)$ in powers of y and he may regard the differential equations hence following as successive approximations:

$$\frac{dy}{dx} = a_0,$$

$$\frac{dy}{dx} = a_0 + a_1y,$$

$$\frac{dy}{dx} = a_0 + a_1y + a_2y^2.$$

Yet the curves satisfying these three equations are of very different nature and the approximation may turn out totally misleading. Fortunately, the physicists rely more on careful judgment than on careful mathematics and so they obtained good results by similar procedures even in cases in which the mathematical fallacy was less obvious and, therefore, more dangerous than in our example.

21. *Optimism rewarded.* The quantities a, b, c, d, e, f, g, and h are given. We investigate whether the system of four equations for the four unknowns x, y, u, and v

$$(S) \qquad \begin{aligned} ax + by + cv + du &= 0, \\ ex + fy + gv + hu &= 0, \\ hx + gy + fv + eu &= 0, \\ dx + cy + bv + au &= 0 \end{aligned}$$

admits any solution different from the trivial solution $x = y = u = v = 0$. This system (S) has, as we know, a non-trivial solution if, and only if, its determinant vanishes, but we wish to avoid the direct computation of this determinant with four rows. The peculiar symmetry of the system (S) may suggest to set

$$u = x, \qquad v = y.$$

Then the first equation of the system (S) coincides with the fourth, and the second with the third, so that the system of four equations reduces to a system of only two distinct equations

$$(a + d)x + (b + c)y = 0,$$
$$(e + h)x + (f + g)y = 0.$$

This system admits a non-trivial solution if, and only if, its determinant vanishes.

Yet we can also reduce the system (S) by setting

$$u = -x, \qquad v = -y.$$

Again, we obtain only two distinct equations

$$(a - d)x + (b - c)y = 0,$$
$$(e - h)x + (f - g)y = 0.$$

The vanishing of the determinant of either system of two equations involves the vanishing of the determinant of the system (S). Hence we may suspect (if we are optimistic enough) that this latter determinant with four rows is the *product* of the two other determinants, each with two rows.

(a) Prove this, and generalize the result to determinants with n rows.

(b) In which respect have we been optimistic?

22. Take the coordinate system as in sect. 9.4. The x-axis is horizontal and the y-axis points downward. Join the origin to the point (a,b)

(1) by a straight line

(2) by a circular arc with center on the x-axis.

A material point starting from rest at the origin attains the point (a,b) in time T_1 or T_2 according as it slides down (without friction) following the path (1) or the path (2). Galileo suggested (as reported in sect. 9.4) that $T_1 > T_2$. After some work this inequality turns out to be equivalent to the following:

$$\int_0^h [x(1 - x)]^{-3/4}\, dx < 4h^{1/4}(1 - h)^{-1/4}$$

if we set

$$a^2(a^2 + b^2)^{-1} = h.$$

We could try to prove the inequality by expanding both sides in powers of h. What would be the simplest (or "most optimistic") possibility?

23. *Numerical computation and the engineer.* The layman is inclined to think that the scientist's numerical computations are infallible, but dull. Actually, the scientist's numerical computation may be exciting adventure, but unreliable. Ancient astronomers tried, and modern engineers try, to obtain numerical results about imperfectly known phenomena with imperfectly known mathematical tools. It is hardly surprising that such attempts may fail; it is more surprising that they often succeed. Here is a typical example. (The technical details, which are suppressed here, will be published elsewhere.)

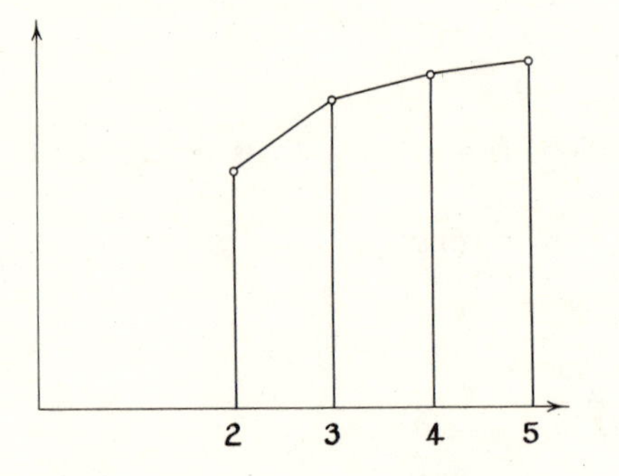

Fig. 11.4. A trial: the abscissa is n.

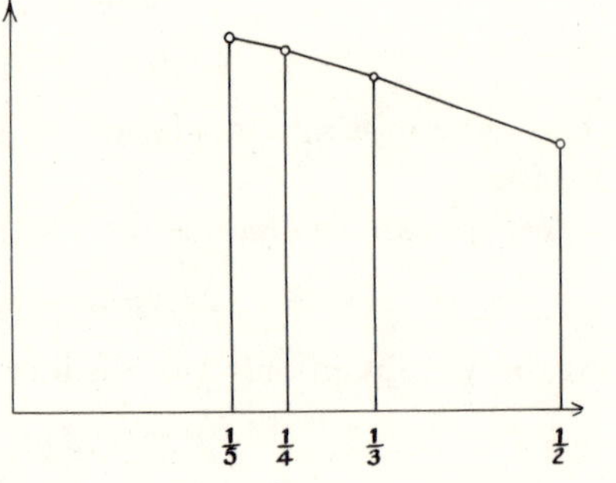

Fig. 11.5. Another trial: the abscissa is $1/n$.

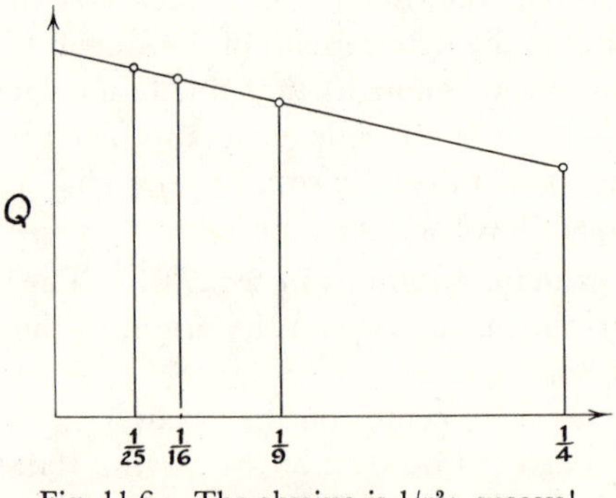

Fig. 11.6. The abscissa is $1/n^2$: success!

An engineer wishes to compute a certain physical quantity Q connected with a square of side 1. (In fact, Q is the torsional rigidity of a beam with square cross-section, but the reader need not know this—in fact, he need not even know what torsional rigidity is.) An exact solution runs into mathematical difficulties, and so the engineer, as engineers often do, resorts to approximations. Following a certain method of approximation, he divides the given square into equal "elements," that is, n^2 smaller squares each of the area $1/n^2$. (In approximating a double integral, we also divide the given area into elements in this way.) It can be reasonably expected that the approximate value tends to the true value as n tends to infinity.

In fact, however, as n increases, the difficulty of the computation also increases, and so rapidly that it soon becomes unmanageable. The engineer considers only the cases $n = 2, 3, 4, 5$ and obtains the corresponding approximate values for Q:

$$0.0937 \qquad 0.1185 \qquad 0.1279 \qquad 0.1324$$

Let us not forget that these numbers correspond to the values

$$1/4 \qquad 1/9 \qquad 1/16 \qquad 1/25$$

of a small square area used in computing, respectively.

The engineer graphs these results. He decides to plot the approximate values obtained for Q as ordinates, but he is hesitant about the choice of the abscissa. He tries first n as abscissa, then $1/n$, and finally $1/n^2$ (which is the numerical value of the area of the small square used in the approximation): see figs. 11.4, 11.5, and 11.6, respectively. The last choice is the best; the four points in fig. 11.6 are *nearly on the same straight line*. Observing this, the engineer produces the line till it intersects the vertical axis and regards the ordinate of the intersection as a "good" approximation to Q.

(a) Why? What is the underlying idea?

(b) Check fig. 11.6 numerically: join each point to the next by a straight line and compute the three slopes.

(c) Choose the two most reliable points in fig. 11.6, use the straight line passing through them in the engineer's construction, compute the resulting approximation to Q, and compare it with 0.1406, the true value of Q.

FINAL REMARK

The reader who went through the foregoing chapters and did some of the foregoing problems had a good opportunity to acquaint himself with some aspects of plausible reasoning. To form a general idea of the nature of plausible reasoning is the aim of the remaining five chapters of this work, collected in Vol. II. This aim deserves, I believe, considerable theoretical interest, but it may have also some practical value: we may perform a concrete task better if we understand more of the underlying abstract idea.

The formulation of certain patterns of plausible reasoning is the principal object of Vol. II. Yet these patterns will be extracted from, and discussed in close contact with, concrete examples. Therefore Vol. II will add several mathematical examples to those treated in the present Vol. I and will treat them in the same manner.

SOLUTIONS

SOLUTIONS, CHAPTER I

1. The primes ending in 1.

2. [Stanford 1948]

$$(n^2 + 1) + (n^2 + 2) + \ldots + (n + 1)^2 = n^3 + (n + 1)^3$$

The terms on the left-hand side are in arithmetic progression.

3. $1 + 3 + \ldots + (2n - 1) = n^2$.

4. 1, 9, 36, 100, . . . are squares. See *How to Solve It*, p. 104.

5. [Stanford 1949] $\left(\frac{n+1}{2}\right)^2$ or $\left(\frac{n+1}{2}\right)^2 - \frac{1}{4}$, according as n is odd or even. A uniform law for both cases: the integer nearest to $(n + 1)^2/4$.

6. First question: Yes. Second question: No; 33 is not a prime.

7. Not for you, if you have some experience with primes [ex. 1, 6, 9]. In fact, (1) can be proved (as particular case of a theorem of Kaluza, *Mathematische Zeitschrift*, vol. 28 (1928) p. 160–170) and (2) disproved: the next coefficient (of x^7) is $-3447 = -3 \cdot 3 \cdot 383$. The "formal computation" has a clear meaning.
Setting

$$\left(\sum_0^\infty n!x^n\right)^{-1} = \sum_0^\infty u_n x^n,$$

we define $u_0 = 1$ and $u_1, u_2, u_3, \ldots$ recurrently by the equations

$$0!u_n + 1!u_{n-1} + 2!u_{n-2} + \ldots + (n-1)!u_1 + n!u_0 = 0$$

for $n = 1, 2, 3, \ldots$.

8. On the basis of the observed data it is quite reasonable to suspect that A_n is positive and increases with n. Yet this conjecture is totally mistaken. By more advanced tools (integral calculus, or theory of analytic functions of a complex variable) we can prove that, for large n, the value of A_n is approximately $(-1)^{n-1}(n-1)!(\log n)^{-2}$.

10. In the case $2n = 60$ we have to make 9 or 7 trials ($p = 3, 5, 7, 11, 13, 17, 19, 23, 29$ or $p' = 31, 37, 41, 43, 47, 53, 59$) according as we follow the first or the second procedure. It is likely that for higher values of n there will be still a greater difference between the number of trials, in favor of the second procedure.

No solution: **9, 11, 12, 13, 14.**

SOLUTIONS, CHAPTER II

1. I think that C or D is the "right generalization" and B "overshoots the mark." B is too general to give any specific suggestion. You may prefer C or D; the choice depends on your background. Yet both C and D suggest to begin with the linear equations and lead eventually to the following plan: express two unknowns in terms of the third unknown from the first two (linear!) equations and, by substituting these expressions in the last equation, obtain a quadratic equation for the third unknown. (In the present case A, can you express *any* two unknowns from the first two equations?) There are two solutions:

$$(x,y,z) = (1, -2, 2), \qquad (29/13, -2/13, 2).$$

2. Rotated 180° about its axis, the pyramid coincides with itself. The right generalization of this pyramid is a solid having an *axis of symmetry* of this kind and the *simplest* solution is a plane passing through the axis and the given point. (There are an infinity of other solutions; by continuity, we can prescribe a straight line through which the bisecting plane should pass.) Note that a regular pyramid with pentagonal base does *not* admit a comparably simple solution. Compare *How to Solve It*, pp. 98–99.

3. A is a special case of B if we allow in B that P may coincide with O, yet the two problems are equivalent: the planes required in A and B are parallel to one another, and so the solution of each problem involves that of the other.

The more general problem B is more accessible, provided that $P \neq O$: choose Q and R, on the two other lines, so that $OP = OQ = OR$. The plane passing through P, Q, and R satisfies the condition of the problem. Therefore, if A is proposed, there is advantage in passing to the more general B.

4. A is a special case of B (for $p = 1$), yet the two problems are equivalent: the substitution $x = yp^{1/2}$ reduces B to A.

The more general problem B is more accessible: differentiate the easy integral

$$\int_{-\infty}^{\infty} (p + x^2)^{-1} dx$$

twice with respect to the parameter p. Therefore, if A is proposed, there is advantage in passing to the more general B.

Observe the parallel situation in ex. 3.

6. The extreme special case in which one of the circles degenerates into a point is more accessilbe and we can reduce to it the general case. In fact, a common outer tangent of two circles remains parallel to itself when both radii decrease by the same amount, and a common inner tangent remains parallel to itself when one of the radii increases, and the other decreases by the same amount. In both cases, we can reduce one of the circles to a point, without changing the direction of the common tangent.

8. The special case in which one of the sides of the angle at the circumference passes through the center of the circle is "leading". From two such special angles, we can combine the general angle at the circumference by addition or subtraction. (This is the gist of the classical proof; Euclid III 20.) For a striking example of a "leading" special case see *How to Solve It*, pp. 166–170.

12. If two straight lines in a plane are cut by three parallel lines, the corresponding segments are proportional. This helps to prove the more difficult analogous theorem in solid geometry; see Euclid XI 17.

13. The diagonals of a parallelogram intersect in their common midpoint.

14. The sum of any two sides of a triangle is greater than the third side. The simpler of the two analogous theorems (Euclid I 20) is used in the proof of the more difficult (Euclid XI 20).

15. Parallelepiped, rectangular parallelepiped (box), cube, bisecting plane of a dihedral angle. *The bisecting planes of the six dihedral angles of a tetrahedron meet in one point which is the center of the sphere inscribed in the tetrahedron.*

16. Prism, right prism, sphere. *The volume of a sphere is equal to the volume of a pyramid the base of which has the same area as the surface of the sphere and the altitude of which is the radius.*

17. Let us call a pyramid an *isosceles pyramid* if all edges starting from its apex are equal. All lateral faces of an isosceles pyramid are isosceles triangles. *The base of an isosceles pyramid is inscribed in a circle and the altitude of the isosceles pyramid passes through the center of this circle.* Cf. ex. 9.26.

22. Yes. Interchanging x and $-x$ we do not change x^2 or the product that represents $(\sin x)/x$ according to E.

23. Prediction: from E follows

$$\left(1-\frac{x^2}{4\pi^2}\right)\left(1-\frac{x^2}{9\pi^2}\right)\left(1-\frac{x^2}{16\pi^2}\right)\cdots = -\frac{\pi^2}{x(x+\pi)}\frac{\sin x-\sin\pi}{x-\pi}$$

and so, for $x \to \pi$, by the definition of the derivative,

$$\left(1-\frac{1}{4}\right)\left(1-\frac{1}{9}\right)\left(1-\frac{1}{16}\right)\cdots = -\frac{\pi^2}{\pi\cdot 2\pi}\cos\pi = \frac{1}{2}.$$

Verification:

$$\left(1-\frac{1}{4}\right)\left(1-\frac{1}{9}\right)\cdots\left(1-\frac{1}{n^2}\right)=\frac{1\cdot 3}{2\cdot 2}\cdot\frac{2\cdot 4}{3\cdot 3}\cdot\frac{3\cdot 5}{4\cdot 4}\cdots\frac{(n-1)(n+1)}{n\cdot n}$$

$$=\frac{1}{2}\frac{n+1}{n}\to\frac{1}{2}.$$

24. 1/6. As ex. 23, or special case $k = 2$ of ex. 25.

25. Prediction: if k is a positive integer

$$\prod_{n=1}^{k-1}\left(1-\frac{k^2}{n^2}\right)\prod_{n=k+1}^{\infty}\left(1-\frac{k^2}{n^2}\right)=\lim_{x\to k\pi}\frac{k^2\pi^2}{x(x+k\pi)}\frac{\sin x}{k\pi-x}$$

$$=(1/2)\,(-\cos k\pi)=(-1)^{k-1}/2.$$

Verification: for $N \geqq k+1$

$$\prod_{n=1}^{k-1}\frac{(n-k)\,(n+k)}{n\ n}\prod_{n=k+1}^{N}\frac{(n-k)\,(n+k)}{n\ n}$$

$$=\frac{(-1)^{k-1}\,(k-1)!(N-k)!\cdot(N+k)!/(k!2k)}{(N!/k)^2}$$

$$=\frac{(-1)^{k-1}}{2}\frac{(N-k)!(N+k)!}{N!\ N!}$$

$$=\frac{(-1)^{k-1}}{2}\frac{(N+1)\,(N+2)\ldots(N+k)}{N\quad(N-1)\ldots(N-k+1)}\to\frac{(-1)^{k-1}}{2}$$

as N tends to ∞.

26. $\pi/4$, the area of the circle with diameter 1. From E, for $x = \pi/2$

$$\frac{2}{\pi}=\left(1-\frac{1}{4}\right)\left(1-\frac{1}{16}\right)\left(1-\frac{1}{36}\right)\left(1-\frac{1}{64}\right)\left(1-\frac{1}{100}\right)\cdots$$

$$=\frac{1\cdot 3}{2\cdot 2}\cdot\frac{3\cdot 5}{4\cdot 4}\cdot\frac{5\cdot 7}{6\cdot 6}\cdot\frac{7\cdot 9}{8\cdot 8}\cdot\frac{9\cdot 11}{10\cdot 10}\cdots.$$

This formula, due to Wallis (1616–1703), was well-known to Euler. There is another way of stating Wallis' formula:

$$\frac{1}{\pi}=\lim_{n\to\infty}\left(\frac{1\cdot 3\cdot 5\ldots(2n-1)}{2\cdot 4\cdot 6\ldots\quad 2n}\right)^2 n.$$

27. $x = \pi z$ in ex. 21 and definition of an infinite product.

28. Yes. From ex. 27

$$\frac{\sin\pi(z+1)}{\pi}=\lim_{n\to\infty}\frac{z+n+1}{z-n}\cdot\frac{(z+n)\ldots(z+1)\,z(z-1)\ldots(z-n)}{(-1)^n\,(n!)^2}.$$

29. From ex. 27 and ex. 26

$$\cos \pi z = \sin \pi(-z + 1/2)$$

$$= \pi \lim \frac{(-z+n+\frac{1}{2})\ldots(-z+\frac{3}{2})(-z+\frac{1}{2})(-z-\frac{1}{2})\ldots(-z+\frac{1}{2}-n)}{(-1)^n(n!)^2}$$

$$= \lim \frac{(2n-1-2z)\ldots(3-2z)(1-2z)(1+2z)(3+2z)\ldots(2n-1+2z)}{(2n-1)\quad\ldots\quad 3 \quad\cdot\quad 1 \quad\cdot\quad 1 \quad\cdot\quad 3 \quad\ldots(2n-1)}$$

$$\cdot \lim \frac{-z+n+\frac{1}{2}}{n} \cdot \pi \lim \left(\frac{1\cdot 3\cdot 5\ldots(2n-1)}{2\cdot 4\cdot 6\ldots\quad 2n}\right)^2 n$$

$$= \left(1-\frac{4z^2}{1}\right)\left(1-\frac{4z^2}{9}\right)\left(1-\frac{4z^2}{25}\right)\cdots.$$

30. Yes. From E and ex. 29

$$\frac{2\sin \pi z/2}{\pi z} \cdot \cos \pi z/2 = \left(1-\frac{z^2}{4}\right)\left(1-\frac{z^2}{16}\right)\left(1-\frac{z^2}{36}\right)\cdots$$

$$\cdot\left(1-\frac{z^2}{1}\right)\left(1-\frac{z^2}{9}\right)\left(1-\frac{z^2}{25}\right)\cdots$$

$$= \left(1-\frac{z^2}{1}\right)\left(1-\frac{z^2}{4}\right)\left(1-\frac{z^2}{9}\right)\left(1-\frac{z^2}{16}\right)\cdots$$

$$= \frac{\sin \pi z}{\pi z}.$$

31. Prediction: for $x = \pi$ ex. 29 yields $\cos \pi = -1$.

Verification: the product of the first n factors

$$\frac{-1\cdot 3}{1\cdot 1}\cdot\frac{1\cdot 5}{3\cdot 3}\cdot\frac{3\cdot 7}{5\cdot 5}\cdot\frac{5\cdot 9}{7\cdot 7}\cdots\frac{(2n-3)(2n+1)}{(2n-1)(2n-1)} = -\frac{2n+1}{2n-1} \to -1.$$

32. Prediction: ex. 29 yields $\cos 2\pi = 1$.

Verification: as ex. 31, or ex. 31 and ex. 35.

33. Prediction: for $x = n\pi$ $(n = 1, 2, 3, \ldots)$ ex. 29 yields

$$\left(1-\frac{4n^2}{1}\right)\left(1-\frac{4n^2}{9}\right)\left(1-\frac{4n^2}{25}\right)\cdots = \cos \pi n = (-1)^n.$$

Verification: from $\cos 0 = 1$ and ex. 35, or directly as ex. 31.

34. Yes. As ex. 22.

35. Yes. By result, or method, of ex. 28.

36. $1 - \sin x = 1 - \cos\left(\frac{\pi}{2} - x\right) = 2\sin^2\left(\frac{\pi}{4} - \frac{x}{2}\right)$

$$= \left(\frac{\sin \pi(1-2z)/4}{\sin \pi/4}\right)^2;$$

we et $x = \pi z$. By ex. 27

$$\frac{\sin \pi(1-2z)/4}{\sin \pi/4}$$

$$= \lim \frac{n+(1-2z)/4}{n+1/4} \cdots \frac{1+(1-2z)/4}{1+1/4}\,\frac{(1-2z)/4}{1/4}\,\frac{-1+(1-2z)/4}{-1+1/4} \cdots \frac{-n+(1-2z)/4}{-n+1/4}$$

$$= \lim \frac{4n+1-2z}{4n+1} \cdots \frac{5-2z}{5}\,\frac{1-2z}{1}\,\frac{3+2z}{3} \cdots \frac{4n-1+2z}{4n-1}$$

$$= \left(1 - \frac{2z}{1}\right)\left(1 + \frac{2z}{3}\right)\left(1 - \frac{2z}{5}\right) \cdots \left(1 + \frac{2z}{4n-1}\right)\left(1 - \frac{2z}{4n+1}\right) \cdots .$$

37. By passing to the logarithms and differentiating in ex. 21 or ex. 27. The precise meaning of the right hand side is

$$\lim_{n\to\infty}\left(\frac{1}{x+n\pi} + \cdots + \frac{1}{x+\pi} + \frac{1}{x} + \frac{1}{x-\pi} + \cdots + \frac{1}{x-n\pi}\right).$$

38. By ex. 37

$$\cot x = \frac{1}{x} + \sum_{n=1}^{\infty}\left(\frac{1}{x+n\pi} + \frac{1}{x-n\pi}\right)$$

$$= \frac{1}{x} + \sum_{n=1}^{\infty}\frac{2x}{x^2 - n^2\pi^2}$$

$$= \frac{1}{x} - 2x\sum_{n=1}^{\infty}\left(\frac{1}{n^2\pi^2} + \frac{x^2}{n^4\pi^4} + \frac{x^4}{n^6\pi^6} + \cdots\right).$$

Let us set

$$y = \cot x = \frac{1}{x} + a_1 x + a_2 x^3 + a_3 x^5 + \cdots .$$

Then, expressing the coefficient of x^{2n-1}, we find

$$S_{2n} = 1 + \frac{1}{2^{2n}} + \frac{1}{3^{2n}} + \frac{1}{4^{2n}} + \frac{1}{5^{2n}} + \cdots = -\frac{a_n \pi^{2n}}{2}$$

for $n = 1, 2, 3, \ldots$. In order to find the coefficients $a_1, a_2, a_3, \ldots$, we use the differential equation

$$y' + y^2 = -1.$$

Substituting for y and y' their expansions and comparing the coefficients of like powers of x, we obtain relations between the coefficients $a_1, a_2, a_3, \dots$ which we can most conveniently survey in the array

	x^{-2}	1	x^2	x^4	x^6	$\dots$
y'	-1	a_1	$3a_2$	$5a_3$	$7a_4$	$\dots$
y^2	1	$2a_1$	$2a_2$	$2a_3$	$2a_4$	$\dots$
			a_1^2	$2a_1a_2$	$2a_1a_3$	$\dots$
					a_2^2	$\dots$
						$\dots$
	0	-1	0	0	0	$\dots$

Cf. ex. 5.1. We obtain so the relations

$$3a_1 = -1, \quad 5a_2 + a_1^2 = 0, \quad 7a_3 + 2a_1a_2 = 0, \ \dots$$

and hence successively

$$S_{2n} = -\frac{a_n\pi^{2n}}{2} = \frac{\pi^2}{6}, \frac{\pi^4}{90}, \frac{\pi^6}{945}, \frac{\pi^8}{9450}, \dots$$

for

$$n = 1, 2, 3, 4, \dots .$$

39. Method of ex. 37 and 38 applied to result of ex. 36. We set now

$$y = \cot\left(\frac{\pi}{4} - \frac{x}{2}\right) = b_1 + b_2x + b_3x^2 + b_4x^3 + \dots .$$

Then

$$T_n = 1 + \frac{(-1)^n}{3^n} + \frac{1}{5^n} + \frac{(-1)^n}{7^n} + \frac{1}{9^n} + \frac{(-1)^n}{11^n} + \dots = \frac{b_n\pi^n}{2^{n+1}}.$$

Now y satisfies the differential equation

$$2y' = 1 + y^2$$

which (observe that $b_1 = 1$) yields the array

	1	x	x^2	x^3	x^4	$\dots$
	1					
y^2	1	$2b_2$	$2b_3$	$2b_4$	$2b_5$	$\dots$
			b_2^2	$2b_2b_3$	$2b_2b_4$	$\dots$
					b_3^2	$\dots$
						$\dots$
$2y'$	$2b_2$	$4b_3$	$6b_4$	$8b_5$	$10b_6$	$\dots$

Hence we obtain first the relations $2b_2 = 2$, $4b_3 = 2b_2$, $6b_4 = 2b_3 + b_2^2, \ldots$ and then the values for

$$n = 1, 2, 3, 4, 5, 6, \ldots$$

$$T_n = \frac{\pi}{4}, \frac{\pi^2}{8}, \frac{\pi^3}{32}, \frac{\pi^4}{96}, \frac{5\pi^5}{1536}, \frac{\pi^6}{960}, \ldots .$$

40. Generally

$$S_{2n}\left(1 - \frac{1}{2^{2n}}\right) = 1 + \frac{1}{2^{2n}} + \frac{1}{3^{2n}} + \frac{1}{4^{2n}} + \frac{1}{5^{2n}} + \ldots - \frac{1}{2^{2n}} - \frac{1}{4^{2n}} - \ldots = T_{2n}.$$

This can be used to check the numerical work in ex. 38 and ex. 39:

$$\frac{1}{6}\cdot\frac{3}{4} = \frac{1}{8}, \quad \frac{1}{90}\cdot\frac{15}{16} = \frac{1}{96}, \quad \frac{1}{945}\cdot\frac{63}{64} = \frac{1}{960}.$$

41.
$$\int_0^1 (1 - x^2)^{-1/2} \arcsin x \cdot dx$$

$$= \int_0^1 (1 - x^2)^{-1/2}\, x\, dx + \frac{1}{2}\frac{1}{3}\int_0^1 (1 - x^2)^{-1/2}\, x^3\, dx + \ldots$$

$$= 1 + \frac{1}{2}\frac{1}{3}\frac{2}{3} + \frac{1}{2}\frac{3}{4}\frac{1}{5}\frac{2\cdot 4}{3\cdot 5} + \frac{1}{2}\frac{3}{4}\frac{5}{6}\frac{1}{7}\frac{2\cdot 4\cdot 6}{3\cdot 5\cdot 7} + \ldots$$

$$= 1 + \frac{1}{3^2} + \frac{1}{5^2} + \frac{1}{7^2} + \ldots .$$

Now evaluate the integral we started from $(=(\pi/2)^2/2)$ and use ex. 40. Cf. Euler, *Opera Omnia*, ser. 1, vol. 14, p. 178–181.

42.
$$\int_0^1 (1 - x^2)^{-1/2} (\arcsin x)^2 dx$$

$$= \int_0^1 (1 - x^2)^{-1/2}\, x^2 dx + \frac{2}{3}\frac{1}{2}\int_0^1 (1 - x^2)^{-1/2}\, x^4 dx + \ldots$$

$$= \frac{1}{2}\frac{\pi}{2} + \frac{2}{3}\frac{1}{2}\cdot\frac{1}{2}\frac{3}{4}\frac{\pi}{2} + \frac{2}{3}\frac{4}{5}\frac{1}{3}\cdot\frac{1}{2}\frac{3}{4}\frac{5}{6}\frac{\pi}{2} + \ldots$$

$$= \frac{\pi}{4}\left(1 + \frac{1}{2^2} + \frac{1}{3^2} + \frac{1}{4^2} + \ldots\right).$$

Now evaluate the integral we started from $(=(\pi/2)^3/3)$. The expansion of

$(\arcsin x)^2$ that we have used will be derived in ex. 5.1. Cf. Euler, *Opera Omnia*, ser. 1, vol. 14, p. 181–184.

43. (a) $$\sum_{n=1}^{\infty} \frac{x^n}{n^2} = \int_0^x \sum_{n=1}^{\infty} \frac{t^{n-1}}{n}\, dt = -\int_0^x t^{-1} \log(1-t)\, dt;$$

integrate by parts and then introduce as new variable of integration $s = 1 - t$.

(b) $x = 1/2$ which renders the greater of the two values, x and $1 - x$, as small as possible.

44. If $P_n(x) = 0$, we have

$$\left(1 + \frac{ix}{n}\right)^n = \left(1 - \frac{ix}{n}\right)^n,$$

$$1 + \frac{ix}{n} = e^{2\pi ki/n}\left(1 - \frac{ix}{n}\right),$$

$$x = \frac{n}{i}\,\frac{e^{\pi ki/n} - e^{-\pi ki/n}}{e^{\pi ki/n} + e^{-\pi ki/n}} = n \tan\frac{k\pi}{n}$$

where we take $k = 0, 1, 2, \ldots n - 1$ if n is odd.

45. If n is odd, we can take in the expression of the roots, see ex. 44,

$$k = 0, \ \pm 1, \ \pm 2, \ \ldots \ \pm (n-1)/2.$$

Therefore,

$$\frac{P_n(x)}{x} = \prod_{k=1}^{(n-1)/2} \left(1 - \frac{x^2}{n^2 \tan^2(k\pi/n)}\right).$$

Observe that, for fixed k,

$$\lim_{n\to\infty} n \tan(k\pi/n) = k\pi.$$

Only a relatively small step is needed to carry us from the point now attained to a proof that is acceptable according to modern standards. A somewhat different arrangement of Euler's argument due to Cauchy seems to have served as a model to Abel as he, led by analogy, discovered the representation of the elliptic functions by infinite products. Cf. A. Cauchy, *Oeuvres complètes*, ser. 2, vol. 3, p. 462–465, and N. H. Abel, *Oeuvres complètes*, vol. 1, p. 335–343.

46. The sum of a finite number of terms is the same in whatever order the terms are taken. The mistake was to extend this statement uncritically to an infinite number of terms, that is, to assume that the sum of an *infinite series* is the same in whatever order the terms are taken. The assumed statement is false; our example shows that it is false. The protection against such a mistake is to go back to the definitions of the terms used and to rely only on rigorous proofs based on these definitions. Thus, the sum of an infinite series is, by definition, the *limit* of a certain sequence (of the sequence of the "partial sums") and interchanging an infinity of terms, as we did, we change essentially the defining sequence. (Under a certain *restrictive condition* a rearrangement of the terms of an infinite series does not change the sum; see Hardy, *Pure Mathematics*, p. 346–347, 374, 378–379. Yet this condition is not satisfied in the present case.)

No solution: **5, 7, 9, 10, 11, 18, 19, 20, 21.**

SOLUTIONS, CHAPTER III

1. Yes: $F = 2n$, $V = n + 2$, $E = 3n$.

2. (1) Yes: $F = m(p + 1)$, $V = pm + 2$, $E = m(p + 1) + pm$; yet there are restrictions: see ex. 21–28. (2) $p = 1$, $m = 4$.

3. (1) Exclude for a moment the tetrahedron; the remaining six polyhedra form three pairs. The two polyhedra in the same pair, as cube and octahedron, are so connected that they have the same E, but the F of one equals the V of the other. The tetrahedron remains alone, but it is connected with itself in this peculiar way. (2) Take the cube. Take any two neighboring faces of the cube and join their centers by a straight line. The 12 straight lines so obtained form the edges of a regular octahedron. This octahedron is inscribed in the cube, its 6 vertices lie in the centers of the 6 faces of the cube. Reciprocally, the centers of the 8 faces of the regular octahedron are the 8 vertices of a cube inscribed in the octahedron. A similar reciprocal relation holds between the polyhedra of the same pair also in the other cases. (Use cardboard models for the dodecahedron and the icosahedron.) The tetrahedron has this peculiar relation to itself: the centers of its 4 faces are the vertices of an inscribed tetrahedron. (3) The passage from one polyhedron of a pair to the other preserves Euler's formula.

4. By E *red* boundary lines, the sphere is divided into F countries; there are V points that belong to the boundary of more than two countries. Choose in each country a point, the "capital" of the country. Connect the capitals of any two neighboring countries by a "road" so that each road crosses just one boundary line and different roads do not cross each other; draw these roads in *blue*. There are precisely E blue lines (roads); they divide the sphere into F' countries with V' points belonging to the boundary of three or more of these countries. Satisfy yourself that $V' = F$ and $F' = V$. The relation between the red and blue subdivisions of the sphere is reciprocal, the passage from one to the other preserves Euler's formula.

5. Euler's formula will hold after "roofing" (sect. 4) if, and only if, it did hold before roofing. Yet by roofing all nontriangular faces of a given polyhedron we obtain another polyhedron with triangular faces only.

6. Analogous to ex. 5: "truncating" introduces vertices with three edges as "roofing" introduces triangular faces. We could also reduce the present case to ex. 5 by using ex. 4.

7. (1) $N_0 = V$, $N_1 = E$, $N_2 = F - 1$. The subscripts 0, 1, 2 indicate the respective dimensionality, see sect. 7. (2) $N_0 - N_1 + N_2 = 1$.

8. (1) Set $l + m = c_1$, $lm = c_2$. Then

$$\begin{aligned} N_0 &= (l+1)(m+1) &&= 1 + c_1 + c_2, \\ N_1 &= (l+1)m + (m+1)l &&= c_1 + 2c_2, \\ N_2 &= lm &&= c_2. \end{aligned}$$

(2) Yes, $N_0 - N_1 + N_2 = 1$, although this simple subdivision of a rectangle cannot be generated exactly in the manner described in ex. 7.

9. $N_2\, 180° = (N_0 - 3)\, 360° + 180°$. In trying to come closer to our goal which is the equation (2) in the solution of ex. 7, we transform this successively into

$$2N_0 - N_2 - 5 = 0,$$

$$2N_0 - 3N_2 + 2N_2 - 3 = 2.$$

By counting the edges in two different ways, we obtain

$$3N_2 = 2N_1 - 3.$$

The last two equations yield

$$N_0 - N_1 + N_2 = 1,$$

which proves Euler's formula, in view of ex. 7 (2).

10. (1) Let $l + m + n = c_1$, $lm + ln + mn = c_2$ and $lmn = c_3$. Then

$$\begin{aligned} N_0 &= (l+1)(m+1)(n+1) = 1 + c_1 + c_2 + c_3, \\ N_1 &= l(m+1)(n+1) + m(l+1)(n+1) + n(l+1)(m+1) \\ &= c_1 + 2c_2 + 3c_3, \\ N_2 &= (l+1)mn + (m+1)ln + (n+1)lm = c_2 + 3c_3, \\ N_3 &= lmn = c_3. \end{aligned}$$

(2) Yes, $N_0 - N_1 + N_2 - N_3 = 1$.

11. We dealt with the case $n = 3$ in sect. 16. In dealing with this case we did not use any simplifying circumstance that would be specific to the particular case $n = 3$. Therefore, this particular case may well "represent" the general case (in the sense of ex. 2.10) as hinted already in sect. 17. The reader should repeat the discussion of sect. 16, saying n for 3, $n + 1$ for 4, P_n for 7, and P_{n+1} for 11, with a little caution. See also ex. 12.

12. Follow the suggestions of sect. 17 and the analogy of ex. 11. Given n planes in general position. They dissect the space into S_n parts. Adjoin one more plane; it is intersected by the foregoing n planes in n straight lines which, *being in a general position*, determine on it P_n regions. Each such plane region acts as a "diaphragm"; it divides an old compartment of space (one of those S_n compartments) into two new compartments, makes one old compartment disappear and two new compartments appear and adds so finally a unit to the previous number S_n of compartments. Hence the relation that we desired to prove.

13. See the third column of the table in sect. 14.

14. The second column of the table in sect. 14 agrees with

$$S_n = 1 + n + \frac{n(n-1)}{1 \cdot 2} + \frac{n(n-1)(n-2)}{1 \cdot 2 \cdot 3}$$

$$= \binom{n}{0} + \binom{n}{1} + \binom{n}{2} + \binom{n}{3};$$

we used the usual notation for binomial coefficients.

15. Finite 3, infinite 8.

16. Let P_n^∞ denote the number of those among the P_n parts defined in ex. 11 which are infinite. By observation, for

$$n = 1, 2, 3$$

$$P_n^\infty = 2, 4, 6.$$

Guess: $P_n^\infty = 2n$. Proof: Take a point in one of the finite parts and imagine an ever-increasing circle with this point as center. When this circle becomes very large, the $P_n - P_n^\infty$ finite parts practically coincide with its center. Now, n different lines through the center of the circle intersect the periphery in $2n$ points and divide it into $2n$ parts. Hence, in fact, $P_n^\infty = 2n$.

$$P_n - P_n^\infty = 1 - n + \frac{n(n-1)}{2}.$$

For instance, the answer to ex. 15 is

$$1 - 4 + 6 = 3.$$

17. Same as ex. 18, by analogy to the solution of ex. 16.

18. Same as ex. 19.

19. See ex. 20.

20. We consider n circles in the plane any two of which intersect in general position. We call S_n^∞ the number of parts into which these circles dissect the plan, in view of ex. 17, 18, and 19. In analogy with sect. 16, notice that the number of parts into which a circle is divided by n circles intersecting it is

$2n$ (general position assumed). Observe (think of all three interpretations of S_n^∞):

$$n = 1 \quad 2 \quad 3 \quad 4$$

$$2n = 2 \quad 4 \quad 6 \quad 8$$

$$S_n^\infty = 2 \quad 4 \quad 8 \quad 14.$$

Guess: $S_{n+1}^\infty = S_n^\infty + 2n$. Proof: as ex. 11, 12. For instance,

$$S_5^\infty = S_4^\infty + 8 = 14 + 8 = 22;$$

This is the solution of ex. 17, 18, and 19. Further guesses:

$$S_n^\infty = 2\binom{n}{0} + 2\binom{n}{2},$$

$$S_n - S_n^\infty = -\binom{n}{0} + \binom{n}{1} - \binom{n}{2} + \binom{n}{3}.$$

21. See ex. 22–30.

22. Wrong: $F = 1$, $V = E = 0$, $1 + 0 \neq 0 + 2$.

23. Wrong: $F = 2$, $V = 0$, $E = 1$, $2 + 0 \neq 1 + 2$.

24. Wrong: $F = 3$, $V = 0$, $E = 2$, $3 + 0 \neq 2 + 2$.

25. Right: $F = 3$, $V = 2$, $E = 3$, $3 + 2 = 3 + 2$.

26. Wrong: $F = p + 1$, $V = 0$, $E = p$, $(p + 1) + 0 \neq p + 2$; see ex. 22, 23, 24 for the cases $p = 0, 1, 2$, respectively. Observe that, in the present case, the solution of ex. 2 (1) becomes inapplicable.

27. The case $m = 3$, $p = 0$ is right, see ex. 25, and so is, more generally, the case $m \geqq 3$: $F = m$, $V = 2$, $E = m$, $m + 2 = m + 2$. The case $m = 0$, $p = 0$ is wrong, see ex. 22. The remaining two cases *can* be so interpreted that they appear right. (1) $m = 1$, $p = 0$: one country with an interior barrier that has two endpoints, $F = 1$, $V = 2$, $E = 1$, $1 + 2 = 1 + 2$. (2) $m = 2$, $p = 0$: two countries separated by two arcs and two corners, $F = 2$, $V = 2$, $E = 2$, $2 + 2 = 2 + 2$. The more obvious interpretation given in ex. 23 yields "wrong." With the present interpretation the solution of ex. 2 (1) remains applicable to the case $m > 0$, $p = 0$.

28. $m \geqq 3$, $p \geqq 1$. The proof uses the fact that, in any convex polyhedron, at least three edges surround a face and at least three edges meet in a vertex.

29. Ex. 22–28 suggest two conditions: (1) A country counted in F, as a face of a convex polyhedron, should be of the "type of a circular region"; a full sphere is not of this type, neither is an annulus of this type. (2) A line counted in E, as an edge of a convex polyhedron, should terminate in corners;

the full periphery of a circle does not terminate so (it does not terminate at all). Ex. 22 fails to satisfy (1), ex. 23 fails to satisfy (2), ex. 24 fails to satisfy either, ex. 25 or, more generally, the case $m > 0, p = 0$, interpreted as in the solution of ex. 27, satisfies both (1) and (2).

30. (1) Take the case (3, 2) of ex. 2 (1), cf. ex. 26, but erase on each meridian the arc between the two parallel circles: $F = 7$, $V = 8$, $E = 12$, $7 + 8 \neq 12 + 2$, there is a spherical zone among the countries and therefore conflict with condition (1), but not with condition (2), of ex. 29. (2) $F = 1$, $V = 1$, $E = 0$ (one country, encompassing the whole globe, except a mathematical point at the north pole); right, $1 + 1 = 0 + 2$, no conflict with (1) or (2) of ex. 29. Etc.

32. $3F_3 + 4F_4 + 5F_5 + \dots = 3V_3 + 4V_4 + 5V_5 + \dots = 2E.$

33. 4π, 12π, 8π, 36π, 20π, respectively.

34. $\Sigma\alpha = \pi F_3 + 2\pi F_4 + 3\pi F_5 + \dots .$

35. By ex. 34, 32, 31

$$\Sigma\alpha = \pi\Sigma(n-2)F_n = 2\pi(E-F).$$

36. A convex spherical polygon with n sides can be dissected into $n - 2$ spherical triangles. Therefore,

$$\begin{aligned} A &= \alpha_1 + \alpha_2 + \dots + \alpha_n - (n-2)\pi \\ &= 2\pi - (\pi - \alpha_1) - (\pi - \alpha_2) - \dots - (\pi - \alpha_n) \\ &= 2\pi - a_1' - a_2' - \dots - a_n' \\ &= 2\pi - P'. \end{aligned}$$

37. The faces of the polyhedron passing through one of the vertices include an interior solid angle; its supplement is called by Descartes the exterior solid angle. Describe a sphere with radius 1 around the vertex as center, but keep only that sector of the sphere that is contained in the exterior solid angle; the sectors so generated at the several vertices of the polyhedron form, when shifted together, a full sphere as the circular sectors in the analogous plane figure (fig. 3.7) form, when shifted together, a full circle. We regard as the measure of a solid angle the area of the corresponding spherical polygon: the joint measure of all the exterior solid angles of the polyhedron is, in fact, 4π.

38. Let $P_1, P_2, \dots P_V$ denote the perimeters of the spherical polygons that correspond to the V interior solid angles of the polyhedron. Then, by ex. 36 and 37,

$$\begin{aligned} \Sigma\alpha &= P_1 + P_2 + \dots + P_V \\ &= 2\pi - A_1' + 2\pi - A_2' + \dots + 2\pi - A_V' \\ &= 2\pi V - 4\pi. \end{aligned}$$

39. By ex. 35 and 38

$$2\pi(E - F) = \Sigma\alpha = 2\pi(V - 2).$$

40. By ex. 31, 32

$$\begin{aligned} 3F &= 3F_3 + 3F_4 + 3F_5 + \dots \\ &\leqq 3F_3 + 4F_4 + 5F_5 + \dots = 2E \end{aligned}$$

which yields the first of the six proposed inequalities. The case of equality is attained when $F = F_3$, that is, when all faces are triangles. Eliminating first E and then F from Euler's theorem and the inequality just proved, we obtain the remaining two inequalities in the first row; they go over into equations if, and only if, all faces are triangles. Interchanging the roles of F and V, as suggested by ex. 3 and 4, we obtain the three proposed inequalities in the second row; they go over into equations if, and only if, all vertices of the polyhedron are three-edged. Some of the inequalities proved are given in Descartes' notes.

41. From Euler's theorem

$$6F - 2E = 12 + 2(2E - 3V)$$

and hence, by ex. 31, 32, and 40,

$$3F_3 + 2F_4 + F_5 = 12 + 2(2E - 3V) + F_7 + 2F_8 + \dots$$

$$3F_3 + 2F_4 + F_5 \geqq 12$$

and so *any convex polyhedron must have some faces with less than six sides.*

No solution: **31.**

SOLUTIONS, CHAPTER IV

1. $R_2(25) = 12$, see sect. 2; $S_3(11) = 3$.

2. $R_2(n)$ denotes the number of the lattice points in a plane that lie on the periphery of a circle with radius $\sqrt{n}$ and center at the origin. (Take the case $n = 25$, ex. 1, and draw the circle.) $R_3(n)$ is the number of lattice points in space on the surface of the sphere with radius $\sqrt{n}$ and center at the origin.

3. If p is an odd prime, $R_2(p^2) = 12$ or 4 according as p divided by 4 leaves the remainder 1 or 3.

4. The comparison of the tables suggests: if p is an odd prime, either both p and p^2 are expressible as a sum of two squares, or neither p nor p^2 is so expressible. A more precise conjecture is also somewhat supported by our observations: if p is an odd prime, $R_2(p) = 8$ or 0, according as p divided by 4 leaves the remainder 1 or 3.

5. If $p = x^2 + y^2$, it follows that

$$p^2 = x^4 + 2x^2y^2 + y^4 = (x^2 - y^2)^2 + (2xy)^2.$$

That is, if $R_2(p) > 0$ also $R_2(p^2) > 0$. This is only one half of our less precise, and only a small part of our more precise, conjecture. (If we know that $R_2(p^2) > 0$, a conclusion concerning $R_2(p)$ is definitely less obvious.) Still, it seems reasonable that such a partial verification greatly strengthens our confidence in the less precise conjecture, and strengthens somewhat our confidence in the more precise conjecture too.

6. $R_3(n) = 0$ for $n = 7, 15, 23$, and 28, and for no other n up to 30; see Table II on pp. 74–75.

7. The respective contributions to $S_4(n)$ are: (1) 24, (2) 12, (3) 6, (4) 4. (5) 1.

8. First, refer to the cases distinguished in ex. 7. If $S_4(4u)$ is odd, the case (5) necessarily arises, and so

$$4u = a^2 + a^2 + a^2 + a^2,$$

$$u = a^2.$$

Second, to any divisor d of u corresponds the divisor u/d and these two divisors are different unless $u = d^2$. Therefore, the number of the divisors of u is odd or even, according as u is or is not a square, and the same holds for the sum of these divisors, since each divisor of u is odd, as u itself. We conjectured in sect. 6 that $S_4(4u)$ and the sum of the divisors of u are equal; we proved now that these two numbers leave the same remainder when divided by 2. Having proved a part of our conjecture, we have, of course, more faith in it.

9.

(1) $24 \times 2^4 = 8 \times 48$	(6) $24 \times 2^3 = 8 \times 24$
(2) $12 \times 2^4 = 8 \times 24$	(7) $12 \times 2^3 = 8 \times 12$
(3) $6 \times 2^4 = 8 \times 12$	(8) $4 \times 2^3 = 8 \times 4$
(4) $4 \times 2^4 = 8 \times 8$	(9) $12 \times 2^2 = 8 \times 6$
(5) $1 \times 2^4 = 8 \times 2$	(10) $6 \times 2^2 = 8 \times 3$

(11) $4 \times 2 = 8 \times 1$.

10. See Table II, p. 74. Check at least a few entries. It follows from ex. 9 that $R_4(n)$ is divisible by 8.

11. Trying to notice at least fragmentary regularities (as we did in sect. 6), you may be led to grouping some more conspicuous cases as follows:

(1)	2	3	5	7	11	13	17	19	23	29
	3	4	6	8	12	14	18	20	24	30
(2)		2		4		8		16		
		3		3		3		3		
(3)	4	8	12	16	20	24	28			
	3	3	12	3	18	12	24.			

In (1), (2), and (3) the first line gives n, the second line $R_4(n)$.

12. Done in solution of ex. 11: (1) primes, (2) powers of 2, (3) numbers divisible by 4.

13. By the analogy of sect. 6 and a little observation, the law is relatively easy to discover when n is not divisible by 4. Therefore, we concentrate on the case (3) in the solution of ex. 11.

$$\begin{array}{lccccccccc} n & = & 4 & 8 & 12 & 16 & 20 & 24 & 28 \\ n/4 & = & 1 & 2 & 3 & 4 & 5 & 6 & 7 \\ R_4(n)/8 & = & 3 & \mathbf{3} & 12 & 3 & 18 & \mathbf{12} & 24. \end{array}$$

A number in heavy print in the third line is the *sum of all divisors* of the corresponding number in the second line—and, therefore, the sum of *some* divisors of the corresponding number in the first line, in which we are really interested. This observation leads to another trial:

$$\begin{array}{lcccc} n & = & 4 & 8 & 12 & 16 \\ R_4(n)/8 & = & 1+2 & 1+2 & 1+2+3+6 & 1+2 \\ n & = & 20 & & 24 & 28 \\ R_4(n)/8 & = & 1+2+5+10 & & 1+2+3+6 & 1+2+7+14. \end{array}$$

Which divisors are added together? Which divisors are omitted?

14. $R_4(n)$, the number of representations of n as a sum of four squares, equals 8 times the sum of those divisors of n which are not divisible by 4. (If n itself is not divisible by 4, none of its divisors is, and hence the rule is simpler in this more frequent case.)

15. Correspondingly to the columns of Table II:

$$\begin{array}{llll} 31 & 25+4+1+1 & 12\times 16 & 32 = 31+1 \\ & 9+9+9+4 & 4\times 16 & \\ 32 & 16+16 & 6\times 4 & 3 = 2+1 \\ 33 & 25+4+4 & 12\times 8 & 48 = 33+11+3+1 \\ & 16+16+1 & 12\times 8 & \\ & 16+9+4+4 & 12\times 16 & \end{array}$$

16. $5 = 1+1+1+1+1 = 4+1$

$$R_8(5) = \binom{8}{5} 2^5 + 8\cdot 7\cdot 2^2 = 2016 = 16\times 126.$$

$$40 = 25+9+1+1+1+1+1+1$$

$$40 = 9+9+9+9+1+1+1+1$$

$$S_8(40) = 8\cdot 7 + \binom{8}{4} = 126.$$

17. Ex. 16. Table III has been actually constructed by a method less laborious than that of ex. 16; cf. ex. 6.17 and 6.23.

18. Within the limits of Table III, both $R_8(n)$ and $S_8(8n)$ increase steadily with n whereas $R_4(n)$ and $S_4(4(2n-1))$ oscillate irregularly.

19. The analogy with $R_4(n)$ and $S_4(4(2n-1))$ points to divisors. One fragmentary regularity is easy to observe: if n is odd, $R_8(n)/16$ and $S_8(8n)$ are exactly equal; if n is even, they are different, although the difference is relatively small in most cases.

20. Odd and even already in ex. 19. Powers of 2:

n	1	2	4	8	16
$S(8n)$	1	8	64	512	4096.

Also the second line consists of powers of 2:

n	2^0	2^1	2^2	2^3	2^4
$S(8n)$	2^0	2^3	2^6	2^9	2^{12}.

What is the law of the exponents?

21. If n is a power of 2, $S(8n) = n^3$. This (and the smooth increase of $R_8(n)$ and $S_8(8n)$) leads to constructing the following table.

n	$R_8(n)/16 - n^3$	$S_8(8n) - n^3$
1	0	0
2	−1	0
3	1	1
4	7	0
5	1	1
6	−20	8
7	1	1
8	71	0
9	28	28
10	−118	8
11	1	1
12	260	64
13	1	1
14	−336	8
15	153	153
16	583	0
17	1	1
18	−533	224
19	1	1
20	946	64

In the column concerned with $R_8(n)$, the $+$ and $-$ signs are regularly distributed.

22. Cubes of divisors!

n	$R_8(n)/16 = S_8(8n)$
1	1^3
3	$3^3 + 1^3$
5	$5^3 + 1^3$
7	$7^3 + 1^3$
9	$9^3 + 3^3 + 1^3$
11	$11^3 + 1^3$
13	$13^3 + 1^3$
15	$15^3 + 5^3 + 3^3 + 1^3$
17	$17^3 + 1^3$
19	$19^3 + 1^3$

n	$R_8(n)/16$	$S_8(8n)$
2	$2^3 - 1^3$	2^3
4	$4^3 + 2^3 - 1^3$	4^3
6	$6^3 - 3^3 + 2^3 - 1^3$	$6^3 + 2^3$
8	$8^3 + 4^3 + 2^3 - 1^3$	8^3
10	$10^3 - 5^3 + 2^3 - 1^3$	$10^3 + 2^3$
12	$12^3 + 6^3 + 4^3 - 3^3 + 2^3 - 1^3$	$12^3 + 4^3$
14	$14^3 - 7^3 + 2^3 - 1^3$	$14^3 + 2^3$
16	$16^3 + 8^3 + 4^3 + 2^3 - 1^3$	16^3
18	$18^3 - 9^3 + 6^3 - 3^3 + 2^3 - 1^3$	$18^3 + 6^3 + 2^3$
20	$20^3 + 10^3 - 5^3 + 4^3 + 2^3 - 1^3$	$20^3 + 4^3$

23. (1) $(-1)^{n-1}R_8(n)/16$ equals the sum of the cubes of all odd divisors of n, less the sum of the cubes of all even divisors of n. (2) $S_8(8n)$ equals the sum of the cubes of those divisors of n whose co-divisors are odd. (If d is a divisor of n, we call n/d the co-divisor of d.) See ex. 6.24 on the history of these theorems and references.

24. Construct the table

0	3	6	9	12
5	8	11	14	
10	13			

which should be imagined as extending without limit to the right and downwards and shows that the only positive integers not expressible in the proposed form are 1, 2, 4, 7.

25. Case $a = 3$, $b = 5$ in ex. 24; a and b are co-prime. Last integer non-expressible $ab - a - b = (a - 1)(b - 1) - 1$. This is incomparably easier to prove than the laws concerned with sums of squares.

26. (1) is generally true. (2) is not generally true, but the first exception is $n = 341$. (See G. H. Hardy and E. M. Wright, *An introduction to the theory of numbers*, Oxford, 1938, p. 69, 72.)

SOLUTIONS, CHAPTER V

1. [Cf. Putnam 1948]

(a) $x + \frac{2}{3}x^3 + \frac{2}{3}\frac{4}{5}x^5 + \ldots + \frac{2}{3}\frac{4}{5}\frac{6}{7}\cdots\frac{2n}{2n+1}x^{2n+1} + \ldots.$

(b) Having verified the differential equation, put

$$y = a_0 x + a_1 x^3 + a_2 x^5 + \ldots + a_n x^{2n+1} + \ldots.$$

To compare the coefficients of like powers you may use the array

	1	x^2	x^4	$\ldots$	x^{2n}	
y'	a_0	$3a_1$	$5a_2$	$\ldots$	$(2n+1)a_n$	$\ldots$
$-x^2y'$		$-a_0$	$-3a_1$	$\ldots$	$-(2n-1)a_{n-1}$	$\ldots$
$-xy$		$-a_0$	$-a_1$	$\ldots$	$-a_{n-1}$	$\ldots$
	1	0	0	$\ldots$	0	$\ldots$

which yields $a_0 = 1$ and $(2n + 1)a_n = 2na_{n-1}$ for $n \geqq 1$.

2. [Cf. Putnam 1950]

(a) $y = \frac{x}{1} + \frac{x^3}{1 \cdot 3} + \frac{x^5}{1 \cdot 3 \cdot 5} + \ldots + \frac{x^{2n-1}}{1 \cdot 3 \cdot 5 \ldots (2n-1)} + \ldots.$

(b) This expansion satisfies

$$y' = 1 + \frac{x^2}{1} + \frac{x^4}{1 \cdot 3} + \frac{x^6}{1 \cdot 3 \cdot 5} + \ldots,$$

$$y' = 1 + xy.$$

The given product y satisfies the same differential equation. Both the expansion and the product vanish when $x = 0$. Hence, they are identical.

3. The relations between the coefficients a_n derived from

$$\frac{1}{1+x} + \frac{4a_1x}{(1+x)^3} + \frac{16a_2x^2}{(1+x)^5} + \dots$$
$$= 1 + a_1x^2 + a_2x^4 + \dots = f(x)$$

are exhibited in the array (see ex. 1)

$$\begin{array}{ccccc}
1 & -1 & 1 & -1 & 1 \\
 & 4a_1 & -4a_1\cdot 3 & 4a_1\cdot 6 & -4a_1\cdot 10 \\
 & & 16a_2 & -16a_2\cdot 5 & 16a_2\cdot 15 \\
 & & & 64a_3 & -64a_3\cdot 7 \\
 & & & & 256a_4 \\
\hline
1 & 0 & a_1 & 0 & a_2
\end{array}$$

They yield

$$f(x) = 1 + \left(\frac{1}{2}\right)^2 x^2 + \left(\frac{1}{2}\frac{3}{4}\right)^2 x^4 + \left(\frac{1}{2}\frac{3}{4}\frac{5}{6}\right)^2 x^6 + \left(\frac{1}{2}\frac{3}{4}\frac{5}{6}\frac{7}{8}\right)^2 x^8 + \dots$$

This example is of historical interest. See Gauss, *Werke*, vol. 3, p. 365–369.

4. Study the arrangement of the following array (see ex. 1, 3):

$$\begin{array}{c|ccccc}
f(x)^3 & a_0^3 & 3a_1a_0^2 & 3a_2a_0^2 & 3a_3a_0^2 & 3a_4a_0^2 \\
 & & & 3a_1^2a_0 & 6a_2a_1a_0 & 6a_3a_1a_0 \\
 & & & & a_1^3 & 3a_2^2a_0 \\
 & & & & & 3a_2a_1^2 \\
3f(x)f(x^2) & 3a_0^2 & 3a_1a_0 & 3a_2a_0 & 3a_3a_0 & 3a_4a_0 \\
 & & & 3a_0a_1 & 3a_1a_1 & 3a_2a_1 \\
 & & & & & 3a_0a_2 \\
2f(x^3) & 2a_0 & & & 2a_1 & \\
\hline
 & 6a_1 & 6a_2 & 6a_3 & 6a_4 & 6a_5
\end{array}$$

Starting from $a_0 = 1$, we obtain recursively a_1, a_2, a_3, a_4, and $a_5 = 8$. See G. Pólya, *Zeitschrift für Kristallographie*, vol. (A) 93 (1936) p. 415–443, and *Acta Mathematica*, vol. 68 (1937) p. 145–252.

5. From comparing the expansions in sect. 1.

6. (a) $\varepsilon^2/15$. (b) $+\infty$. In both extreme cases, the error is positive, the approximate value larger than the true value.

7. (a) $\varepsilon^2/15$. (b) $1/3$. In both extreme cases, the approximate value is larger than the true value.

8. $4\pi(a^2 + b^2 + c^2)/3$. There is some reason to suspect that this approximation yields values in excess of the true values. See G. Pólya, *Publicaciones del Instituto di Matematica*, Rosario, vol. 5 (1943).

9. In passing from the integral to the series, use the binomial expansion and the integral formulas ex. 2.42.

$$P = 2\pi a \left[1 - \frac{1}{2}\sum_1^\infty \frac{1}{2}\frac{3}{4}\cdots\frac{2n-1}{2n}\,\frac{\varepsilon^{2n}}{2n-1}\right],$$

$$P' = 2\pi a \left[1 - \sum_1^\infty \frac{3}{4}\frac{7}{8}\cdots\frac{4n-1}{4n}\,\frac{\varepsilon^{2n}}{4n-1}\right].$$

10. Use the solution of ex. 9 and put $\frac{1}{2}\frac{3}{4}\frac{5}{6}\cdots\frac{2n-1}{2n} = g_n$. Then $g_1 > g_n$ for $n \geqq 2$ and, for $\varepsilon > 0$,

$$E - P = 2\pi a \sum_2^\infty (g_1 g_n - g_n^2)\, \varepsilon^{2n}/(2n-1) > 0.$$

11. The initial term of the relative error of P'' is

$$-\,[\alpha + 3(1-\alpha)]\, \varepsilon^4/64 + \ldots$$

and so it is of order 4 unless $\alpha = 3/2$ and $P'' = P + (P - P')/2$.

12. $(P'' - E)/E = 3 \cdot 2^{-14}\varepsilon^8 + \ldots$ when ε small

$= (3\pi - 8)/8 = .1781$ when $\varepsilon = 1$

$= .00019$ when $\varepsilon = 4/5$.

Hence the conjecture $P'' > E$. See G. Peano, *Applicazioni geometriche del calcolo infinitesimale*, p. 231–236.

13.

$$e^p = \lim_{n\to\infty}\left(1 + \frac{p}{n}\right)^n.$$

Therefore, the desired conclusion is equivalent to

$$\limsup_{n\to\infty}\left(\frac{n(a_1 + a_{n+p})}{(n+p)a_n}\right)^n \geqq 1.$$

The opposite assumption implies

$$\frac{n(a_1 + a_{n+p})}{(n+p)a_n} < 1$$

for $n > N$, N fixed, or, which is the same,

$$\frac{a_{n+p}}{n+p} - \frac{a_n}{n} < -\frac{a_1}{n+p}.$$

Consider the values $n = (m - 1)p$:

$$\frac{a_{mp}}{mp} - \frac{a_{(m-1)p}}{(m-1)p} < -\frac{a_1}{mp},$$

$$\frac{a_{(m-1)p}}{(m-1)p} - \frac{a_{(m-2)p}}{(m-2)p} < -\frac{a_1}{(m-1)p},$$

$$\cdots\cdots\cdots\cdots\cdots\cdots$$

As in sect. 5, we conclude that, with a suitable constant C,

$$\frac{a_{mp}}{m} < C - a_1\left(1 + \frac{1}{2} + \ldots + \frac{1}{m}\right)$$

and this leads for $m \to \infty$ to a contradiction to the hypothesis $a_n > 0$.

14. The example $a_n = n^c$ of sect. 4 suggests

$$a_1 = 1,$$
$$a_n = n \log n \qquad \text{for } n = 2, 3, 4, \ldots .$$

With this choice

$$\left(\frac{a_1 + a_{n+p}}{a_n}\right)^n = \left\{\frac{1 + (n+p)\,[\log n + \log(1 + (p/n))]}{n \log n}\right\}^n$$

$$= \left\{\frac{(n+p)\log n + 1 + (n+p)\left[\frac{p}{n} - \frac{p^2}{2n^2} + \ldots\right]}{n \log n}\right\}^n$$

$$= \left\{1 + \frac{p + \alpha_n}{n}\right\}^n \to e^p$$

since $\alpha_n \to 0$.

15. The mantissas in question are the 900 ordinates of the slowly rising curve $y = \log x - 2$ that correspond to the abscissas $x = 100, 101, \ldots, 999$; log denotes the common logarithm. Table I specifies how many among these 900 points on the curve fall in certain horizontal strips of width 1/10. Let us consider the points at which the curve enters and leaves such a strip. If x_n is the abscissa of such a point

$$\log x_n - 2 = n/10,$$

$$x_n = 100 \cdot 10^{n/10},$$

where $n = 0, 1, 2, \ldots, 10$. The number of integers in any interval is approximately equal to the length of the interval: the difference is less than one unit. Therefore, the number of the mantissas in question with first figure n is $x_{n+1} - x_n$, with an error less than 1. Now,

$$x_{n+1} - x_n = 100(10^{1/10} - 1)10^{n/10}$$

is the nth term of a geometric progression with ratio

$$10^{1/10} = 1.25893.$$

Predict and observe the analogous phenomenon in a six-place table of common logarithms.

16. The periodical repetition can (and should) be regarded as a kind of symmetry; but it is present in all cases, and so we shall not mention it again. The following kinds of symmetry play a role in our classification.

(1) Center of symmetry. Notation: c, c'.

(2) Line of symmetry. Notation: h if the line is horizontal, v or v' if it is vertical.

(3) Gliding symmetry: the frieze shifted horizontally *and* reflected in the central horizontal line *simultaneously*, coincides with itself (in friezes 5, 7, *a*, *b*). Notation: g.

The following types of symmetry are represented in fig. 5.2. (The dash ′ is used to distinguish two elements of symmetry of the same kind, as c and c', or v and v', when their situation in the figure is essentially different.)

1, d: no symmetry (except periodicity)

2, g: $c, c', c, c', \ldots$

3, f: $v, v', v, v', \ldots$

4, e: h

5, a: g

6, c: h; $(v, c), (v', c'), (v, c), (v', c'), \ldots$

7, b: g; $v, c, v, c, \ldots$.

All possible kinds of symmetry are represented in fig. 5.2, as you may convince yourself inductively.

17. Three different kinds of symmetry are represented: 1 is of the same type as 2, 3 as 4. Try to find all types. Cf. G. Pólya, *Zeitschrift für Kristallographie*, vol. 60 (1924) p. 278–282, P. Niggli, *ibid.*, p. 283–298, and H. Weyl, *Symmetry*, Princeton, 1952.

18. Disregard certain details, depending on the style of the print. Then (1) vertical line of symmetry, (2) horizontal line of symmetry, (3) center of symmetry, (4) all three preceding kinds of symmetry jointly, (5) no symmetry. The same for the five curves representing the five equations in rectangular coordinates. Some variant of this problem can be used to enliven a class of analytic geometry.

SOLUTIONS, CHAPTER VI

2. $x(1-x)^{-2}$. Particular case of ex. 3, with $f(x) = (1-x)^{-1}$; obtain it also by combining ex. 4 and 5.

3. $xf'(x) = \sum_{n=0}^{\infty} na_n x^n$.

4. $xf(x) = \sum_{n=1}^{\infty} a_{n-1}x^n.$

5. $(1-x)^{-1}f(x) = \sum_{n=0}^{\infty} (a_0 + a_1 + \ldots + a_n)x^n$; particular case of ex. 6.

6. $f(x)g(x) = \sum_{n=0}^{\infty} (a_0b_n + a_1b_{n-1} + \ldots + a_nb_0)x^n.$

7. $D_3 = 1, D_4 = 2, D_5 = 5, D_6 = 14.$ For D_6, refer to fig. 6.1; there are 2 different dissections of type I, 6 of type II, and 6 of type III.

8. The recursion formula is verified for $n = 6$:

$$14 = 1 \times 5 + 1 \times 2 + 2 \times 1 + 5 \times 1.$$

Choose a certain side as the base of the polygon (the horizontal side in fig. 6.2) and start the dissection by drawing the second and third sides of the triangle Δ whose first side is the base. Having chosen Δ, you still have to dissect a polygon with k sides to the left of Δ and another polygon with $n + 1 - k$ sides to the right; both polygons jointly yield $D_k\,D_{n+1-k}$ possibilities. Choose $k = 2, 3, 4, \ldots n - 1$; of course, you have to interpret suitably the case $k = 2$.

9. By ex. 4 and 6, the recursion formula for D_n yields

$$xg(x) = D_2x^3 + [g(x)]^2.$$

Choose the solution of this quadratic equation the expansion of which begins with x^2:

$$g(x) = (x/2)[1 - (1 - 4x)^{1/2}].$$

Expanding and using the notation for binomial coefficients, you find:

$$D_n = -\frac{1}{2}\binom{1/2}{n-1}(-4)^{n-1}.$$

10. Better so

$$\sum_{u=-\infty}^{\infty} x^{u^2} \sum_{v=-\infty}^{\infty} x^{v^2} \sum_{w=-\infty}^{\infty} x^{w^2} = \sum_{-\infty}^{\infty} \sum_{-\infty}^{\infty} \sum_{-\infty}^{\infty} x^{u^2+v^2+w^2}$$

where u, v, and w range over all integers (from $-\infty$ to $+\infty$) independently, so that the triple sum is extended over all the lattice points of space (see ex. 4.2). To see that this *is* the generating function of $R_3(n)$, you just collect those terms in which the exponent $u^2 + v^2 + w^2$ has the same value n.

11. $$\sum_{n=0}^{\infty} R_k(n)x^n = \left[\sum_{n=0}^{\infty} R_1(n)x^n\right]^k.$$

12. $$\sum_{n=1}^{\infty} S_k(n)x^n = \left[\sum_{n=1}^{\infty} x^{(2n-1)^2}\right]^k.$$

13. Let I, J, K, and L denote certain power series all coefficients of which are integers. Then the generating functions of $R_1(n)$, $R_2(n)$, $R_4(n)$, and $R_8(n)$ are of the form

$$1 + 2I,$$
$$(1 + 2I)^2 = 1 + 4J,$$
$$(1 + 4J)^2 = 1 + 8K,$$
$$(1 + 8K)^2 = 1 + 16L,$$

respectively.

14. $$x + x^9 + x^{25} + x^{49} + x^{81} + \dots$$
$$= x(1 + x^8 + x^{24} + x^{48} + x^{80} + \dots) = xP$$

where P denotes a power series in which the coefficient of x^n vanishes when n is not divisible by 8. The generating functions of

$$S_1(n), \quad S_2(n), \quad S_4(n), \quad S_8(n)$$

are

$$xP, \quad x^2P^2, \quad x^4P^4, \quad x^8P^8,$$

respectively.

15. From ex. 6 and 11
$$G^{k+l} = G^k G^l$$

where G stands for the generating function of $R_1(n)$.

16. Analogous to ex. 15, from ex. 6 and 12.

17. Use ex. 15 and 16 with $k = l = 4$. The actual computation was performed by this method, with occasional checks from other sources, as ex. 4.16 and ex. 23.

18. (1) From ex. 14 and ex. 16 follows

$$S_4(4)S_4(8n-4) + S_4(12)S_4(8n-12) + \cdots + S_4(8n-4)S_4(4) = S_8(8n).$$

It was conjectured in sect. 4.6 that $S_4(4(2n-1)) = \sigma(2n-1)$ and in ex. 4.23 that $S_8(8u) = \sigma_3(u)$ if u is an odd number.

(2) $$\sigma(1)\sigma(9) + \sigma(3)\sigma(7) + \sigma(5)\sigma(5) + \sigma(7)\sigma(3) + \sigma(9)\sigma(1)$$
$$= 2(1 \times 13 + 4 \times 8) + 6 \times 6$$
$$= 126 = 5^3 + 1^3 = \sigma_3(5).$$

(3) It seems reasonable that such a verification increases our confidence in both conjectures, to some degree.

19. $$\sum \left[\frac{u-1}{2} - 5\,\frac{k(k+1)}{2} \right] s_{u-k(k+1)} = 0$$

for $u = 1, 3, 5, \ldots$; the summation is extended over all non-negative integers k satisfying the inequality

$$0 \leqq k(k+1) < u.$$

20. $\sigma(3) = 4\sigma(1)$

$$2\sigma(5) = 3\sigma(3)$$

$$3\sigma(7) = 2\sigma(5) + 12\sigma(1)$$

$$4\sigma(9) = \sigma(7) + 11\sigma(3).$$

The last is true, since

$$4 \times 13 = 8 + 11 \times 4.$$

21. The recursion formula has been proved for $S_4(4(2n-1))$ in ex. 19. This recursion formula means, in fact, an infinite system of relations which determine $S_4(4(2n-1))$ unambiguously if $S_4(4)$ is given. Now, we know that

$$S_4(4) = \sigma(1) = 1.$$

If $\sigma(2n-1)$ satisfies the same system of recursive relations as $S_4(4(2n-1))$,

$$S_4(4(2n-1)) = \sigma(2n-1)$$

for $n = 1, 2, 3, \ldots$ because the system is unambiguous. If, conversely, the last equation holds, $\sigma(2n-1)$ satisfies those recursive relations.

22. Assume that

$$G = a_0 + a_1x + a_2x^2 + a_3x^3 + \ldots,$$

$$H = u_0 + u_1x + u_2x^2 + u_3x^3 + \ldots,$$

$$G^k = H.$$

It follows, as in ex. 19, that

$$GxH' - kxG'H = 0.$$

Equate to 0 the coefficient of x^n:

$$\sum_{m=0}^{n} [n - (k+1)m]a_m u_{n-m} = 0.$$

Consider $a_0, a_1, a_2, \ldots$ as given. From the last equation you can express u_n in terms of $u_{n-1}, u_{n-2}, \ldots u_1, u_0$ provided that $a_0 \neq 0$. Observe that $u_0 = a_0^k$.

23. Apply ex. 22 to the case $k = 8$,

$$G = 1 + 2x + 2x^4 + 2x^9 + 2x^{16} + \ldots.$$

By ex. 11 the result of ex. 22 yields

$$nR_8(n) = 2(9-n)R_8(n-1) + 2(36-n)R_8(n-4) + 2(81-n)R_8(n-9) + \ldots.$$

Set $R_8(n)/16 = r_n$. Then $r_0 = 1/16$ and we find $r_1, r_2, r_3, \ldots r_{10}$ successively from

$$\begin{aligned}
r_1 &= 16r_0 \\
2r_2 &= 14r_1 \\
3r_3 &= 12r_2 \\
4r_4 &= 10r_3 + 64r_0 \\
5r_5 &= 8r_4 + 62r_1 \\
6r_6 &= 6r_5 + 60r_2 \\
7r_7 &= 4r_6 + 58r_3 \\
8r_8 &= 2r_7 + 56r_4 \\
9r_9 &= 54r_5 + 144r_0 \\
10r_{10} &= -2r_9 + 52r_6 + 142r_1
\end{aligned}$$

In using these formulas numerically, we have an important check: the right hand side of the equation that yields r_n must be divisible by n.

The same method yields a recursion formula for $R_k(n)$, k being any given integer ≥ 2, and also for $S_k(n)$.

25. Call s the infinite product. Computing $-xd \log s/dx$ and using No. 10 of Euler's memoir, you obtain

$$k\, \Sigma\sigma(l)x^l = \frac{\Sigma n a_n x^n}{1 - \Sigma a_m x^m};$$

the limits for all three sums are 1 and ∞. Multiply with the denominator of the right hand side and focus the coefficient of x^n.

Euler's case is $k = 1$. Also the case $k = 3$ yields a relatively simple result (see the work of Hardy and Wright, quoted in ex. 24, p. 282 and 283, theorems 353 and 357). In the other cases we do not know enough about the law of a_n.

No solution: **1, 24.**

SOLUTIONS, CHAPTER VII

1. [Stanford 1950]

$$1 - 4 + 9 - 16 + \ldots + (-1)^{n-1} n^2 = (-1)^{n-1} \frac{n(n+1)}{2}.$$

The step from n to $n + 1$ requires to verify that

$$(-1)^n (n+1)^2 = (-1)^n \frac{(n+1)(n+2)}{2} - (-1)^{n-1} \frac{n(n+1)}{2}.$$

2. To prove

$$P_n = \binom{n}{0} + \binom{n}{1} + \binom{n}{2},$$

$$S_n = \binom{n}{0} + \binom{n}{1} + \binom{n}{2} + \binom{n}{3},$$

$$S_n^\infty = 2\binom{n}{0} + 2\binom{n}{2}$$

we use ex. 3.11, ex. 3.12 combined with the expression of P_n, and ex. 3.20, respectively. Then, supposing the above expressions, we are required to verify

$$P_{n+1} - P_n = \binom{n}{0} + \binom{n}{1}$$

$$S_{n+1} - S_n = \binom{n}{0} + \binom{n}{1} + \binom{n}{2}$$

$$S_{n+1}^\infty - S_n^\infty = 2\binom{n}{1}.$$

All three follow from the well known fact (the basic relation in the Pascal triangle) that

$$\binom{n+1}{k+1} - \binom{n}{k+1} = \binom{n}{k}.$$

3. *How to Solve It*, p. 103–110.

4. $\frac{3}{4}, \frac{2}{3} = \frac{4}{6}, \frac{5}{8}, \frac{3}{5} = \frac{6}{10}, \ldots \frac{n+1}{2n}.$

The step from n to $n + 1$ requires to verify that

$$1 - \frac{1}{(n+1)^2} = \frac{n+2}{2n+2}\,\frac{2n}{n+1}.$$

Cf. ex. 2.23.

5. $-\frac{3}{1}, -\frac{5}{3}, -\frac{7}{5}, -\frac{9}{7}, \ldots -\frac{2n+1}{2n-1}.$

The step from n to $n + 1$ requires to verify that

$$1 - \frac{4}{(2n+1)^2} = \frac{2n+3}{2n+1}\,\frac{2n-1}{2n+1}.$$

Cf. ex. 2.31.

6. The general case is, in fact, equivalent to the limiting case

$$\frac{x}{1-x} = \frac{x}{1+x} + \frac{2x^2}{1+x^2} + \frac{4x^4}{1+x^4} + \frac{8x^8}{1+x^8} + \dots$$

from which the particular case proposed is derived as follows: substitute x^{16} for x and multiply by 16, obtaining

$$\frac{16x^{16}}{1-x^{16}} = \frac{16x^{16}}{1+x^{16}} + \frac{32x^{32}}{1+x^{32}} + \dots ;$$

then subtract from the original equation. If we set $2^{n+1} = m$, the step from n to $n+1$ requires

$$\frac{mx^m}{1+x^m} = -\frac{2mx^{2m}}{1-x^{2m}} + \frac{mx^m}{1-x^m}.$$

7. To prove

$$1 + 3 + 5 + 7 + \dots + 2n - 1 = n^2.$$

The step from n to $n+1$ requires to verify that

$$2n + 1 = (n+1)^2 - n^2.$$

8. The nth term in the fourth row of the table is

$$(1+2) + (4+5) + \dots + (3n-5+3n-4) + 3n-2$$
$$= 3 + 9 + \dots + [6(n-1) - 3] + 3n - 2$$
$$= 6\frac{n(n-1)}{2} - 3(n-1) + 3n - 2 = 3n^2 - 3n + 1.$$

The step from $n-1$ to n requires, in fact,

$$n^3 - (n-1)^3 = 3n^2 - 3n + 1.$$

9. After n^2, n^3 and n^4, the generalization concerned with n^k is obvious. The simple case of n^2 was known since antiquity; Alfred Moessner discovered the rest quite recently by empirical induction, and Oskar Perron proved it by mathematical induction. See *Sitzungsberichte der Bayerischen Akademie der Wissenschaften*, Math.-naturwissenschaftliche Klasse, 1951, p. 29–43.

10. For $k = 1$ the theorem reduces to the obvious identity

$$1 - n = -(n-1).$$

The step from k to $k+1$ requires to verify that

$$(-1)^{k+1}\binom{n}{k+1} = (-1)^{k+1}\binom{n-1}{k+1} - (-1)^k\binom{n-1}{k}$$

which is the basic relation in the Pascal triangle, already encountered in ex. 2.

11. [Stanford 1946]. Call the required number of pairings of $2n$ players P_n. The nth player can be matched with any one of the other $2n - 1$ players. Once his antagonist is chosen

$$2n - 2 = 2(n - 1)$$

players remain who can be paired in P_{n-1} ways. Hence

$$P_n = (2n - 1)P_{n-1}.$$

12. Call A_n the statement proposed to prove, concerned with $f_n(x)$. Instead of A_n we shall prove A_n'.

A_n'. The function $f_n(x)$ is a quotient *the denominator of which is* $(1 - x)^{n+1}$ and the numerator a polynomial *of degree n* the constant term of which is 0 and the other coefficients positive integers.

Observe that A_n' asserts more than A_n; the points in which A_n' goes beyond A_n are emphasized by italics. Assuming A_n', set

$$(1 - x)^{n+1} f_n(x) = P_n(x) = a_1 x + a_2 x^2 + \dots + a_n x^n$$

where $a_1, a_2, \dots a_n$ are supposed to be positive integers. From the recursive definition we derive the recursive formula

$$P_{n+1}(x) = x[(1 - x)P_n'(x) + (n + 1)P_n(x)]$$

and this shows that the coefficients of $x, x^2, x^3, \dots x^n$ and x^{n+1} in $P_{n+1}(x)$ are a_1, $na_1 + 2a_2$, $(n - 1)\, a_2 + 3a_3, \dots 2a_{n-1} + na_n$, a_n, respectively, which makes the assertion obvious.

13. (1) The sum of all the coefficients of $P_n(x)$ is $n!$ In fact, this sum is $P_n(1)$ and the recursive formula yields

$$P_{n+1}(1) = (n + 1)P_n(1).$$

(2) $P_n(x)/x$ is a reciprocal polynomial or

$$P_n(1/x)x^{n+1} = P_n(x).$$

In fact, assume that $a_1 = a_n$, $a_2 = a_{n-1}, \dots$; the corresponding relations for the coefficients of $P_{n+1}(x)$ follow from their expression given at the end of the solution of ex. 12.

16. $Q_1 = 1, \quad Q_2 = 3, \quad Q_3 = 45, \quad Q_4 = 4725$

$$Q_2/Q_1 = 3, \quad Q_3/Q_2 = 15, \quad Q_4/Q_3 = 105$$

suggest

$$Q_n = 1^n 3^{n-1} 5^{n-2} \dots (2n - 3)^2 (2n - 1)^1.$$

In fact, the definition yields

$$\frac{Q_{n+1}}{Q_{n-1}} = \frac{Q_n Q_{n+1}}{Q_{n-1} Q_n} = \frac{(2n)!(2n + 1)!}{(n!2^n)^2} = [1 \cdot 3 \cdot 5 \dots (2n - 1)]^2 (2n + 1)$$

and hence you prove the general law by inference from $n - 1$ to $n + 1$.

Observe that

$$\frac{2n!}{n!2^n} = \frac{1 \cdot 2 \cdot 3 \cdot 4 \cdot 5 \cdot 6 \ldots (2n-1)2n}{2 \cdot 4 \cdot 6 \ldots \cdot 2n}.$$

17. The reasoning that carries us from 3 to 4 applies to the passage from n to $n + 1$ with one exception: it breaks down, as it must, in the passage from 1 to 2.

18. From $n = 3$ to $n + 1 = 4$: consider the lines a, b, c, and d. Consider first the case that there are two different lines among these four lines, for example b and c. Then the point of intersection of b and c is uniquely determined and must also lie on a (because, allegedly, the statement holds for $n = 3$) and also on d (for the same reason). Therefore, the statement holds for $n + 1 = 4$. If, however, no two among the four given lines are different, the statement is obvious. This reasoning breaks down, as it must, in the passage from 2 to 3.

No solution: **14**, **15**.

SOLUTIONS, CHAPTER VIII

1. (1) Straight line, (2) perpendicular, (3) common perpendicular, (4) segments of line through given point and center (Euclid III, 7, 8), (5) segment of perpendicular through center, no maximum, (6) segments of line joining centers. The case in which the minimum distance is 0 has been consistently discarded as obvious, although it may be important.

2. (1) straight line, (2) perpendicular, (3) common perpendicular, (4) perpendicular, (5) common perpendicular, (6) see sect. 4, (7) segments of line joining point to center, (8) segment of perpendicular through center, no maximum, (9) segment of perpendicular through center, no maximum, (10) segments of line joining centers. The case in which the minimum distance is 0 has been discarded.

3. (1) concentric circles, (2) parallel lines, (3) concentric circles.

4. (1) concentric spheres, (2) parallel planes, (3) coaxial cylinders, (4) concentric spheres.

5. (2) See sect. 3. Others similar.

6. (6) There is just one cylinder that has the first given line as axis and the second given line as tangent. The point of contact is an endpoint of the shortest distance. Others similar.

7. Call one of the given sides the base. Keep the base in a fixed position, let the other side rotate about its fixed endpoint and call its other endpoint X. The prescribed path of X is a circle, the level lines are parallel to the base, the triangle with maximum area is a right triangle (which is obvious).

8. Call the given side the base, keep it in a fixed position, call the opposite vertex X, and let X vary. The prescribed path of X is an ellipse, the level lines are parallel to the base, the triangle with maximum area is isosceles.

9. The level lines are straight lines $x + y =$ const., the prescribed path is (one branch of) an equilateral hyperbola with equation $xy = A$, where A is the given area. It is clear by symmetry that there is a point of contact with $x = y$.

10. Consider expanding concentric circles with the given point as center. It seems intuitive that there is a first circle hitting the given line; its radius is the shortest distance. This is certainly so in the cases ex. 1 (2) and (4).

11. Crossing means passing from one side of the level line to the other, and on one side f takes higher, on the other lower, values than at the point of crossing.

12. You may, but you need not. The highest point may be the peak P (you may wish to see the view) or the pass S (you may cross it hiking from one valley to the other) or the initial point of your path, or its final point, or an angular point of it.

13. (1) The level line for 180° is the segment AB, the level line for 0° consists of the straight line passing through A and B, except the segment AB. Any other level line consists of two circular arcs both with endpoints A and B, and symmetrical to each other with respect to the straight line passing through A and B. (2) If two level lines are different, one lies inside the other; $\angle AXB$ takes a higher value on the inner, a lower value on the outer, level line. With suitable interpretation, this applies also to 0°.

14. The minimum is attained at the point where the line l crosses the line through A and B, that is, a level line. This does not contradict the principle laid down in ex. 11; on both sides of this particular level line $\angle AXB$ takes higher values than on it.

15. Notation as in sect. 6. Keep c constant for a moment. Then, since $V = abc/3$ is given, also ab is constant, and

$$S = 2ab + 2(a + b)c$$

is a minimum, when $2(a + b)$, the perimeter of a rectangle with given area ab, is a minimum. This happens, when $a = b$. Now, change your standpoint and keep another edge constant.

16. Keep one of the sides constant. Then you have the case of ex. 8, and the two other sides must be equal (the triangle is isosceles). Any two sides are accessible to this argument, and the triangle should be equilateral.

17. Keep the plane of the base and the opposite vertex fixed, but let vary the base which is a triangle inscribed in a given circle. The altitude is constant; the area of the base (and with it the volume) becomes a maximum when the base is equilateral, by sect. 4 (2). We can choose any face as base, and so each face must be equilateral and, therefore, the tetrahedron regular when the maximum of the volume is attained.

18. Regard the triangle between a and b as the base. Without changing the corresponding altitude, change the base into a right triangle; this change increases the area of the base (ex. 7) and, therefore, the volume. You could treat now another pair of sides similarly, yet it is better to make c perpendicular both to a and to b right away.

19. Fixing the endpoint on the cylinder, you find, by ex. 2 (7), that the shortest distance is perpendicular to the sphere. Fixing the endpoint on the sphere, you may convince yourself that the shortest distance is also perpendicular to the cylinder. Therefore, it should be perpendicular to both. This can be shown also directly.

20. The procedure of ex. 19 shows that the shortest distance should be perpendicular to both cylinders. In fact, it falls in the same line as the shortest distance between the axes of the cylinders; see sect. 4 (1).

21. The procedure of ex. 19 and an analogue of ex. 10 in space.

22. By hypothesis

$$f(X,Y,Z,\dots) \leqq f(A,B,C,\dots)$$

for all admissible values of $X,Y,Z,\dots$. Therefore, in particular

$$f(X,B,C,\dots) \leqq f(A,B,C,\dots)$$

$$f(X,Y,C,\dots) \leqq f(A,B,C,\dots)$$

and so on; $X,Y,Z,\dots$ may be variable numbers, lengths, angles, points, ...

24. Either $x_1 = y_1 = z_1$ (exceptional case) or, for $n \geqq 1$, of the three values x_n, y_n, z_n just two are different. Call d_n the absolute value of the difference; for example

$$d_1 = |x_1 - z_1|, \qquad d_2 = |y_2 - x_2|.$$

By definition

$$\pm d_2 = x_2 - y_2 = \frac{z_1 + x_1}{2} - y_1 = \frac{z_1 + x_1}{2} - z_1$$

$$= \frac{x_1 - z_1}{2} = \pm\frac{d_1}{2}$$

or $d_2 = d_1/2$. In the same way

$$d_n = d_{n-1}/2 = d_{n-2}/4 = \dots = d_1/2^{n-1}$$

and so

$$\left|x_n - \frac{l}{3}\right| = \left|x_n - \frac{x_n + y_n + z_n}{3}\right| = \frac{|x_n - y_n + x_n - z_n|}{3}$$

$$\leqq 2d_n/3 \to 0.$$

25. We consider n positive numbers $x_1, x_2, \ldots x_n$ with given arithmetic mean A,

$$x_1 + x_2 + \ldots + x_n = nA.$$

If $x_1, x_2, \ldots x_n$ are not all equal, one of them, say x_1, is the smallest and another, say x_2, is the largest. (The choice of the subscripts is a harmless simplification, just a matter of convenient notation. We did *not* make the unwarranted assumption that *only* x_1 takes the smallest value.) Then

$$x_1 < A < x_2.$$

Let us put now

$$x_1' = A, \qquad x_2' = x_1 + x_2 - A, \qquad x_3' = x_3, \ldots x_n' = x_n.$$

Then

$$x_1 + x_2 + \ldots + x_n = x_1' + x_2' + \ldots + x_n'$$

and

$$x_1'x_2' - x_1x_2 = Ax_1 + Ax_2 - A^2 - x_1x_2 = (A - x_1)(x_2 - A) > 0.$$

Therefore

$$x_1x_2x_3 \ldots x_n < x_1'x_2'x_3' \ldots x_n'.$$

If $x_1', x_2', \ldots x_n'$ are not all equal, we repeat the process obtaining another set of n numbers $x_1'', x_2'', \ldots x_n''$ such that

$$x_1' + x_2' + \ldots + x_n' = x_1'' + x_2'' + \ldots + x_n''$$

$$x_1'x_2'x_3' \ldots x_n' < x_1''x_2''x_3'' \ldots x_n''.$$

The set $x_1', x_2', \ldots x_n'$ has at least one term equal to A, the set $x_1'', x_2'', \ldots x_n''$ has at least two terms equal to A. At the latest, the set $x_1^{(n-1)}, x_2^{(n-1)}, \ldots x_n^{(n-1)}$ will contain $n - 1$ and, therefore, n terms equal to A, and so

$$x_1x_2 \ldots x_n < x_1^{(n-1)}x_2^{(n-1)} \ldots x_n^{(n-1)}$$

$$= A^n = \left(\frac{x_1 + x_2 + \ldots + x_n}{n}\right)^n.$$

26. Connecting the common initial point of the perpendiculars x,y,z with the three vertices of the triangle, we subdivide the latter into three smaller triangles. Expressing that the sum of the areas of these three parts is equal to the area of the whole, we obtain $x + y + z = l$. The equation $x =$ const. is represented by a line parallel to the base of the equilateral triangle, the equation $y = z$ by the altitude. The first segment of the broken line in fig. 8.9 is parallel to the base and ends on the altitude. The first step of ex. 25 is represented by a segment parallel to the base and ends on the line with equation $y = l/3$, which is parallel to another side and passes through the center. The second step is represented by a segment along the line $y = l/3$ and ends at the center.

27. For the solution modelled after ex. 25 see Rademacher-Toeplitz, pp. 11–14, 114–117.

28. Partial variation.

29. At the point where the extremum is attained, the equations

$$\frac{\partial f}{\partial x} + \lambda \frac{\partial g}{\partial x} = 0, \qquad \frac{\partial f}{\partial y} + \lambda \frac{\partial g}{\partial y} = 0$$

hold with a suitable value of λ. This condition is derived under the assumption that $\partial g/\partial x$ and $\partial g/\partial y$ do not both vanish. Under the further assumption that $\partial f/\partial x$ and $\partial f/\partial y$ do not both vanish, the equations with λ express that the curve $g = 0$ (the prescribed path) is tangent to the curve $f =$ const. (a level line) that passes through the point of extremum, at this point.

30. At a peak, or at a pass, $\partial f/\partial x = \partial f/\partial y = 0$. At an angular point of the prescribed path, $\partial g/\partial x$ and $\partial g/\partial y$ (if they exist) are both $= 0$. An extremum at the initial (or final) point of the prescribed path does not fall at all under the analytic condition quoted in ex. 29 which is concerned with an extremum relative to *all* points (x,y) in a certain neighborhood, satisfying $g(x,y) = 0$.

31. The condition is

$$\frac{\partial f}{\partial x} + \lambda \frac{\partial g}{\partial x} = \frac{\partial f}{\partial y} + \lambda \frac{\partial g}{\partial y} = \frac{\partial f}{\partial z} + \lambda \frac{\partial g}{\partial z} = 0.$$

It assumes that the three partial derivatives of g are not all 0. Under the further assumption that the three partial derivatives of f are not all 0, the three equations express that the surface $g = 0$ and the surface $f =$ const. passing through the point of extremum are tangent to each other at that point.

32. The condition consists of three equations of which the first, relative to the x-axis, is

$$\frac{\partial f}{\partial x} + \lambda \frac{\partial g}{\partial x} + \mu \frac{\partial h}{\partial x} = 0.$$

It assumes that three determinants of which the first is

$$\frac{\partial g}{\partial y}\frac{\partial h}{\partial z} - \frac{\partial g}{\partial z}\frac{\partial h}{\partial y}$$

do not all vanish. Under the further assumption that the three partial derivatives of f are not all 0, the three equations express that the curve which is the intersection of the two surfaces $g = 0$ and $h = 0$, is tangent to the surface $f =$ const., passing through the point of extremum, at that point.

34. The desired conclusion is: the cube alone attains the minimum. Therefore, when the inequality becomes an equality, the cube should appear; that is, we should have $x = y$, or (looking at the areas) $x^2 = xy$. Yet, the inequality used (without success) becomes an equality when $2x^2 = 4xy$: we could have predicted hence that it will fail.

With, or without, peeking at sect. 6, we divide S into three pairs of opposite faces

$$S = 2x^2 + 2xy + 2xy$$

and apply the theorem of the means:

$$(S/3)^3 \geqq 2x^2 \cdot 2xy \cdot 2xy = 8x^4y^2 = 8V^2.$$

The equality is attained if, and only if, $2x^2 = 2xy$, or $x = y$; that is, only for the cube.

Draw the moral: foreseeing the case of equality may guide your choice, may yield a cue.

35. Let V, S, x, and y stand for volume, surface, radius, and altitude of the cylinder, respectively, so that

$$V = \pi x^2 y, \qquad S = 2\pi x^2 + 2\pi xy.$$

The desired conclusion, $y = 2x$, guides our choice: with

$$S = 2\pi x^2 + \pi xy + \pi xy$$

the theorem of the means yields

$$(S/3)^3 \geqq 2\pi x^2 \cdot \pi xy \cdot \pi xy = 2\pi^3 x^4 y^2 = 2\pi V^2,$$

with equality just for $y = 2x$.

36. Ex. 34 is a particular case, ex. 35 a limiting case. Let V, S, y, and x stand for volume, surface, altitude of the prism, and for the length of a certain side of its base, respectively. Let a and l denote the area and the perimeter, respectively, of a polygon similar to the base in which the side corresponding to the side of length x of the base is of length 1. Then

$$V = ax^2 y, \qquad S = 2ax^2 + lxy.$$

In ex. 34 and 35, the maximum of S is attained, when the area of the base (now ax^2) is $S/6$. Expecting that this holds also in the present general case, we have a cue; we set

$$S = 2ax^2 + lxy/2 + lxy/2$$

and obtain, using the theorem of the means,

$$(S/3)^3 \geqq 2ax^2 \cdot (lxy)^2/4 = [l^2/(2a)]\,V^2$$

with equality if $ax^2 = lxy/4 = S/6$.

37. Let V, S, x, and y stand for the volume of the double pyramid, its surface, a side of its base, and the altitude of one of the constituent pyramids, respectively. Then

$$V = 2x^2y/3, \qquad S = 8x[(x/2)^2 + y^2]^{1/2}/2.$$

In the case of the regular octahedron, the double altitude of a constituent pyramid equals the diameter of the base, or

$$2y = 2^{1/2}x, \quad \text{or} \quad 2y^2 = x^2.$$

Having obtained this cue, we set

$$S^2 = 4x^2(x^2 + 2y^2 + 2y^2),$$
$$(S^2/3)^3 \geqq 4^3x^6x^22y^22y^2 = 4^4x^8y^4 = (6V)^4.$$

Equality occurs only if $x^2 = 2y^2$. Note that in this case $S = 3^{1/2}2x^2$.

38. Let V, S, x, and y denote the volume of the double cone, its surface, the radius of its base, and the altitude of one of the constituent cones, respectively. Then

$$V = 2\pi x^2y/3, \qquad S = 2 \cdot 2\pi x(x^2 + y^2)^{1/2}/2.$$

Consider the right triangle with legs x, y and hypotenuse $(x^2 + y^2)^{1/2}$. If the projection of x on the hypotenuse is $1/3$ of the latter (as we hope that it will be in the case of the minimum),

$$x^2 = (x^2 + y^2)/3$$

or $2x^2 = y^2$. Having obtained this cue, we set

$$S^2 = 2\pi^2x^2(2x^2 + y^2 + y^2),$$
$$(S^2/3)^3 \geqq 8\pi^6x^6 \cdot 2x^2y^2y^2 = \pi^2(3V)^4.$$

Equality occurs only if $2x^2 = y^2$. Note that in this case $S = 3^{1/2}2 \cdot \pi x^2$.

39. Let V, S, and y denote the volume of the double pyramid, its surface, and the altitude of one of the constituent pyramids, respectively. Let x, a, and l be connected with the base of the double pyramid in the same way as in the solution of ex. 36 with the base of the prism. Let p stand for the radius of the circle inscribed in the base. Then

$$V = 2ax^2y/3,$$
$$ax^2 = lxp/2,$$
$$S = 2lx(p^2 + y^2)^{1/2}/2 = (4a^2x^4 + l^2x^2y^2)^{1/2}.$$

In ex. 37 and 38, S is a minimum when $S = 3^{1/2}2ax^2$ which yields

$$l^2x^2y^2 = 8a^2x^4.$$

Noticing this cue, we set

$$S^2 = 4a^2x^4 + l^2x^2y^2/2 + l^2x^2y^2/2,$$

$$(S^2/3)^3 \geqq 4a^2x^4 \cdot (l^2x^2y^2)^2/4 = (l^4/a^2)(3V/2)^4.$$

Equality occurs if, and only if, the base $ax^2 = S/(3^{1/2}2)$.

40. There is a plausible conjecture: the equilateral triangle has the minimum perimeter for a given area, or the maximum area for a given perimeter. Let a, b, c, A, and $L = 2p$ stand for the three sides of the triangle, its area, and the length of its perimeter, respectively. By Heron's formula,

$$A^2 = p(p - a)(p - b)(p - c).$$

The use of the theorem of the means is strongly suggested: A should not be too large, when p is given; the right hand side is a product. How should we apply the theorem? There is a cue: *if* the triangle is equilateral, $a = b = c$, or $p - a = p - b = p - c$. Therefore, we proceed as follows:

$$\begin{aligned} A^2/p &= (p - a)(p - b)(p - c) \\ &\leqq \left(\frac{p - a + p - b + p - c}{3}\right)^3 \\ &= (p/3)^3. \end{aligned}$$

That is, $A^2 \leqq L^4/(2^4 3^3)$, and there is equality only in the case of the equilateral triangle. Cf. ex. 16.

41. There is a plausible conjecture: the square.

Let a and b include the angle ϕ, c and d the angle ψ, and $\phi + \psi = \varepsilon$. We obtain

$$2A = ab \sin \phi + cd \sin \psi.$$

Expressing in two different ways the diagonal of the square that separates ϕ and ψ, we obtain

$$a^2 + b^2 - 2ab \cos \phi = c^2 + d^2 - 2cd \cos \psi.$$

We have now three relations to eliminate ϕ and ψ. Adding

$$(a^2 + b^2 - c^2 - d^2)^2 = 4a^2b^2 \cos^2 \phi + 4c^2d^2 \cos^2 \psi - 8abcd \cos \phi \cos \psi$$

$$16A^2 = 4a^2b^2 \sin^2 \phi + 4c^2d^2 \sin^2 \psi + 8abcd \sin \phi \sin \psi$$

we obtain

$$\begin{aligned} 16A^2 + (a^2 + b^2 - c^2 - d^2)^2 &= 4a^2b^2 + 4c^2d^2 - 8abcd \cos \varepsilon \\ &= 4(ab + cd)^2 - 16abcd (\cos \varepsilon/2)^2. \end{aligned}$$

Finally, noticing differences of squares and setting

$$a + b + c + d = 2p = L,$$

we find

$$A^2 = (p - a)(p - b)(p - c)(p - d) - abcd\,(\cos \varepsilon/2)^2.$$

In the probable case of equality (the square) the sides are equal, and so are the quantities $p - a$, $p - b$, $p - c$, $p - d$. With this cue, we obtain

$$\begin{aligned} A^2 &\leqq (p - a)(p - b)(p - c)(p - d) \\ &\leqq \left(\frac{p - a + p - b + p - c + p - d}{4}\right)^4 \\ &= (p/2)^4 = (L/4)^4. \end{aligned}$$

In order that both inequalities encountered should become equalities, we must have $\varepsilon = 180°$ and $a = b = c = d$.

42. The prism is much more accessible than the two other solids which we shall tackle, after careful preparation, in ex. 46 and 47. Let L denote the perimeter of the base and h the altitude of the prism. Any lateral face is a parallelogram; its base is a side of the base of the solid, and its altitude $\geqq h$. Therefore, the lateral surface of the prism is $\geqq Lh$, and equality is attained if, and only if, all lateral faces are perpendicular to the base and so the prism a right prism.

43. Let x_j, y_j be the coordinates of P_j, for $j = 0, 1, 2, \ldots n$, and put

$$x_j = x_{j-1} + u_j, \qquad y_j = y_{j-1} + v_j$$

for $j = 1, 2, \ldots n$. Then the left-hand side of the desired inequality is the length of the broken line $P_0P_1P_2 \ldots P_n$ and the right-hand side the length of the straight line P_0P_n, which is the shortest distance between its endpoints.

44. In the case $n = 2$, we examine (the notation is slightly changed) the assertion

$$(u^2 + v^2)^{1/2} + (U^2 + V^2)^{1/2} \geqq [(u + U)^2 + (v + V)^2]^{1/2}.$$

We transform it into equivalent forms by squaring and other algebraic manipulations:

$$(u^2 + v^2)^{1/2}(U^2 + V^2)^{1/2} \geqq uU + vV,$$

$$u^2V^2 + v^2U^2 \geqq 2uvUV,$$

$$(uV - vU)^2 \geqq 0.$$

In its last form, the assertion is obviously true. Equality is attained if, and only if,

$$u : v = U : V.$$

We handle the case $n = 3$, by applying repeatedly the case $n = 2$:

$$\begin{aligned}&(u_1^2 + v_1^2)^{1/2} + (u_2^2 + v_2^2)^{1/2} + (u_3^2 + v_3^2)^{1/2}\\ &\geqq [(u_1 + u_2)^2 + (v_1 + v_2)^2]^{1/2} + (u_3^2 + v_3^2)^{1/2}\\ &\geqq [(u_1 + u_2 + u_3)^2 + (v_1 + v_2 + v_3)^2]^{1/2}.\end{aligned}$$

And so on, for $n = 4, 5, \ldots$. In fact, we use mathematical induction.

45. Let h be the altitude and let the base be divided by the foot of the altitude into two segments, of lengths p and q, respectively. We have to prove that

$$\begin{aligned}(p^2 + h^2)^{1/2} + (q^2 + h^2)^{1/2} &\geqq 2\left[\left(\frac{p+q}{2}\right)^2 + h^2\right]^{1/2}\\ &= [(p+q)^2 + (h+h)^2]^{1/2}\end{aligned}$$

which is a case of ex. 43. For equality, we must have

$$p : h = q : h,$$

or $p = q$, that is, an isosceles triangle.

46. Let h be the altitude of P, $a_1, a_2, \ldots a_n$ the sides of the base of P, and $p_1, p_2, \ldots p_n$ the perpendiculars from the foot of the altitude on the respective sides. Let Σ denote a summation with j ranging from $j = 1$ to $j = n$. Then

$$A = \Sigma a_j p_j / 2$$

$$S = A + \Sigma a_j (p_j^2 + h^2)^{1/2}/2.$$

These expressions become simpler for the right pyramid P_0, since all perpendiculars from the foot of the altitude on the sides have a common value p_0. Therefore,

$$\begin{aligned}A_0 &= L_0 p_0/2\\ S_0 &= A_0 + L_0(p_0^2 + h^2)^{1/2}/2\\ &= A_0 + (4A_0^2 + h^2L_0^2)^{1/2}/2;\end{aligned}$$

P and P_0 have the same altitude $3V/A = 3V_0/A_0 = h$. Using ex. 43 and our assumptions, we obtain

$$\begin{aligned}2(S - A) &= \Sigma[(a_j p_j)^2 + (a_j h)^2]^{1/2}\\ &\geqq [(\Sigma a_j p_j)^2 + (\Sigma a_j h)^2]^{1/2}\\ &= [4A^2 + h^2L^2]^{1/2}\\ &\geqq [4A^2 + h^2L_0^2]^{1/2}\\ &= 2(S_0 - A).\end{aligned}$$

Therefore, $S \geqq S_0$. For equality, both inequalities encountered must become equalities and so two conditions must be satisfied. First,

$$p_1 : h = p_2 : h = \ldots = p_n : h,$$

that is, P is a right pyramid. Second, $L = L_0$.

47. We take two steps: (1) We change the base of D into that of D_0, and both pyramids, of which D consists, into right pyramids, leaving, however, their altitudes unchanged. We obtain so a double pyramid D' which is not necessarily a right double pyramid. (Its two constituent pyramids are right pyramids, but perhaps of different altitudes.) (2) We change D' into D_0. Step (1) can only diminish the surface, by ex. 46. The altitudes of the two constituent pyramids of D', of lengths h_1 and h_2, fall in the same straight line. Let p_0 denote the radius of the circle inscribed in the base of D_0. Then the surface of D' is

$$S' = [(p_0^2 + h_1^2)^{1/2} + (p_0^2 + h_2^2)^{1/2}]L_0/2$$

$$\geqq 2\left[p_0^2 + \left(\frac{h_1 + h_2}{2}\right)^2\right]^{1/2} \frac{L_0}{2} = S_0$$

by ex. 45.

48. Leaving the volume V unchanged all the time, we take three steps: (1) Leaving the base unchanged, in shape and size, we transform the given prism into a right prism. (2) Leaving its area A unchanged, we transform the base into a square. (3) We transform the right prism with square base into a cube. Steps (1) and (3) can only diminish the surface S, by exs. 42 and 34, respectively. Step (2) leaves the altitude $h = V/A$ unchanged, and can only diminish L, the perimeter of the base, by ex. 41; but $S = 2A + Lh$. Unless the prism is a cube from the outset, at least one of the three steps actually diminishes S. The weaker theorem 34 served as a stepping stone.

49. It follows from ex. 47, 41, and 37 as the foregoing ex. 48 follows from ex. 42, 41, and 34. Yet we can combine with advantage the two steps corresponding to (1) and (2) of ex. 48 into one step, thanks to the sharp formulation of ex. 47.

50. We start from any pyramid with triangular base (any tetrahedron, not necessarily regular). We transform it into a right pyramid, leaving unchanged the volume V and the area of the base A, but changing (if necessary) the base into an equilateral triangle. This diminishes the perimeter of the base L, by ex. 40, and, therefore, the surface S, by ex. 46. The lateral faces of the new pyramid are isosceles triangles. Unless they happen to be equilateral, we take one of them as base, and repeat the process, diminishing again S. By the principle of partial variation (ex. 22), if there is at all a tetrahedron with a given V and minimum S, it must be the regular tetrahedron.

51. See ex. 53.

52. See ex. 53.

53. Let V, S, and y denote the volume, the surface, and the altitude of the pyramid, and let x, a, and l be connected with the base of the pyramid in the same way as in the solution of ex. 36 with the base of the prism. Let p stand for the radius of the circle inscribed in the base. Then

$$\begin{aligned} V &= ax^2y/3 \\ ax^2 &= lxp/2 \\ S &= ax^2 + lx(p^2 + y^2)^{1/2}/2 \\ &= ax^2 + (4a^2x^4 + l^2x^2y^2)^{1/2}/2. \end{aligned}$$

Trying to introduce expressions which depend on the shape, but not on the size, we are led to consider

$$\frac{S}{ax^2} = 1 + \left[1 + \left(\frac{ly}{2ax}\right)^2\right]^{1/2} = 1 + (1 + t)^{1/2}$$

(we introduced an abbreviation, setting $[ly/(2ax)]^2 = t$) and

$$\frac{S^3}{(3V)^2} = \frac{l^2}{4a} \frac{[1 + (1 + t)^{1/2}]^3}{t}.$$

As V is given, and S should be a minimum, the left-hand side should be a minimum. Therefore, the right-hand side should be a minimum. Yet the shape is given and so l^2/a is given. Therefore, all that remains to do is to find the value of t that makes the right hand side a minimum: this value is *independent of the shape*. It fits all special shapes equally, for example, the shapes mentioned in ex. 51 and 52. Yet there is a special shape, for which we know the result: if the base is an equilateral triangle, the best ratio $S : ax^2$, or total surface to base, is 4 : 1, by ex. 50. This remains true for all shapes, since $S/(ax^2)$ depends only on t, and yields

$$1 + (1 + t)^{1/2} = 4, \qquad t = 8.$$

54. The reader should copy the following table, substituting for the number of each problem a suitable figure.

	(1)	(2)	(3)	(4)	(5)	(6)	(7)
(a)	34	35	36	42	u	48	x
(b)	51	52	53	46	50	v	y
(c)	37	38	39	47	w	49	z
(d)	—	—	—	—	40	41	n

The rows: (a) prisms, (b) pyramids, (c) double pyramids, (d) polygons (relevant only for the last three columns).

The columns: (1) right with square base, (2) right with circle as base, (3) right with base of given shape, (4) transition from oblique to right, (5) arbitrary triangular base, (6) arbitrary quadrilateral base, (7) arbitrary polygonal base with a given number n of sides.

Analogy may suggest the theorems which can be expected to fill the gaps marked by the letters u, v, w, x, y, z, and n. Here are some:

(u) The prism that, of all triangular prisms with a given volume, has the minimum surface, has the following properties: its base is an equilateral triangle, the area of its base is 1/6 of the total area of its surface, it is circumscribed about a sphere which touches each face at the center of the face.

(y) The pyramid that, of all pyramids with n-sided polygonal base and a given volume, has the minimum surface, has the following properties: its base is a regular polygon, the area of its base is 1/4 of the total area of its surface, it is circumscribed about a sphere which touches each face at the centroid of the face.

On the basis of the foregoing we can easily prove (u), (v), and (w), but (x), (y), and (z) depend on (n), which we shall discuss later; see sect. 10.7 (1).

55. (1) Using the *method* and the notation of sect. 6, we have

$$V = abc, \qquad S_5 = ab + 2ac + 2bc.$$

$$(S_5/3)^3 \geqq ab \cdot 2ac \cdot 2bc = 4V^2$$

with equality if, and only if,

$$ab = 2ac = 2bc$$

or $a = b = 2c$: the box is one half of a cube.

(2) Using the *result* of sect. 6, we regard the plane of the face not counted in S_5 as a mirror. The box together with its mirror image forms a new box of volume $2V$ the *whole* surface of which is given, $= 2S_5$. By sect. 6, the new (double) box must be a cube when the maximum of the volume is attained.

56. Following ex. 55, regard the plane of the missing face as a mirror. Maximum for the triangular prism which is one half of a cube, halved by a diagonal plane. Apply ex. 48, with an additional remark.

57. [Putnam 1950] Following ex. 55 and 56, regard the planes of both missing faces as mirrors. Maximum for the triangular prism which is one quarter of a cube; the cube is divided into four congruent fragments by two planes of symmetry, one a diagonal plane, the other perpendicular to the first and parallel to two faces.

58. Let A, L, r, and s stand for the area, the perimeter, the radius, and the arc of the sector, respectively. Then

$$A = rs/2, \qquad L = 2r + s$$

$$(L/2)^2 \geqq 2r \cdot s = 4A;$$

we use the theorem of the means. Equality is attained when $s = 2r$ and the angle of the sector equals two radians.

59. Let u, v, and w denote the sides of the triangle, A the area, γ the given angle opposite w. Then

$$2A = uv \sin \gamma.$$

(1)
$$[(u+v)/2]^2 \geqq uv = 2A/\sin \gamma,$$

by the theorem of the means. Equality is attained when $u = v$, that is, the triangle is isosceles.

(2)
$$w^2 = u^2 + v^2 - 2uv \cos \gamma$$
$$= u^2 + v^2 - 4A \cot \gamma$$
$$(u^2 + v^2)/2 \geqq uv = 2A/\sin \gamma.$$

The equality is attained, and so w a minimum, when $u^2 = v^2$ and the triangle is isosceles.

(3) As both $u + v$ and w, also $u + v + w$ is a minimum, when the triangle is isosceles.

60. Use the notation of ex. 59. The given point lies on the side w. Draw from the given point parallels to u and v, terminating on v and u, and call them a and b, respectively; a and b are given (they are, in fact, oblique coordinates). From similar triangles

$$\frac{v-b}{a} = \frac{b}{u-a} \quad \text{or} \quad \frac{a}{u} + \frac{b}{v} = 1$$

$$\frac{1}{4} = \left(\left[\frac{a}{u} + \frac{b}{v}\right]/2\right)^2 \geqq \frac{ab}{uv} = \frac{ab \sin \gamma}{2A}$$

$$A \geqq 2ab \sin \gamma.$$

There is equality if, and only if,

$$\frac{a}{u} = \frac{b}{v} = \frac{1}{2}, \qquad u = 2a, \qquad v = 2b$$

and the given point is the midpoint of w.

61. Use the notation of sect. 6 and the theorem of the means.

(1)
$$V = abc \leqq [(a+b+c)/3]^3 = [E/12]^3$$

(2)
$$S = 2ab + 2ac + 2bc$$
$$\leqq a^2 + b^2 + a^2 + c^2 + b^2 + c^2$$
$$= 2(a+b+c)^2 - 4(ab+ac+bc)$$

that is,

$$3S \leqq 2(E/4)^2.$$

In both cases, equality is attained only for $a = b = c$, that is, for the cube.

62. Use the notation of sect. 6 and the theorem of the means. The length is c, the girth $2(a + b)$, and

$$V = (2a \cdot 2b \cdot c)/4 \leqq [(2a + 2b + c)/3]^3/4 \leqq l^3/108.$$

Equality is attained only for $2a = 2b = c = l/3$.

63. Use the notation in the solution of ex. 35. Then

$$d^2 = (2x)^2 + (y/2)^2 = 2(x^2 + x^2 + y^2/8)$$

and, therefore, by the theorem of the means

$$V^2 = \pi^2 x^4 y^2 = 8\pi^2 x^2 \cdot x^2 \cdot y^2/8 \leqq 8\pi^2(d^2/6)^3$$

with equality only if

$$x^2 = y^2/8 = d^2/6.$$

For the historical background cf. O. Toeplitz, *The Calculus*, p. 82–83.

No solution: **23**, **33**.

SOLUTIONS, CHAPTER IX

1. (1) Imagine two mirrors perpendicular to the plane of the drawing, the one through l and the other through m. A person at P looks at m and sees himself from the side: the light coming from P returns to P after a first reflection in l and a second in m. The light, choosing the shortest path, describes the desired $\triangle PYZ$ with minimum perimeter; the sides of $\triangle PYZ$ include equal angles with l and m at the points Y and Z, respectively. (2) Let P' and P'' be mirror images of P with respect to l and m, respectively. The straight line joining P' and P'' intersects l and m in the required Y and Z, respectively, and its length is that of the desired minimum perimeter. (By the idea of fig. 9.3, applied twice.)

2. (1) Light, after three successive reflections on three circular mirrors, returns to its source from the opposite direction. (2) A closed rubber band connects three rigid rings. Both interpretations suggest that the two sides of the required triangle that meet in a vertex on a given circle include equal angles with the radius.

3. Roundtrip of light or closed rubber band, as in ex. 2; XY and XZ are equally inclined to BC, etc.

4. A polygon with n sides and minimum perimeter inscribed in a given polygon with n sides has the following property: those two sides of the minimum polygon of which the common vertex lies on a certain side s of the given polygon, are equally inclined to s. See, however, ex. 6 and 13.

5. Call A the point of intersection of l and m. Take a point B on m and a point C on l so that $\angle BAC$ (less than 180°) contains the point P in its interior. Then, by reflection,

$$\angle P''AB = \angle BAP, \qquad \angle PAC = \angle CAP'$$

and hence

$$\angle P''AP' = 2\angle BAC.$$

The solution fails when $\angle P''AP' \geqq 180°$ or, which is the same, when the given $\angle BAC \geqq 90°$.

6. The solution cannot apply when the given triangle has an angle $\geqq 90°$, see ex. 5. The solution of ex. 4 is, of course, *a fortiori* liable to exception.

7. Fix for the moment X in the position P on the side BC. Then the solution (2) of ex. 1 applies (since $\angle BAC$ is acute, see ex. 5); the minimum perimeter is $P'P''$. Now the length $P'P''$ depends on P; it remains to find the minimum of $P'P''$. (As $P'P''$ itself was obtained as a minimum, we seek a minimum of the minima or a "minimum minimorum.") Now, by reflection, $P''A = PA = P'A$. Therefore, $\triangle P''AP'$ is isosceles; its angle at A is independent of P (see ex. 5) and so its *shape is independent of* P. Therefore, $P'P''$ becomes a minimum when $P'A = PA$ becomes a minimum, and this is visibly the case when $PA \perp BC$; cf. sect. 8.3. *The vertices of the triangle with minimum perimeter inscribed in a given acute triangle are the feet of the three altitudes of the given triangle.* Comparing this with the solution of ex. 3, we see that *the altitudes of an acute triangle bisect the respective angles of the inscribed triangle of which their feet are the vertices.* The latter result is, of course, quite elementary. The present solution is due to L. Fejér. Cf. Courant-Robbins, p. 346–353.

8. No. If $\triangle ABC$ has an angle $\geqq 120°$, the vertex of that angle is the traffic center. This is strongly suggested by the mechanical solution in sect. 2 (2).

9. [Putnam 1949] This is closely analogous to the simpler plane problem treated in sect. 1 (4), sect. 2 (2) and ex. 8. Which method should we adopt? (1) *Mechanical interpretation*, modelled on fig. 9.7. There are four pulleys, one at each of the four given points A, B, C, and D. Four strings are attached together at the point X; each string passes over one of the pulleys and carries a weight of one pound at its other endpoint. As in sect. 2 (2), a first consideration (of the potential energy) shows that the equilibrium position of this mechanical system corresponds to the proposed minimum problem. A second consideration deals with the forces acting on the point X. There are four such forces; they are equal in magnitude and in the direction of the four taut strings, going to A, B, C, and D, respectively. The resultant of the first two forces must counterbalance the resultant of the last two forces. Therefore, these resultants are in the same line which bisects both $\angle AXB$ and $\angle CXD$. The equality of these angles follows from the congruence of the two parallelograms of forces (both are rhombi). Similarly related pairs are: $\angle AXC$ and $\angle BXD$, $\angle AXD$ and $\angle BXC$. (2) *Partial variation and optical interpretation*, modelled on fig. 9.4. Keep constant (for a moment) $CX + DX$, the sum of two distances. Then the point X has to vary on the surface of a prolate spheroid (ellipsoid of revolution) with foci at C and D. We conceive this surface as a mirror. The light starting from A, reflected at our spheroidal mirror and arriving at B renders $AX + XB$ a minimum; along its path $\angle AXB$ is bisected by the normal to the mirror at the point X. The same normal bisects $\angle CXD$ by sect. 1 (3) or sect. 2 (1). Yet the equality of $\angle AXB$ and $\angle CXD$ is not so easily obtained by this method: although both methods work equally well in the simpler analogous case, they do not equally well apply to the present theorem.

10. Yes. Wherever the point X may be

$$AX + XC \geqq AC, \qquad BX + XD \geqq BD$$

since the straight path is the shortest between two points, and both inequalities become equations if, and only if, X is the point of intersection of the diagonals AC and BD: this is the traffic center. The statement of ex. 9 remains fully correct in view of the fact that the normal to the plane of the quadrilateral is a common bisector of $\angle AXC$ and $\angle BXD$, which are both straight angles.

11. Follows by partial variation from the result of sect. 1 (4). Cf. Courant-Robbins, p. 354–361.

13. Drive the ball parallel to a diagonal of the table. Fig. 9.14 applies ex. 12 four times in succession. Imagine fig. 9.14 drawn on transparent paper and fold it along the reflecting lines; then the several segments of the straight line PP just cover the parallelogrammatic path of the billiard ball. By the way, we see here a case in which ex. 4 has an infinity of solutions.

14. By fig. 9.15 (which should be drawn on transparent paper)

$$n2\alpha < 180^\circ < (n + 1)2\alpha,$$

$$90^\circ/(n + 1) < \alpha < 90^\circ/n.$$

Draw figures illustrating the cases $n = 1, 2, 3$. Consider the case $n = \infty$.

15. The particular case treated in sect. 1 and ex. 12 yields several suggestions; see ex. 16, 17, and 18.

16. If A, B, and $AX + XB$ are given, the locus of X is the surface of a prolate spheroid (ellipsoid of revolution) with foci A and B; such spheroids are the level surfaces. The spheroid to which the given line l is tangent at the point X yields the solution. The normal to this spheroid at the point X is *perpendicular to* l, and *bisects* $\angle AXB$ by a property of the ellipse proved in sect. 1 (3) and sect. 2 (1).

17. Place a sheet of paper folded in two so that the crease coincides with l and one half of the sheet (a half-plane) passes through A and the other half through B. The desired shortest line is certainly described on this folded sheet. If the sheet is unfolded, the shortest line becomes straight. On the folded as on the unfolded sheet the lines *XA and XB include the same angle with l.*

18. Fasten one end of a rubber band of suitable length to A, pass the band over the rigid rod l at X and fasten the other end to B so that the band is stretched: it forms so the shortest line required by ex. 15 (if the friction is negligible). Three forces act at the point X: two tensions equal in magnitude, one directed toward A and the other toward B, and the reaction of the rod which is perpendicular to l (since the friction is negligible). The parallelogram of forces is a rhombus and so a normal to l bisects $\angle AXB$, as found in ex. 16. The reaction of the rod has no component parallel to the rod, and so the components of the tensions parallel to l must be equal in amount (and opposite in direction). Therefore, XA and XB are equally inclined to l, as found in ex. 17. By the way, the equivalence of the results of ex. 16 and 17 can be shown by a little solid geometry. (Trihedral angles are congruent if they have three appropriate data in common.)

19. Closed rubber band around three knitting needles held rigidly. Partial variation and ex. 16, or ex. 18; the bisectors of the three angles of a triangle meet in the center of the inscribed circle. Ex. 3 is a limiting case.

20. Each vertex of the triangle is the midpoint of an edge of the cube. The triangle is equilateral; its center is the center of the cube; its perimeter is $3\sqrt{6}a$.

21. By partial variation, sect. 8.3, and sect. 1 (4) or 2 (2), TX, TY, and TZ are perpendicular to a, b, and c, respectively, and equally inclined to each other (120°). We could call T the "traffic center of three skew lines." The problem of sect. 1 (4) is an extreme case: a, b, and c become parallel. There is an obvious generalization and there are some obvious analogous problems: the traffic center of three spheres, the traffic center of a point, a straight line and a plane, and so on.

22. The traffic center of three skew edges of a cube is, of course, the center of the cube. Represent clearly the rotation of the cube through 120° that interchanges the three given skew edges, and the situation of the triangle found in ex. 20.

23. In order to find the shortest line between two given points A and B on the surface of a polyhedron, imagine the polyhedral surface made of cardboard, of plane polygons hinged together and folded up suitably. Unfold the polyhedral surface in one plane (lay the cardboard flat on the desk): the shortest line required becomes the *straight line from A to B.* Before unfolding it, however, we have to cut the polyhedral surface along suitable edges which the shortest line *does not cross*. As we do not know in advance which faces and which edges the shortest line will cross, we have to examine all admissible combinations. We turn now to the proposed problem, list the essential sequences of faces, and note after each sequence the square of the rectilinear distance from the spider to the fly along that sequence.

(1) End wall, ceiling, end wall:

$$(1 + 20 + 7)^2 = 784;$$

(2) End wall, ceiling, side wall, end wall:

$$(1 + 20 + 4)^2 + (4 + 7)^2 = 746;$$

(3) End wall, ceiling, side wall, floor, end wall:

$$(1 + 20 + 1)^2 + (4 + 8 + 4)^2 = 740.$$

24. An arc of a great circle is a geodesic on the sphere. A great circle is a plane curve; the plane in which the great circle lies is its osculating plane at all its points. This plane passes through the center of the sphere, and therefore it contains all the normals to the sphere (all the radii) that pass through points of the great circle. A small circle is not a geodesic; in fact, the plane of the small circle contains none of the normals to the sphere that pass through the points of the small circle.

25. By the conservation of energy, the magnitude of the velocity of the point is constant although, of course, the direction of the velocity varies. The difference of the velocity vectors at the two endpoints of a short arc of the trajectory is due to the normal reactions of the surface and is, therefore, almost normal to the surface. This is the characteristic property of a geodesic; see ex. 24 (2). Another version of the same argument: reinterpret the tensions along the rubber band of ex. 24 (2) as velocities along the trajectory; all vectors are of the same magnitude and the variation in the direction is due to normal reactions in both cases.

26. Push the n free edges gently against a plane (your desk), forming a pyramid with n isosceles lateral faces the base of which is the desired polygon. In fact, the base is inscribed in a circle the center of which is the foot of the altitude of the pyramid. The radius of the circle is the third side of a right triangle of which the hypotenuse is the radius of the great circle described on the cardboard and the second side the altitude of the pyramid.

27. If the center of gravity is as close to the floor as possible, there is equilibrium. As little mechanics as this is enough to suggest the desired solution: take a point D on the surface of P such that the distance CD is a minimum. An easy discussion shows that D can neither be a vertex of P nor lie on an edge of P, and that CD is perpendicular to the face F of P on which D lies. See Pólya-Szegö, *Analysis*, vol. 2, p. 162, problem 1.

28. (a) Imagine the globe completely dried up, so that all peaks, passes, and deeps are exposed. Now cover just one deep with some water. The remaining part of the globe has P peaks, S passes, and $D - 1$ deeps, and can be regarded as an island. By the result proven in the text,

$$P + (D - 1) = S + 1.$$

(b) The level lines and the lines of steepest descent subdivide the globe into F "countries"; this is the terminology of ex. 3.2. Take so many lines that each remarkable point, peak, deep, or pass becomes a vertex, as on figs. 9.16 and 9.17, and that no country has more than one remarkable point on its boundary.

We "distribute" each edge, or boundary line, equally between the two countries that it separates, giving 1/2 of the edge to each country. Similarly, we distribute each vertex equally among the countries of which it is a vertex. In return, each country will contribute to the left-hand side of Euler's equation

$$V - E + F = 2;$$

it will contribute one unit to F and a suitable fraction to V and to $-E$. Let us compute this contribution for the various kinds of countries.

I. If there is no remarkable point on its boundary, the country is a quadrilateral, contained between two level lines and two lines of steepest descent. Its contribution to $V - E + F$ is

$$4 \times \frac{1}{4} - 4 \times \frac{1}{2} + 1 = 0.$$

II. If there is a peak or a deep on its boundary, the country is a triangle; see fig. 9.16. If the peak, or the deep, is a common vertex to n countries, the contribution of each country to $V - E + F$ is

$$\left(2 \times \frac{1}{4} + \frac{1}{n}\right) - 3 \times \frac{1}{2} + 1 = \frac{1}{n}$$

and the joint contribution of all n countries is $n \cdot 1/n = 1$.

III. If there is a pass on its boundary, the country is a quadrilateral; see fig. 9.17. Its contribution to $V - E + F$ is

$$\left(3 \times \frac{1}{4} + \frac{1}{8}\right) - 4 \times \frac{1}{2} + 1 = -\frac{1}{8}$$

and the joint contribution of the 8 countries of which the pass is the common vertex is $8 \cdot (-1/8) = -1$.

The grand total of all contributions is, by Euler's theorem,

$$P + D - S = 2.$$

(c) The proof using the idea of the "Deluge" is not properly an example of "physical mathematics": it uses ideas in touch with everyday experience, but not with any specific physical theory. The hint in part (b) of the question was misleading: it appeared to suggest that P, D, and S are somehow analogous to F, V, and E, which is by no means the case. Still, it was a useful hint: it directed our attention to Euler's theorem. This is, however, quite natural: the ideas that guide us in the solution of problems are quite often mistaken but still useful.

29. (a) Let t_1 be the time of descent of the stone and t_2 the time of ascent of the sound. Then

$$t = t_1 + t_2, \qquad d = gt_1^2/2, \qquad d = ct_2.$$

Eliminating t_1 and t_2 and solving a quadratic equation, we find

$$d^{1/2} = -c(2g)^{-1/2} \pm [c^2(2g)^{-1} + ct]^{1/2}.$$

Since $t = 0$ should give $d = 0$, we have to choose the sign $+$. Then

$$d = c^2g^{-1} + ct - c^2g^{-1}[1 + 2gc^{-1}t]^{1/2}.$$

(b)
$$d = gt^2/2 - g^2t^3/(2c) + \dots .$$

Neglecting the terms not written here, we can use the two terms retained as a suitable approximate formula.

(c) It is typical that we can foresee the principal term of the expansion and even the sign of the correction on the basis of physical considerations. Also the mathematical procedure used to obtain a suitable approximate formula is typical: we *expanded* (the expression for d) *in powers of a small quantity* (the time t). Cf. sect. 5.2.

30. The elliptic mirror becomes a parabolic mirror which collects all rays of light that fall on it parallel to its axis into its focus. Such a parabolic mirror is the most essential part of a reflecting telescope.

31. The equation is separable. We obtain

$$dx = \left(\frac{y}{c - y}\right)^{1/2} dy$$

by obvious transformations. We set

$$\left(\frac{y}{c - y}\right)^{1/2} = \tan\varphi,$$

introducing the auxiliary variable φ, and obtain

$$y = c\sin^2\varphi, \qquad x = c(\varphi - (1/2)\sin 2\varphi).$$

We find x by integration and have to choose the constant of integration so that the curve passes through the origin: $\varphi = 0$ implies $x = y = 0$. Setting $2\varphi = t$, $c = 2a$, we obtain the usual equations of the cycloid:

$$x = a(t - \sin t), \qquad y = a(1 - \cos t).$$

The cycloid passes through the point A, which is the origin. There is just one value of a that makes the first branch of the cycloid (corresponding to $0 < t < 2\pi$) pass through the given point B. In order to see this, let a vary from 0 to ∞; this "inflates" the cycloid which, sweeping over a quadrant of the plane, hits B when inflated to the right size.

33. Let a be the radius of the sphere (as in sect. 5), h the height of the segment, V its volume, and C the volume of the cone with the same base as the segment and the same height h. The origin (the point O in fig. 9.13) is the common vertex of the segment and the cone. From elementary geometry and the equation of the circle given in sect. 5

$$C = \frac{\pi(2ah - h^2)h}{3}.$$

Use fig. 9.13 but consider now only the cross-sections at the distance x from 0 with $0 < x < h$. Passing from the equilibrium of the cross-sections expressed by equation (A) to the equilibrium of the solids (segment, cone—but not that with volume C—and cylinder) we find

$$\text{(B)} \qquad 2a(V + \pi h^2 \cdot h/3) = (h/2)\pi(2a)^2 h.$$

Hence

$$V = \frac{\pi h^2(3a - h)}{3} = \frac{a + (2a - h)}{2a - h} C;$$

$2a - h$ is the height of the complementary segment.

34. Write the equation of the circle considered in sect. 5 in the form

$$\text{(A)} \qquad 2a\pi x^2 = x\pi y^2 + x\pi x^2.$$

Only πx^2, the cross-section of a cone, hangs now from the point H of fig. 9.13; the cross-section πy^2 of the sphere and the cross-section πx^2 of another cone (congruent with the first) remain in their original position (with abscissa x). Consider $0 < x < a$, pass to the equilibrium of the three solids, introduce $\bar{x}$, the abscissa of the center of gravity of the hemisphere, and remember the position of the center of gravity of a cone (its distance from the vertex is 3/4 of the altitude):

$$\text{(B)} \qquad 2a \cdot \pi a^2 \cdot a/3 = \bar{x} \cdot 2\pi a^3/3 + (3a/4)\pi a^2 \cdot a/3,$$

$$\bar{x} = 5a/8.$$

35. Keep the notation of ex. 33, but change that of ex. 34 in one respect: $\bar{x}$ denotes now the abscissa of the center of gravity of a segment with height h. Considering $0 < x < h$, pass from (A) of ex. 34 to

$$\text{(B)} \qquad 2a \cdot \pi h^2 \cdot h/3 = \bar{x}V + (3h/4)\pi h^2 \cdot h/3$$

which yields in view of the value of V found in ex. 33

$$\frac{\bar{x}}{h - \bar{x}} = \frac{h + 4(2a - h)}{h + 2(2a - h)}.$$

36. Let h denote the height and V the volume of the segment. Write the usual equation of the parabola in the form

$$\text{(A)} \qquad 2p \cdot \pi y^2 = x \cdot \pi(2p)^2.$$

Notice the cross-section πy^2 of the paraboloid and the cross section $\pi(2p)^2$ of a cylinder. Considering $0 < x < h$ and passing from the equilibrium of the cross-sections to that of the solids, we find

$$\text{(B)} \qquad 2p \cdot V = (h/2)\pi(2p)^2 h,$$

$$V = \pi p h^2 = (3/2)\pi 2ph(h/3).$$

Notice that, by the equation of the parabola, $\pi 2ph$ is the base of the segment.

37. Keep the notation of ex. 36 and let $\bar{x}$ denote the abscissa of the center of gravity of the segment. Now write the equation of the parabola in the form

$$\text{(A)} \qquad x \cdot \pi y^2 = 2p \cdot \pi x^2.$$

Notice πx^2, the cross-section of a cone. Considering $0 < x < h$ and passing from the cross-sections to the solids, we find

$$\bar{x}V = 2p \cdot \pi h^2(h/3)$$

and hence, by ex. 36,

$$\bar{x} = 2h/3.$$

38. $n = 0$: volume of prism, area of parallelogram; $n = 1$: area of triangle, center of gravity of parallelogram or prism; $n = 2$: volume of cone or pyramid, center of gravity of triangle. $n = 3$: center of gravity of cone or pyramid.

Observe that the method of Archimedes as presented here in sect. 5 and ex. 33–38 would be suitable for a class of Analytic Geometry and could lend a new interest to this subject, which may so easily become dry and boring in the usual presentation. The propositions of the "Method" that we have not discussed can be similarly treated and could be similarly used.

No solution: **12, 32.**

SOLUTIONS, CHAPTER X

1. No. The gap is not too bad: the existence of the maximum can be established with the help of the general theorem quoted in ch. VIII, footnote 3.

2. The explicit formula given in the solution of ex. 8.41 shows that $A^2 \leqq (p-a)(p-b)(p-c)(p-d)$, and equality is attained if, and only if, $\varepsilon = 180°$ in which case the quadrilateral is inscribed in a circle.

3. Let A, B, and C be consecutive vertices of the regular polygon with n sides, and M the midpoint of the side BC. Replace $\triangle ABM$ by the isosceles triangle $\triangle AB'M$ (in which $AB' = B'M$) that has the same base AM and the same perimeter and, therefore, a larger area; see ex. 8.8.

4. If we express both areas in terms of r, the radius of the circle, and n, the number of the sides of the polygon, it remains to prove the inequality:

$$\frac{\pi^2 r^2}{n \tan (\pi/n)} < \pi r^2.$$

It is more elegant to observe that a regular polygon is circumscribable about a circle: the desired result is a particular case of ex. 5.

5. A polygon with area A and perimeter L is circumscribed about a circle with radius r. Then, obviously, $\pi r^2 < A$. Lines drawn from the center of the circle to the vertices of the polygon divide it into triangles with the common altitude r; hence $A = Lr/2$. Combining both results obtained, we find

$$A = \frac{L^2 r^2}{4A} < \frac{L^2}{4\pi}.$$

Now, $L^2/(4\pi)$ is precisely the area of the circle that has the perimeter L.

6. Let A denote the area and L the perimeter of a given curve, and r the radius of the circle with the same perimeter so that $L = 2\pi r$. Let A_n denote the area and L_n the perimeter of a polygon P_n that tends to the given curve as $n \to \infty$. Then A_n tends to A and L_n to L. Consider the polygon P'_n that is similar to P_n and has the perimeter L; the area of P'_n is $A_n(L/L_n)^2$. Since P'_n has the same perimeter as the circle with radius r, we conclude from sect. 7 (4) that

$$A_n(L/L_n)^2 < \pi r^2.$$

Passing to the limit, we find that

$$\lim_{n \to \infty} A_n(L/L_n)^2 = A \leqq \pi r^2.$$

This justifies statement I of sect. 8. Yet the text of sect. 7 (5) is objectionable: we definitely did *not* prove that $A < \pi r^2$, as that text appears to suggest. In fact, the relation expressed by $<$ can go over into that expressed by $\leqq$ as we pass to the limit.

7. Both statements are equivalent to the inequality

$$\frac{216V^2}{S^3} \leqq 1;$$

V denotes the volume and S the area of the surface of the box. In sect. 8.6 we proved this inequality directly.

8. The equivalence of I′, II′, and III′ is shown by the same method as that of I, II, and III in sect. 8. Yet I′ is not equivalent to I. In fact, I′ explicitly denies the possibility left unsettled by I that a curve which is not a circle could have the same perimeter and also the same area as a circle.

The argument of sect. 7 (5) as amplified in ex. 6 proves I, but does not prove I′: it proves $\leqq$ which is enough for I, but not $<$ which would be needed for I′.

9. The solution of the proposed problem would add to I of sect. 8 what is still needed to obtain I′ of ex. 8. So much for the importance; concerning the other points see ex. 10–13.

10. Call C'' the smallest triangle containing C, L'' its perimeter and A'' its area. Then, obviously, $L'' < L$ and $A'' > A$. Take as C' the triangle similar to C'' with perimeter L; the area of C' is $A' = A''(L/L'')^2 > A'' > A$.

11. If C is any curve, but *not convex*, we consider first C'', the least convex curve containing C, and then C', similar to C'', but having a perimeter equal to that of C. The whole argument of ex. 10, down to the final inequality, can be repeated in the more general situation.

12. Take two different points P and Q on the closed curve C. There must be on C a third point R that is not on the straight line through P and Q, since C cannot be wholly contained in a straight line. Consider the circle through P, Q, and R. If this circle does not coincide with C, there is a fourth point S on C which is not on the circle: the problem of ex. 9 is equivalent to that of ex. 13.

13. If C is not convex, ex. 11 yields the desired construction. If C is convex, P, Q, R, and S are, in some order, the vertices of a convex quadrilateral. The region surrounded by C consists of this quadrilateral and of four segments. Each segment is bounded by a side of the quadrilateral and by one of the four arcs into which P, Q, R, and S divide C. Following Steiner's idea (see sect. 5 (2), figs. 10.3 and 10.4) we consider the four segments as rigid (of cardboard) and rigidly attached to the respective sides of the quadrilateral which we consider as articulated (with flexible joints at the four vertices). We adopt the notation of ex. 8.41. Then, by our main condition, $\varepsilon \neq 180°$. Let a slight motion of the articulated quadrilateral change ε into ε'. We choose ε' so close to ε that the four arcs rigidly attached to the sides *still form a not self-intersecting* curve C'. Moreover, we choose ε' so that

$$|\varepsilon' - 180°| < |\varepsilon - 180°|.$$

This implies that the area of C' is larger than that of C, in virtue of the formula for A^2 given in the solution of ex. 8.41. Yet the C', consisting of the same four arcs as C, has the same perimeter.

14. Both inferences have the same logical form. Yet the second inference, that leads to an obviously false result, must be incorrect. Therefore, also the first inference must be incorrect, although it aims at a result that might be true. The second inference is, in fact, an ingenious parody of the first, due to O. Perron.

The difference between the two cases must be some outside circumstance not mentioned in the proposed text. There is no greatest integer. Yet, among all isoperimetric curves there is one with the greatest area. This, however, we did not learn from ex. 10–13.

15. The curve C is not a circle, but it has the same perimeter as a certain circle. The area of C cannot be larger than that of the circle, by ex. 6. I say that the area of C cannot be equal to that of the circle. Otherwise there would be, as we know from ex. 10–13, another curve C' still with the same perimeter as the circle, but with a larger area, which is impossible in virtue of what we proved in ex. 6.

16. Given two points, A and B in fig. 10.13, joined by a straight line and a variable curve which include a region together. We consider the length of the curve and the area of the region. In the text we regarded the including length as given and sought the maximum of the included area. Here we regard the included area as given and seek the minimum of the including length. In both cases, the solution is the same: an arc of a circle. Even the proof is essentially the same. We may use fig. 10.14 here as there. Of course, there are obvious differences; the (unshaded) segment of the circle in fig. 10.14, I is constructed now from a given area, not from a given length, and we use now theorem II′ of ex. 8, not theorem I′.

17. Use ex. 16: identify the points X and Y of fig. 10.11 with the points A and B of fig. 10.13, respectively, and add the invariable $\triangle XYC$ to fig. 10.13. There is maximum when the line of given length is an arc of circle.

18. In fig. 10.11 regard the line CY as a mirror, let X' be the mirror image of X, and apply ex. 17 to $\angle XCX'$ and the two given points X and X' on its two sides. There is maximum when the line of given length is an arc of circle perpendicular to CY at the point Y.

19. Use partial variation. Regard X as fixed: the solution is an arc of circle perpendicular to CY, by ex. 18. Regard Y as fixed: the arc of circle is also perpendicular to CX. Finally, the solution is an arc of circle perpendicular both to CX and to CY, and so its center is at C, as conjectured in sect. 9.

20. There is maximum when the straight line is perpendicular to the bisector of the angle. This would follow from symmetry, if we knew in advance that there is just one solution. The result follows without any such assumption from ex. 8.59 (2).

21. By the idea of fig. 10.14, there is maximum when the strings BC and DA are arcs of the *same* circle of which the sticks AB and CD are chords.

22. A closed line consisting of $2n$ pieces, n sticks alternating with n strings, surrounds a maximum area, when all the sticks are chords and all the strings are arcs of the same circle.

23. When all the strings of ex. 22 are of length 0, we obtain sect. 5 (2) and figs. 10.3 and 10.4.

24. Analogous to ex. 16: the rigid disk, the variable surface with given area and the circle that forms the rim of both correspond to the stick, the string, and the pair of points A and B, respectively. The method of ex. 16 applies. (In fig. 10.14 rotate the circle I about its vertical diameter and do the same to the segment at the base of fig. II, but change its upper part in a more arbitrary manner.) Assuming the isoperimetric theorem in space, we obtain: the included volume is a maximum, when the surface with given area is a portion (a zone with one base) of a sphere.

25. Take the three planes perpendicular to each other and take for granted the isoperimetric theorem in space. Then the trihedral angle becomes an octant and you can use the analogue of fig. 10.12 in space. By successive reflections on the three planes, the surface cutting the octant becomes a *closed* surface; its area and the volume surrounded are eight times the given area and the volume cut off by the original surface, respectively. The closed surface with given area that surrounds the maximum volume is the sphere. Therefore, in our special case of the proposed problem there is maximum when the surface with given area is a portion (1/8) of a sphere with center at the vertex of the trihedral angle.

26. The configuration considered in the solution of ex. 25 is the special case $n = 2$ of the following general situation. There are $n + 1$ planes; n planes pass through the same straight line and divide the space into $2n$ equal wedges (dihedral angles) and the last plane is perpendicular to the n foregoing. These $n + 1$ planes divide the space into $4n$ equal trihedral angles to any one of which the method of repeated reflections, used in ex. 25, applies and yields the same result: *the volume cut off is a maximum when the surface of a given area is a portion of a sphere with center at the vertex of the trihedral angle.*

(There are three more configurations containing trihedral angles to which the method applies and yields the same result. These configurations are connected with the regular solids, the first with the tetrahedron, the second with the cube and the octahedron, and the third with the dodecahedron and the icosahedron. Their study requires more effort or more preliminary knowledge and so we just list them in the following table which starts with the simple configuration described above.

Planes	Parts of space	Angles		
$n + 1$	$4n$	90°	90°	$180°/n$
6	24	90°	60°	60°
9	48	90°	60°	45°
15	120	90°	60°	36°

The "planes" are planes of symmetry, the "parts of space" trihedral angles, and the "angles" are included by the three planes bounding the trihedral angle.)

It is natural to conjecture that the result remains valid for *any* trihedral angle. This conjecture is supported inductively by the cases listed and also by analogy; the similarly obtained similar conjecture about angles in a plane (sect. 9) has been proved (ex. 19).

It is even natural to extend the conjecture to polyhedral angles and there we can find at least a limiting case accessible to verification. We call here "cone" the infinite part of space described by an acute plane angle rotating about one of its sides. We seek the surface with given area that cuts off the maximum volume from the cone. It can be proved that this surface is (1) a surface of revolution, (2) a portion of a sphere, and (3) that the center of this sphere is the vertex of the cone. We cannot go into detail here, but it should be observed that part (2) of the proof results in the same way from ex. 24 as the solution of ex. 17 results from ex. 16. The problem of ex. 25, raised by Steiner, still awaits a complete solution.

27. If a region with area A had two bisectors without any common point, it would be divided by them into three sub-regions, two with area $A/2$ and a third with a non-vanishing area, which is obviously impossible.

28. The straight line is shorter: $1 < (\pi/2)^{1/2}$.

29. See ex. 30.

30. Assume that the endpoints of a given bisector lie on two different sides which meet in the vertex O, but none of the endpoints coincides with O. By suitable reflections (idea of fig. 10.12) we obtain six equal triangles one of which is the original triangle and six equal arcs one of which is the given bisector. The six triangles form a regular hexagon with center O. The six arcs form a closed curve which surrounds one-half of the area of the hexagon and especially the point O in which three axes of symmetry of the curve meet. If the length of the bisector is a minimum, the closed curve must be a circle or a regular hexagon, according as all bisectors are admitted (the present ex. 30) or only straight bisectors are admitted (ex. 29); we have to use theorem II′ of ex. 8 or the theorem conjugate to that of sect. 7 (1), respectively. The solution of ex. 30 is the sixth part of a circle with center in one of the vertices, the solution of ex. 29 is a line parallel to one of the sides; in each case there are three solutions. The given bisector may have some other situation (both endpoints on the same side, or at the same vertex, and so on) but the discussion of these situations corroborates the result obtained.

31. [Cf. Putnam 1946] Let O be the center of the circle. If the straight line segment PP' is bisected by O, we call the points P and P' *opposite* to each other. We call two curves opposite to each other if one of them consists of the points opposite to the points of the other. Let now A and B be the endpoints of bisector which we call shortly AB. Let the points A', B' and the arc $A'B'$ be opposite to A, B, and AB, respectively. Then $A'B'$ is a bisector. Let P be a common point of AB and $A'B'$ (ex. 27) and P' opposite to P. Then also P' is a common point of AB and $A'B'$. Let A, P, P', and B follow each other in this order on AB and let PB' be the shorter (not longer) of the two arcs PA and PB'. (This choice is possible; it is, in fact, just a matter of notation.) Consider the curve consisting of two pieces: the arc $B'P$ (of $A'B'$) and the arc PB (of AB). This curve is (1) shorter (not longer) than AB, and (2) longer (not shorter) than the diameter BB' which is the straight path from B to B'. It follows from (1) and (2) that AB is longer (not shorter) than the diameter BB', and this is the theorem.

32. The minor axis. See ex. 33.

33. *The shortest bisector of any region is either a straight line or an arc of a circle.* See ex. 16. *If the region has a center of symmetry* (as the square, the circle and the ellipse have, but not the equilateral triangle) *the shortest bisector is a straight line.* The proof is almost the same as for the circle (ex. 31).

34. Practically the same as ex. 27.

35. Ex. 16 and "unfolding" as in ex. 9.23.

36. In all five cases, the plane of the shortest bisector passes through the center of the circumscribed sphere.

Tetrahedron: square in a plane parallel to two opposite edges; 3 solutions.

Cube: square parallel to one of the faces; 3 solutions.

Octahedron: hexagon in a plane parallel to one of the faces; 4 solutions.

Dodecahedron: decagon in a plane parallel to one of the faces; 6 solutions.

Icosahedron: decagon in a plane perpendicular to an axis that joins two opposite vertices; 6 solutions.

The proof is greatly facilitated in the last four cases by a general remark; see ex. 38.

37. Let O be the center of the sphere. Define opposite points and curves as in ex. 31. Let b be a bisector. Then also b', the curve opposite to b, is a bisector and b and b' have a common point P (ex. 34). Also P', the point opposite to P, is a common point. The points P and P' divide b into two arcs none of which can be shorter than the shortest line joining P and P' which is one half of a great circle.

38. Four of the five regular solids (all except the tetrahedron) have a center of symmetry. *A closed surface which has a center of symmetry has a bisector which is a geodesic.* The proof is almost the same as for the sphere (ex. 37).

39. (See *Elemente der Mathematik*, vol. 4 (1949), p. 93 and vol. 5 (1950), p. 65, problem 65.) Call d the distance of the rim of the diaphragm from its vertex. The area of the diaphragm is πd^2; this proposition is due to Archimedes. Cf. ex. 11.4.

(1) If the center of S is the vertex of the diaphragm, $d = a$, $\pi d^2 = \pi a^2$.

(2) Let l be the straight line that joins the center C of the sphere S and the center C' of the other sphere of which the diaphragm is a portion. Let A be the intersection of l with S that lies on the same side of C as C', and D and B the intersections of l with the diaphragm and with the plane that passes through the rim of the diaphragm, respectively. If the diaphragm bisects the volume of S, the points A, B, C, and D follow each other *in this order* along l. The point of l nearest to the rim of the diaphragm is B, and D is farther from this rim than C. Therefore, $d > a$, $\pi d^2 > \pi a^2$.

(3) Conjecture: No surface bisecting the volume of the sphere with radius a has an area less than πa^2. The proof may be difficult.

41. (1) Maximum $f = n^2$, attained when

$$x_1 = x_2 = \dots = x_n = 1 \text{ or } -1.$$

Minimum $f = 0$, attained for infinitely many different systems $x_1, \dots x_n$ when $n \geq 3$.

(2) Maximum same as before and unique. Minimum $f = n$, attained when

$$x_1 = n^{1/2}, \qquad x_2 = \dots = x_n = 0$$

and in $n - 1$ similar cases.

42. The conjecture is correct for the regular solids with three-edged vertices, but it is incorrect if a vertex has more than three edges. See M. Goldberg, *Tôhoku Mathematical Journal*, vol. 40 (1935) p. 226–236.

No solution: **40, 43.**

SOLUTIONS, CHAPTER XI

1. (a) yes (b) no: α is unnecessary, the area is $ah/2$.

2. (a) yes (b) no: α and β are unnecessary, the area is mh.

3. $2\pi rh$, *independent* of d. Solution by the method of Archimedes or by integral calculus: from $x^2 + y^2 = r^2$, follows

$$y^2 \left(\frac{dy}{dx}\right)^2 + y^2 = r^2, \qquad \int_d^{d+h} 2\pi y \left[1 + \left(\frac{dy}{dx}\right)^2\right]^{1/2} dx = 2\pi rh.$$

4. Let h be the altitude of the zone of which the area is required. From similar right triangles $h : a = a : 2b$, and so the area required is $2\pi bh = \pi a^2$, *independent* of b. The zone becomes a full sphere when $b = a/2$ and a circle when $b = \infty$. Cf. ex. 10.39.

5. Have you observed the analogy with ex. 1–4? The volume of the perforated sphere can be obtained elementarily, or by the use of analytic geometry and integral calculus as

$$\int_{-h/2}^{h/2} \pi y^2 dx - \pi y_1^2 h = \pi h^3/6;$$

$x^2 + y^2 = r^2$ and y_1 is the ordinate corresponding to $x = h/2$.

6. $\pi h^3/12$, *independent* of a and b. Solution similar to that of ex. 5, connected with that of ex. 7. In the extreme case $a = b = 0$ the segment becomes a full sphere with diameter h. If h is small, the difference between Mh and V is intuitively seen to be very small.

7. $\pi c^2 h/6$, *independent* of r. If $c = h$, the cone degenerates into a cylinder and we have the case of sect. 2 and ex. 5. Solution similar to that of ex. 8.

8. With O as origin and OX as x-axis, the equations of the circle and the parabola in fig. 11.3 are

$$(x - d)^2 + y^2 = r^2, \qquad 2px = y^2,$$

respectively. Let x_1 and x_2 denote the abscissas of the points of intersection of the two curves, $x_1 < x_2$. Then $x_2 - x_1 = h$ and the volume required is

$$\pi \int_{x_1}^{x_2} [r^2 - (x - d)^2 - 2px]\, dx = \pi \int_{x_1}^{x_2} (x_2 - x)\,(x - x_1)\, dx = \pi h^3/6,$$

independent of r and d; substitute $x - x_1 = t$. We used the decomposition in factors of a polynomial of the second degree when the two roots and the coefficient of x^2 are known.

9. (a) yes; the volume is $\pi h^2(a + 2b)/3$. (b) no.

10. Yes. As u_1 and u_2 can be arbitrarily given, there are an infinity of possible systems $u_1, u_2, \ldots u_{10}$ satisfying the recursive relation $u_n = u_{n-1} + u_{n-2}$. We examine two special systems:

$u_1', u_2', u_3', \ldots u_{10}'$ with $u_1' = 0$, $u_2' = 1$

$u_1'', u_2'', u_3'', \ldots u_{10}''$ with $u_1'' = 1$, $u_2'' = 1$.

We find

$$u_7' = 8, \quad u_1' + u_2' + \ldots + u_{10}' = 88,$$

$$u_7'' = 13, \quad u_1'' + u_2'' + \ldots + u_{10}'' = 143.$$

With a little luck, we may observe that

$$(*) \qquad u_1' + u_2' + \dots + u_{10}' = 11u_7', \quad u_1'' + u_2'' + \dots + u_{10}'' = 11u_7''$$

and then guess, and finally prove, that

$$(**) \qquad u_1 + u_2 + \dots + u_{10} = 11u_7.$$

The proof is so: we verify directly that

$$(***) \qquad u_n = (u_2 - u_1)u_n' + u_1 u_n''$$

holds for $n = 1$ and $n = 2$ and we conclude hence, using the recursive relation, that it holds also for $n = 3, 4, 5, \dots 10$. Adding the two observed equations (*) after having multiplied the first by $u_2 - u_1$ and the second by u_1, we conclude from (***) the desired (**). Main idea of the proof: the general solution u_n of our recursive relation (more aptly called a linear homogeneous difference equation of the second order) is a linear combination of two independent particular solutions u_n' and u_n'' (as the general integral of a linear homogeneous differential equation of the second order is a linear combination of two particular integrals).

11.
$$\int_0^\infty \frac{1}{1+x^\alpha} \frac{dx}{1+x^2} = \int_0^\infty \frac{x^{-\alpha}}{x^{-\alpha}+1} \frac{1}{x^{-1}+x} \frac{dx}{x}$$

$$= \int_0^\infty \frac{x^\alpha}{1+x^\alpha} \frac{dx}{1+x^2}$$

$$= \frac{1}{2} \int_0^\infty \frac{1+x^\alpha}{1+x^\alpha} \cdot \frac{dx}{1+x^2} = \frac{\pi}{4}$$

independent of α. In passing from the second form to the third we introduced x^{-1} as new variable of integration. For $\alpha = 0, \infty, -\infty$ the given integral reduces to

$$\int_0^\infty \frac{1}{2} \frac{dx}{1+x^2}, \qquad \int_0^1 \frac{dx}{1+x^2}, \qquad \int_1^\infty \frac{dx}{1+x^2},$$

respectively. These cases could suggest the above solution.

12. The most obvious fact of this kind is:

$$\int_{-\infty}^{\infty} f(u)du = 0 \qquad \text{if} \qquad f(-u) = -f(u).$$

Set $$u = \log x, \qquad f(\log x) = F(x):$$

$$\int_{0}^{\infty} F(x)x^{-1}dx = 0 \qquad \text{if} \qquad F(x^{-1}) = -F(x).$$

This suggests the following generalization:

$$\int_{0}^{\infty} g(x)[1 + h(x)]x^{-1}dx = \int_{0}^{\infty} g(x)x^{-1}dx \text{ if } g(x^{-1}) = g(x),\ h(x^{-1}) = -h(x).$$

Ex. 11 is the particular case:

$$g(x) = \frac{x}{2(1 + x^2)}, \qquad h(x) = \frac{1 - x^\alpha}{1 + x^\alpha}.$$

13. $0x = 0$, or $x^2 - 4 = (x - 2)(x + 2)$, etc.

14. $x = y = 8$; it is enough to try $x = 8, 9, 10, 11$.

15. $x = y = z = w = 4$, by trials.

17. [Cf. Stanford 1948] The planes of symmetry of a regular solid pass through its center and divide a sphere with the same center into spherical triangles. The three radii through the three vertices of this spherical triangle pass through a vertex, the center of a face and the midpoint of an edge, respectively. The corresponding angles of the spherical triangle are π/v, π/f, and $\pi/2$. Let us call c the side (hypotenuse) of the spherical triangle opposite the angle $\pi/2$. The ratio of the radius of the inscribed sphere to that of the circumscribed sphere is $\cos c$ and, by spherical trigonometry,

$$\cos c = \cot(\pi/f)\cot(\pi/v).$$

The numbers f and v, and the resulting value of $\cos c$, are displayed in the following table for the Tetrahedron, Hexahedron (cube), Octahedron, Dodecahedron, and Icosahedron.

	T	H	O	D	I
$f =$	3	4	3	5	3
$v =$	3	3	4	3	5
$\cos c =$	$\frac{1}{3}$	$\frac{1}{\sqrt{3}}$		$\sqrt{\frac{5 + 2\sqrt{5}}{15}}$	

H. Weyl, *Symmetry*, Princeton, 1952, reproduces Kepler's original figure; see p. 76, fig. 46.

18. See ex. 10.42.

21. (a) We call a determinant with n rows central-symmetric if its elements $a_{j,k}$ satisfy the condition

$$a_{j,k} = a_{n+1-j,n+1-k} \text{ for } j, k = 1, 2, 3, \ldots n.$$

A central-symmetric determinant with n rows is the product of two determinants. Either both factors have n/2 rows, or one factor has $(n + 1)/2$ *rows and the other* $(n - 1)/2$ *rows, according as n is even or odd. Examples:*

$$\begin{vmatrix} a & b \\ b & a \end{vmatrix} = (a+b)(a-b), \quad \begin{vmatrix} a & b & c \\ d & c & d \\ c & b & a \end{vmatrix} = \begin{vmatrix} a+c & b \\ 2d & c \end{vmatrix} (a-c),$$

$$\begin{vmatrix} a & b & c & d \\ e & f & g & h \\ h & g & f & e \\ d & c & b & a \end{vmatrix} = \begin{vmatrix} a+d & b+c \\ e+h & f+g \end{vmatrix} \begin{vmatrix} f-g & e-h \\ b-c & a-d \end{vmatrix}.$$

Proof: Put $n = 2m$ or $n = 2m + 1$, according as n is even or odd. Add the last column to the first, then the column preceding the last to the second, and so on, till the first m columns are changed. After that, subtract the first row from the last, then the second row from the row preceding the last, and so on, till the last m rows are changed. These operations introduce either a rectangle $m \times (m + 1)$, or a square $m \times m$, consisting of vanishing elements in the south-west corner.

(b) The determinant with four rows could be divisible by both determinants with two rows, *without* being their product, namely, if these determinants with two rows had a common factor. Optimistically, we assumed that there is no such common factor: we tried the simplest assumption and succeeded.

22. Most optimistic: the coefficient of any power of h on the left-hand side is less than, or equal to, the coefficient of the same power on the right-hand side. This is really the case: after division by $4h^{1/4}$, the constant term is 1 on both sides and, for $n \geqq 1$, the coefficients of h^n are

$$\frac{3}{4}\frac{7}{8}\frac{11}{12}\cdots\frac{4n-1}{4n}\frac{1}{4n+1}, \qquad \frac{1}{4}\frac{5}{8}\frac{9}{12}\cdots\frac{4n-3}{4n}$$

on the respective sides. Obviously

$$3 \cdot 7 \cdot 11 \cdots (4n - 1) < 5 \cdot 9 \cdots (4n - 3)(4n + 1).$$

23. (a) Call P_n the proximate value obtained with the method in question when the square is subdivided into n^2 smaller squares. Assume that P_n can be expanded in powers of n^{-1}:

$$P_n = Q_0 + Q_1 n^{-1} + Q_2 n^{-2} + \dots .$$

("In general, a function can be expanded in a power series." Cf. ex. 20.) As $n \to \infty$, $P_n \to Q_0$, and we infer that $Q_0 = Q$. Now, the four points in fig. 11.6 are closer to a straight line than those in fig. 11.5. This circumstance *suggests* that $Q_1 = 0$ and the terms $n^{-3}, n^{-4}, \dots$ are negligible even for small n. This leads to

$$P_n \sim Q + Q_2 n^{-2}$$

which represents, if we take n^{-2} as abscissa and P_n as ordinate, a straight line (approximately). In some more or less similar cases it has been proved that the error of approximation is of the order $1/n^2$, and in the light of such analogy the guess appears less wild.

(b) The columns of the following table contain: (1) values of n, (2) ordinates, (3) differences of ordinates, (4) abscissas, (5) differences of abscissas, (6) slopes computed as the ratio of (3) to (5), except that in (5) and (6) the sign $-$ is omitted.

(1)	(2)	(3)	(4)	(5)	(6)
2	0.0937		0.2500		
		0.0248		0.1389	0.1785
3	0.1185		0.1111		
		0.0094		0.0486	0.1934
4	0.1279		0.0625		
		0.0045		0.0225	0.2000
5	0.1324		0.0400		

(c) It is natural to regard $n = 5$ as the most reliable computation and $n = 4$ as the next best. If the points (x_1, y_1) and (x_2, y_2) lie on the straight line with equation $y = mx + b$, we easily find (from a system of two equations for m and b) that

$$b = \frac{y_1/x_1 - y_2/x_2}{1/x_1 - 1/x_2},$$

which, in the present case yields

$$Q \sim \frac{25 \times 0.1324 - 16 \times 0.1279}{25 - 16} = 0.1404.$$

If you have expected anything better than that, you are too sanguine.

No solution: **16, 19, 20.**

BIBLIOGRAPHY

I. CLASSICS

EUCLID, *Elements.* The inexpensive shortened edition in Everyman's Library is sufficient here. "EUCLID III 7" refers to Proposition 7 of Book III of the Elements.

DESCARTES, *Oeuvres*, edited by Charles Adam and Paul Tannery. The work "Regulae ad Directionem Ingenii," vol. 10, pp. 359–469, is of especial interest.

EULER, *Opera Omnia*, edited by the "Societas scientiarum naturalium Helvetica."

LAPLACE, *Oeuvres complètes.* The "Introduction" of vol. 7, pp. V–CLIII, also separately printed (and better known) under the title "Essai philosophique sur les probabilités" is of especial interest.

II. SOME BOOKS OF SIMILAR TENDENCY

R. COURANT and H. ROBBINS, *What is mathematics?*

H. RADEMACHER and O. TOEPLITZ, *The Enjoyment of Mathematics*, Princeton, 1957.

O. TOEPLITZ, *The Calculus*, Chicago, 1963; of especial interest.

III. RELATED FORMER WORK OF THE AUTHOR

Books

1. *Aufgaben und Lehrsätze aus der Analysis*, 2 volumes, Berlin, 1925. Jointly with G. Szegö. Reprinted New York, 1945.

2. *How to Solve It*, Princeton, 1945. The 5th printing, 1948, is slightly enlarged.

3. Wahrscheinlichkeitsrechnung, Fehlerausgleichung, Statistik. From *Abderhalden's Handbuch der biologischen Arbeitsmethoden*, Abt. V, Teil 2, pp. 669–758.

Papers

1. Geometrische Darstellung einer Gedankenkette. *Schweizerische Pädagogische Zeitschrift*, 1919, 11 pp.

2. Wie sucht man die Lösung mathematischer Aufgaben? *Zeitschrift für mathematischen und naturwissenschaftlichen Unterricht*, v. 63, 1932, pp. 159–169.

3. Wie sucht man die Lösung mathematischer Aufgaben? *Acta Psychologica*, v. 4, 1938, pp. 113–170.

4. Heuristic reasoning and the theory of probability. *American Mathematical Monthly*, v. 48, 1941, pp. 450–465.

5. On Patterns of Plausible Inference. *Courant Anniversary Volume*, 1948, pp. 277–288.

6. Generalization, Specialization, Analogy. *American Mathematical Monthly*, v. 55, 1948, pp. 241–243.

7. Preliminary remarks on a logic of plausible inference. *Dialectica*, v. 3, 1949, pp. 28–35.

8. With, or without, motivation? *American Mathematical Monthly*, v. 56, 1949, pp. 684–691.

9. Let us teach guessing. *Etudes de Philosophie des Sciences, en hommage à Ferdinand Gonseth*, 1950, pp. 147–154. Editions du Griffon, Neuchatel, Switzerland.

10. On plausible reasoning. *Proceedings of the International Congress of Mathematicians*, 1950, v. 1, pp. 739–747.

IV. PROBLEMS

Among the examples proposed for solution there are some taken from the *William Lowell Putnam Mathematical Competition* or the *Stanford University Competitive Examination in Mathematics*. This fact is indicated at the beginning of the solution with the year in which the problem was proposed as "Putnam 1948" or "Stanford 1946." The problems of the Putnam Examination are published yearly in the *American Mathematical Monthly* and most Stanford examinations have been published there too.